Deep Reinforcement Learning and Its Industrial Use Cases

Scrivener Publishing
100 Cummings Center, Suite 541J
Beverly, MA 01915-6106

Publishers at Scrivener
Martin Scrivener (martin@scrivenerpublishing.com)
Phillip Carmical (pcarmical@scrivenerpublishing.com)

Deep Reinforcement Learning and Its Industrial Use Cases

AI for Real-World Applications

Edited by
Shubham Mahajan
CHRIST (Deemed to be University), Delhi NCR, India
Pethuru Raj
Reliance Jio Platforms Ltd., Bangalore, India
and
Amit Kant Pandit
School of Electronics & Communication Engineering, Shri Mata Vaishno Devi University, Katra, India

Scrivener Publishing

WILEY

This edition first published 2024 by John Wiley & Sons, Inc., 111 River Street, Hoboken, NJ 07030, USA and Scrivener Publishing LLC, 100 Cummings Center, Suite 541J, Beverly, MA 01915, USA

For more information about Scrivener publications please visit www.scrivenerpublishing.com.

Wiley Global Headquarters
111 River Street, Hoboken, NJ 07030, USA

For details of our global editorial offices, customer services, and more information about Wiley products visit us at www.wiley.com.

Library of Congress Cataloging-in-Publication Data

ISBN 978-1-394-27255-6

Cover image: Pixabay.Com
Cover design by Russell Richardson

Set in size of 11pt and Minion Pro by Manila Typesetting Company, Makati, Philippines

Printed in the USA

10 9 8 7 6 5 4 3 2 1

Contents

Preface

In recent years, the field of artificial intelligence (AI) has undergone a remarkable transformation, with deep reinforcement learning (DRL) emerging as a powerful paradigm for solving complex decision-making problems. From optimizing supply chains to personalizing healthcare treatments, DRL holds the promise of revolutionizing industries across the globe.

As the demand for AI solutions continues to grow, it becomes increasingly vital for professionals and enthusiasts alike to understand not just the theoretical underpinnings of DRL but also its practical applications in real-world settings. This book is born out of a passion for exploring the intersection of theoretical research and practical applications in the realm of AI. It aims to bridge the gap between academic insights and industrial implementations, providing readers with a comprehensive understanding of how DRL can be leveraged to tackle complex challenges and unlock new opportunities.

Throughout the pages of this book, we seek to discover the inner workings of DRL and its real-world applications. We start by laying down the foundational principles of reinforcement learning, building up to advanced DRL algorithms and techniques. Along the way, we delve into diverse case studies, examining how leading organizations are harnessing the power of DRL to drive innovation and gain a competitive edge. From financial trading to autonomous manufacturing systems, each case study offers valuable insights into the practical considerations and challenges involved in deploying DRL solutions.

Whether you're a seasoned AI professional looking to expand your toolkit, a business leader seeking to leverage AI for strategic advantage, or a curious enthusiast eager to understand the cutting-edge technology shaping our future, this book is designed to meet your needs. We hope that through the knowledge and insights shared within these pages, you will be empowered to navigate the complex landscape of DRL with confidence and clarity.

We are deeply grateful to everyone who helped with this book and greatly appreciate the dedicated support and valuable assistance rendered by Martin Scrivener and the Scrivener Publishing team during its publication.

We invite you to join us on this exciting expedition into the realm of deep reinforcement learning. Together, let us explore how AI is reshaping industries, transforming businesses, and shaping the future of our world.

Happy reading!

The Editors
July 2024

1

Deep Reinforcement Learning Applications in Real-World Scenarios: Challenges and Opportunities

Sunilkumar Ketineni and Sheela J.*

Department of School of Computer Science and Engineering VIT-AP University, Amaravathi, Andhra Pradesh, India

Abstract

Deep reinforcement learning (DRL) has proven to be incredibly effective at resolving complicated issues in a variety of fields, from game play to robotic control. Its seamless transfer from controlled surroundings to practical applications, meanwhile, poses a variety of difficulties and chances. This paper comprehensively examines the opportunities and challenges in applying DRL in real-world settings, offering a comprehensive exploration of the challenges and opportunities within this dynamic field. It highlights the pressing issues of data scarcity and safety concerns in critical domains like autonomous driving and medical diagnostics, emphasizing the need for sample-efficient learning and risk-aware decision-making techniques. Additionally, the chapter uncovers the immense potential of DRL to transform industries, optimizing complex processes in finance, energy management, and industrial operations, leading to increased efficiency and reduced costs. This chapter serves as a valuable resource for researchers, practitioners, and decision-makers seeking insights into the evolving landscape of DRL in practical settings.

***Keywords*:** Deep reinforcement learning, decision-making, transfer learning, meta-learning, domain adaptation

1.1 Introduction

The deep reinforcement learning (DRL) paradigm has become a potent tool for teaching agents to make successive judgments in challenging situations. Its uses can be found in a wide range of industries, including robotics, autonomous vehicles, banking, and healthcare [1]. DRL algorithm translation from controlled laboratory conditions to real-world scenarios is not without its difficulties and potential though. This chapter explores the challenging landscape of implementing DRL in real-world settings. For the purpose of addressing difficult decision-making issues across a variety of areas, deep reinforcement learning (DRL) has proven to be a powerful paradigm. The potential of DRL has been astounding, from gaming to robotics and autonomous systems.

**Corresponding author*: sheela.j@vitap.ac.in

Shubham Mahajan, Pethuru Raj and Amit Kant Pandit (eds.) Deep Reinforcement Learning and Its Industrial Use Cases: AI for Real-World Applications, (1–28) © 2024 Scrivener Publishing LLC

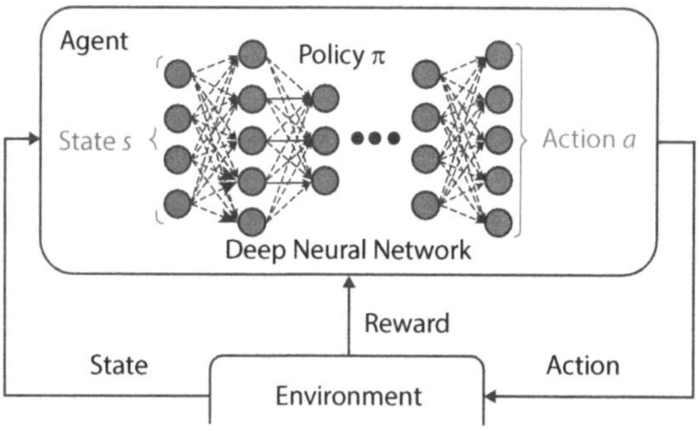

Figure 1.1 Deep reinforcement learning agent with an updated structure.

Deep reinforcement learning (DRL) has received a lot of interest recently as a potentially effective method for handling challenging decision-making problems in a variety of contexts. DRL has a wide range of possible applications, from robots and autonomous driving to healthcare and finance. The deep reinforcement learning agent with an updated structure is displayed in Figure 1.1.

1.1.1 Problems with Real-World Implementation

- Sample effectiveness: Sample efficiency is the ability of a learning algorithm to perform well with a small number of training instances or samples. It is frequently necessary to train models on a lot of data in machine learning and reinforcement learning to achieve excellent performance [2]. However, gathering data can frequently be expensive, time-consuming, or even impractical in real-world situations. Sample effectiveness is crucial in industries like healthcare, manufacturing, and marketing, impacting quality control, market research, and product development.
- Reinforcement learning (RL): Agents study how to interact with their surroundings in real life to optimize reward signals. Less contact with the environment is needed for sample-efficient RL algorithms to develop efficient rules. In circumstances where engaging with the environment is expensive, risky, or time-consuming, this is crucial.
- Supervised learning: To make predictions or categorize data, models undergo supervised learning from labeled instances. Using fewer labeled examples, sample-efficient algorithms can perform well and eliminate the need for labor-intensive manual labeling.
- Transfer learning: Transfer learning entails developing a model for one activity or domain and then applying it to another task or domain that is related to the original. Even with little data, domain-specific transfer learning techniques can use what is learned there to enhance performance in a different domain.
- Active learning: A model actively chooses the most educational examples for labeling in a process known as active learning, which aims to enhance the model's performance [3]. The most useful instances may be swiftly identified

and labeled using sample-efficient active learning procedures, which will save time and effort overall.

- Meta-learning: Through the process of meta-learning, models are trained on a range of activities to enhance their capacity to pick up new skills fast and with little input. Since models must generalize from a limited number of examples, sample efficiency is a crucial component of effective meta-learning.

Contributions of the Book Chapter

In this chapter, we provide an in-depth exploration of the challenges and opportunities in applying deep reinforcement learning (DRL) in practical situations, offering valuable insights into how innovative techniques are addressing sample efficiency, data scarcity, and safety concerns while also highlighting the immense potential for DRL to transform industries by streamlining complex processes and enhancing decision-making across various domains.

1.2 Application to the Real World

Data may be sparse or challenging to obtain in many real-world scenarios, just like with robotics or medical applications. Due to sample-efficient algorithms, these circumstances lend themselves to the application of machine learning techniques.

1.2.1 Security and Robustness

Deep reinforcement learning (DRL) agents must be deployed in realistic situations while taking safety and robustness into account. Deep reinforcement learning includes educating agents to choose actions that will maximize a cumulative reward signal as they interact with their environment [4]. To avoid unintended effects and unanticipated actions in complex and dynamic real-world contexts, it is crucial to guarantee the safety and robustness of these agents. A summary of the main ideas and issues around safety and robustness in DRL is given below:

a) Safety: When discussing safety in DRL, it is important to note that it refers to an agent's capacity to work within predetermined boundaries and refrain from doing any activities that might result in disastrous consequences or safety rule breaches. Safety must be ensured through the following:

- Constraint enforcement: Agents should be built to adhere to safety constraints, which are states or behaviors that an agent must not cross. This may entail punishing or refraining from behavior that violates these limitations.
- Handling uncertainty: Dynamic and unpredictable settings exist in real life. Agents should be able to manage uncertainty in their observations and make judgments that are resistant to changes in the environment.
- Learning from human feedback: Including feedback from people during training can assist agents in acquiring safe behaviors and giving priority to taking actions that are consistent with human preferences and values.

b) Robustness: In DRL, robustness refers to an agent's capacity to function successfully in a variety of settings and circumstances, even when there is noise, disturbance, or variation. Developing robustness entails the following:

- ➢ Domain adaptation: It can be difficult to train an agent in one environment and then hope that it will transfer to another. Agents can adapt to new surroundings more successfully by using transfer learning and domain adaptation techniques. Agents may be susceptible to adversarial attacks, in which minor changes in the input can cause significant modifications in behavior. To fend off such attacks, robust agents ought to be created [5]. Real-world settings may undergo distributional alterations over time, necessitating the constant adaptation and learning of agents to new data distributions.

c) Challenges: There are various difficulties in creating reliable and secure DRL agents for real-world situations, namely:

- ➢ Sample efficiency: In the actual world, training DRL agents can take a lot of time and data. To cut down on the amount of interactions needed for learning, effective exploration tactics are needed.
- ➢ Exploration vs. exploitation: Finding safe and efficient methods depends on striking a balance between exploration (trying new activities) and exploitation (making choices based on knowledge acquired).
- ➢ Incentive engineering: It is difficult to create incentive functions that will lead agents to the desired behaviors while avoiding undesired side effects.
- ➢ A fundamental issue, especially in highly dynamic contexts, is ensuring that agents can transfer their acquired actions to novel and unanticipated situations.
- ➢ Ethics: DRL agents should follow human-defined ideals, observe ethical norms, and refrain from bias.

d) Mitigation methods:

- ➢ Researchers and professionals are looking into different mitigation measures to solve these problems, such as clearly implementing safety limitations into the agent's learning process will guarantee that it never behaves in a dangerous manner. Imitation learning is a technique for teaching agents how to behave safely and to avoid exploring harmful situations.
- ➢ Risk-sensitive learning: This refers to decision-making algorithms that consider risk and uncertainty to prevent taking unnecessary risks. Agents are trained to withstand adversarial perturbations in order to increase their resistance to attacks.
- ➢ Multi-agent training: Teaching agents to communicate with one another can increase safety and produce emergent behaviors that are in line with goals.
- ➢ Human-in-the-loop approaches: Including humans in the decision-making process to give monitoring and intervention as needed.

1.2.2 Generalization

Generalization in deep reinforcement learning (DRL) refers to an agent's capacity to adapt newly learnt skills and behaviors to novel circumstances or contexts. It is an essential component of DRL since real-world scenarios are frequently varied and dynamic and demand that agents perform effectively even in circumstances that they have not directly faced during training [6]. Here is a description of the generalization idea in DRL, along with some tips for improving generalization. It is not possible to prepare DRL agents for every situation they might face in real-world applications. Agents may generalize to new contexts and adapt and decide appropriately without having to undergo substantial retraining. The following reasons make generalization particularly crucial.

The settings in real life are unpredictable and complex. Agents must eventually adapt to the constantly changing environment.

a) Challenges: Several factors make it difficult to achieve good generalization in DRL, including the following:

- Distributional shift: There may be a discrepancy between the distribution of data seen during training and the distribution observed in the real world, resulting in an improper alignment of the training and deployment settings.
- Sparse reward: In some circumstances, incentives may be scarce or delayed, which makes it more difficult for agents to develop useful behavior in a training environment.
- Trade-off between exploration and exploitation: Agents must both investigate novel actions and states to learn new things, but they must also utilize what they have discovered to their advantage in order to maximize rewards [7]. It is essential to strike the proper balance. The curse of dimensionality prevents agents from properly exploring and generalizing in real-world contexts because high-dimensional state and action fields are frequently present.

b) Generalization strategies: Researchers have suggested a number of strategies to enhance the generalize capacities of DRL agents. Transfer learning is the process of pre-training an agent in one environment and then optimizing it for or migrating it to a different environment. With this, learning in the target environment is accelerated by using the knowledge acquired in the source environment [8]. By reducing the discrepancies between the source and target domains, domain modification allows an agent to adapt its learned policy to a new environment. Agents are trained in a range of tasks or contexts through meta-learning, which helps them develop their learning capabilities. As a result, new situations can be adapted to more quickly.

- Addition: By introducing noise or changes to the training data, agents can learn more resilient strategies that can deal with uncertainty in the real world.

c) Evaluation of generalization: To make sure an agent function successfully in unanticipated settings, it is crucial to assess its generalization capabilities. Typical evaluation methods include the following:

- ➢ Zero-shot testing: Testing an agent in a setting it has never encountered before to judge its capacity for generalization.
- ➢ Evaluation of transfer learning: Assessing an agent's capacity to transition successfully from one setting to another. During testing, domain randomization is used to provide random variations to imitate novel situations and gauge an agent's adaptation.

d) Exploration of space in high dimensions: A significant issue in reinforcement learning (RL) is navigating multidimensional landscapes, especially when using real-world examples. In highly dimensional landscapes, the difficulty of effective exploration and exploitation of actions becomes a substantial barrier to RL algorithms' goal of finding optimum policies that maximize cumulative rewards. Here are some tactics and things to think about when exploring high-dimensional places in real life, especially in realistic situations [9]. Curiosity-based techniques motivate the agent to investigate new or uncharted territory by awarding it when it comes across circumstances that result in unforeseen results. By enabling the agent to try out various behaviors surrounding the present best estimation, noise injection encourages research. Gradient-based exploring is possible because of algorithms like Trust Region Policy Optimization (TRPO) and Proximal Policy Optimization (PPO), which improve changes to policies while limiting the change.

Some algorithms instead examine the parameter space of the policy or value functions rather than simply investigating the action space.

- ➢ Sparse rewards and shaping: The agent's investigation can be guided by designing suitable reward mechanisms. When prizes are scarce and only available in certain circumstances, the agent may be forced to travel to several state areas in order to collect awards.
- ➢ Exploration that is hierarchy and skill-based: In extremely complex environments, mastering intricate techniques or sub-policies can help in an investigation. The agent acquires a variety of talents that can be used in various contexts to efficiently handle various scenarios.
- ➢ Model-based exploration: By learning a model of the environment, the agent can simulate several action pathways in advance of taking them, allowing it to better investigate the effects of various decisions.
- ➢ Meta-learning: By exposing the agent to a variety of tasks or situations, we can train it to become more adept at understanding how to learn, which could improve its exploration tactics.
- ➢ Online discovery techniques: Investigation may be hampered in real-world situations by a lack of time, resources, or security concerns. It can be useful to create online exploration tactics that change according to the agent's knowledge.
- ➢ Transfer learning: If the agent has experience with related activities, it may be able to use that knowledge to direct its exploration of the current multidimensional domain.
- ➢ Self-play and competitive agents: In some circumstances, agents are able to investigate by interacting with other representations of themselves, fostering a variety of exploration tactics.

Due to considerations including safety concerns, expensive data collecting, and computing constraints, implementing these tactics in real-world situations might be difficult.

e) Designing rewards: Reinforcement learning (RL), particularly when used in real-world contexts, heavily relies on reward design. By giving suggestions for improving its activities, the reward function specifies the objective and directs an RL the agent's approach to learning [10]. To guarantee that the agent learns the intended behavior successfully and rapidly, it is essential to design an adequate incentive mechanism. Here are some ideas and tactics for designing rewards in real-life situations:

- Clarity of objectives: Outline the task's goals in detail. What actions or results are you hoping the agent will learn? Making an award system that promotes the intended behavior requires a clearly stated purpose.
- Sparse vs. dense rewards: Decide whether to use sparse rewards (given only in specific situations) or dense rewards (given more frequently). Shape the reward function to offer extra interim incentives that direct the agent toward the desired behavior. This could speed up learning and improve the effectiveness of research [11]. Using expert presentations or human likes, inverse reinforcement learning (IRL) systems learn the reward function rather than manually developing it. This method may be useful for capturing intricate actions that are challenging to clearly explain.
- Proxy incentives: Make use of proxy rewards that are simpler to model or quantify. As an example, in robotic activities, the learning process can be made simpler by employing measurements such as the distance to the goal or motion regularity as proxy incentives.
- Learning through curriculum: To aid the agent's gradual comprehension, gradually improve the assignment's difficulty. To progressively increase reward function difficulty, start with smaller goals.
- Normalization of incentives: Make incentives consistent with a fair range by normalizing them. RL algorithms may experience convergence problems because to extremely large or small rewards. Scaling rewards will enable a fair trade-off between short-term gains and long-term objectives. In order to choose the jobs with higher accumulated incentives, the agent may use this information. Put restrictions or sanctions in place to deter undesirable behaviors or acts. The agent's actions may be influenced by, for instance, fines for crashes or transgressions of safety restrictions.
- Behavioral cloning: Before employing reinforced learning to adjust the agent's policy, pre-train the agent using behavioral cloning so that it can learn to imitate expert conduct. Initiating the RL process in this way may be beneficial. Combining various incentive kinds will enable you to encourage behavior in a variety of ways. In order to help the agent learn, this can offer a more complete signal. Implement domain expertise or insights into the design of the rewards function. On what qualifies as good conduct, specialists in the industry might offer insightful advice. Add normalization terms or exploration bonuses to the reward function to promote discovery and prevent quick convergence to inferior solutions. To improve the reward function, seek out

human feedback. Based on their subjective assessments, human evaluators can assist in improving the reward design.

- Designing rewards iteratively is a common practice: Maintain a constant eye on the agent's actions, assess its effectiveness, and make any necessary adjustments to the reward function. An in-depth knowledge of the task, the agent's skills, and the desired behavior is necessary to design an incentive system that works. It is critical to experiment with several reward structures and refine them until you find the one that encourages the required behavior in practical contexts.

f) Considering the law and ethics: There are various moral and legal issues that deep reinforcement learning (DRL) in real-world situations brings up. It is crucial to negotiate these challenges appropriately because the use of DRL agents in the real world can have a big impact on society. Here are some significant moral and legal criteria for DRL applications, namely:

- Protection and risk reduction: The primary concern should always be safety. DRL agents have the ability to act in ways that have immediate effects; thus, it is crucial that they take care to avoid doing anything that can endanger people or the natural world.
- Equity and disparity: In data used for training, biases can be picked up by DRL agents. Ensure that training data is accurate and devoid of biases that can result in unjust or biased outputs [12]. Watch out for biases in decision-making all the time and rectify them as necessary.
- Openness and comprehensibility: DRL models—in particular, deep neural networks—are frequently referred to as "black boxes." The openness and explanation of processes should be improved. Utilize ways to comprehend and rationalize the agent's choices, such as SHAP values, attention processes, and model-independent accessibility strategies.
- Data protection: Observe all rules and legislation pertaining to confidentiality of data. Use privacy-compliant data collection and usage practices, such as the GDPR in Europe. As much as feasible, anonymize or pseudonymize data to safeguard people's identity.
- Accounting and responsibility: Follow all laws and regulations relevant to data protection. Use data collecting and usage methods that respect privacy, such as those required by the GDPR in Europe. To protect people's privacy, anonymize or pseudonymize data as much as you can.
- Human-in-the-loop: Think about placing a human in the loop in crucial applications so that they may step in and make changes to the DRL agent's judgments as needed. To avoid disagreements or abuse, create precise rules for human interaction.
- Guidelines and certification: There might be rules or standards for certification that must be met based on the application. In order to ensure adherence to current regulations and requirements, work with the appropriate regulatory agencies.

- Normative structures: Adopt moral guidelines for AI and DRL, such as the concepts of openness, responsibility, justice, and society benefit. Before implementation, conduct moral impact analyses to assess any possible ethical problems.
- Information security: Defend against unauthorized access to the data that DRL agents use. Encrypting data, access restriction, and recurring checks for safety are all appropriate security methods.
- Consent and knowledge-based decisions: When DRL agents contact with people, they should always get their knowledge and consent before doing so, and they should also be upfront about how their data will be utilized [13]. Make sure people are aware of the DRL system's strengths and restrictions.
- Constant observation and assessment: Constantly keep an eye on how DRL agents behave in the real world. Recognize and correct variations from intended behavior as soon as possible. Analyze the system's effectiveness and moral ramifications on a regular basis.
- Public participation: Involve the public and partners in conversations about DRL implementations, particularly if the innovation has societal implications. Solicit public input and take into account public concerns and desires.

1.2.2.1 Overcoming Challenges in DRL

Overcoming challenges in deep reinforcement learning (DRL) is an ongoing and dynamic process that involves a combination of research, engineering, and problem-specific strategies. Here is a more detailed guide on how to address these challenges:

1. Sample efficiency
Experience replay: Implement a replay buffer to store and reuse past experiences, helping to break temporal correlations in the data and enhance learning efficiency.
Prioritized experience replay: Assign different priorities to experiences and sample them based on their importance, focusing on high-impact experiences.
Off-policy learning: Utilize off-policy algorithms like DDPG, SAC, or TD3, which can improve sample efficiency by reusing data effectively.

For example, in autonomous robotics, the cost and time required to collect real-world data for training agents can be prohibitive. This challenge drives the development of innovative techniques such as experience replay and off-policy learning, which allow DRL models to make the most of limited data. Overcoming this challenge not only accelerates the deployment of DRL in practical applications but also reduces the environmental and financial costs associated with data collection.

2. Data scarcity
Transfer learning: Pre-train your agent in simulation environments or on related tasks to provide a head start, and then fine-tune the agent in the target environment using limited real-world data.
Domain adaptation: Implement domain adaptation techniques to reduce the distribution gap between the training and target data, making the agent more adaptable.

In healthcare, for instance, a DRL model trained on a large dataset of medical images from one domain can be fine-tuned for a specific medical facility or patient population, mitigating data scarcity issues. This approach extends the reach of DRL in domains where data collection is challenging.

3. Safety concerns
Reward engineering: Carefully design reward functions that promote safe and desired behavior. Penalize unsafe actions to guide the agent towards safety.

In autonomous driving, DRL agents need to navigate complex, dynamic environments while ensuring the safety of passengers and pedestrians. By addressing safety concerns, we can unlock the potential of DRL to revolutionize critical domains without compromising human well-being.

Constraint optimization: Augment DRL algorithms with constraints that prevent undesirable actions, ensuring safety.

Mimic learning: Combine DRL with imitation learning, training the agent to mimic human demonstrations to ensure safe exploration.

4. Exploration–exploitation balance
Exploration strategies: Experiment with various exploration techniques, such as ε-greedy, Boltzmann exploration, or noise injection, to balance the agent's exploration and exploitation.

Intrinsic rewards: Develop intrinsic motivation signals that encourage the agent to explore areas of the environment where it lacks knowledge.

5. Interpretable models
Model interpretability: Use techniques like attention mechanisms, saliency maps, or surrogate models to make DRL models more interpretable and provide insights into their decision-making process.

Explainable AI: Combine DRL with explainable AI techniques to enhance transparency and interpretability.

6. Transfer learning
Domain randomization: Train the agent in simulation environments with varying parameters to enhance its adaptability to real-world scenarios.

Multi-task learning: Train the agent on multiple related tasks to transfer knowledge across domains.

7. Continuous learning
Develop agents that can continually learn from new experiences and adapt to changing environments, possibly using techniques like online reinforcement learning.

8. Human feedback
Incorporate human demonstrations and feedback into the training process to accelerate learning and ensure safety, a process known as imitation learning.

9. Simulation environments
Create high-fidelity and realistic simulation environments that closely mimic real-world conditions for a more effective pre-training.

10. Parallelization and hardware acceleration
Utilize powerful hardware such as GPUs, TPUs, or distributed computing to speed up training, enabling more experimentation and faster convergence.

11. Ensemble learning
Combine multiple DRL models to improve robustness and decision-making.

12. Regulatory and ethical considerations
Comply with regulations and ethical guidelines, especially in critical applications like healthcare and autonomous vehicles to ensure responsible deployment.

Overcoming DRL challenges is a multidisciplinary endeavor that involves continuous research, experimentation, and adaptation of strategies to the specific problem at hand. Collaboration with experts in machine learning, reinforcement learning, and domain-specific knowledge is often crucial to successfully tackle these challenges and advance the field.

1.3 Possibilities for Making a Difference in the Real World

1.3.1 Transfer Learning and Domain Adaptation

Transfer learning and area adjustment are two related ideas in machine learning and artificial intelligence that entail utilizing information obtained from one task or area to boost performance on another. The systematic diagram of transfer learning is shown in Figure 1.2.

- Transfer learning: Transfer learning is the process of using knowledge obtained from one task (source task) to improve perform on a related but distinct task (target task). Models are often trained for specific tasks in traditional

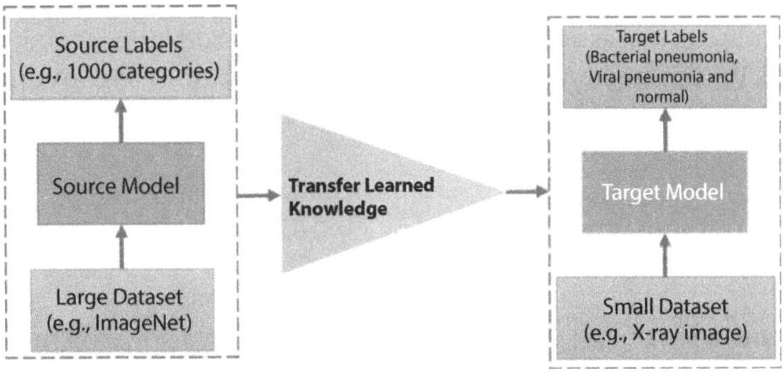

Figure 1.2 Concept of transfer learning.

machine learning using a defined dataset. However, the goal behind transfer learning is to transfer previously learned knowledge from a source activity to a target task, frequently with the premise that certain underlying elements of the tasks are connected [14]. When there is limited data for the target task, transfer learning can be advantageous because the model can use the information learnt from the source task to produce better predictions on the target task.

- Fine-tuning: It entails taking a model that has been trained (typically on a huge dataset) and training it on the dataset for the goal job. The goal is to tweak the model's weights to better fit the target task while maintaining some of the information learned from the original task.
- Feature extraction: In this method, the learnt features of the pre-trained model are extracted and used as input for a new model that is particularly trained for the target task. This is especially beneficial when the output formats of the source and destination jobs are different.
- Pre-training on a domain-specific dataset: Rather than training on a general dataset, models can be pre-trained on a dataset that is more closely related to the domain. Transfer learning and area adaptation are two related ideas in machine learning and artificial intelligence that entail utilizing information obtained from one task or area to boost effectiveness on another.
- Domain adaptation: Domain adaptation is concerned with the case in which the source and target domains are similar but not identical. The key problem is coping with the disparities in distribution of information across the source and destination domains [15]. In other words, the goal is to modify the model's information from the source domain to perform well in the target domain.
- Instance-based adaptation: It entails selecting or re-weighting instances from the source domain to make them more similar to the targeted domain.
- Feature-based adaptation: Techniques such as domain adversarial modeling enable the model to learn features that are domain-invariant, decreasing the gap in feature space between the source and target domains.
- Parameter-based adaptation: These techniques change the model's parameters to match the target domain by reducing the distribution difference within the domains.

Transfer learning as well as adaptation to the domain are both critical in situations where data with labels for the target task or domain is few, costly, or time-consuming to obtain. They allow models to make good use of existing information while also generalizing well to new scenarios.

1.4 Meta-Learning

Deep reinforcement learning (DRL) and meta-learning are two powerful machine learning methods. They form a foundation for teaching agents how to react quickly to novel and dynamic real-world events when coupled. Let us investigate how meta-learning techniques

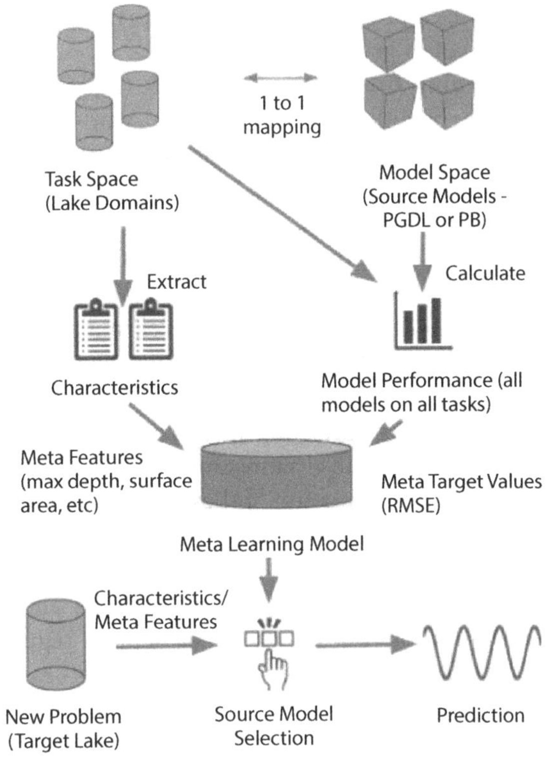

Figure 1.3 Meta learning framework.

facilitate DRL, particularly in dynamic real-world contexts. Meta-learning, commonly referred to as "learning to learn," entails teaching models how to learn. Meta-learning in the context of reinforcement learning seeks to create agents capable of swiftly adapting to new tasks or settings with minimum data and interaction [16]. Meta-learning is concerned with developing a generalizable policy that can be swiftly fine-tuned for new tasks is shows in Figure 1.3. The primary concept is to expose the agent to a range of tasks during the meta-training phase so that it can learn to extract important knowledge or methods that can be applied efficiently to new, unknown problems.

1.5 Deep Reinforcement Learning (DRL)

Deep reinforcement learning is a machine learning subfield that deals with teaching agents to make successive decisions in a given environment in order to maximize a cumulative reward. DRL algorithms approximation rules or function values with neural networks, allowing agents to learn sophisticated behaviors directly from raw sensory input.

The approaches to meta-learning in DRL for dynamic scenarios are as follows:

➢ Model-agnostic meta-learning (MAML): MAML is a well-known meta-learning method that can be used to improve DRL. It entails training a model in such a way that it can be fine-tuned to perform effectively on new tasks with a limited number of gradient changes.

- Reinforcement learning from human feedback (RLHF): Acquiring correct reward signals might be difficult in dynamic real-world circumstances. Meta-learning can aid in this situation by teaching an agent with human feedback. The agent learns by analyzing feedback signals provided by humans interacting with the agent in various tasks.
- Contextual meta-reinforcement learning: The context of the environment frequently changes in dynamic circumstances. Contextual meta-reinforcement learning is concerned with teaching agents to learn a policy that is reliant on the situation. This means that the agent adapts its behavior based on the present context, allowing it to successfully deal with dynamic changes.
- Hierarchical meta-learning: Learning hierarchies of rules can be advantageous in complex and dynamic circumstances. Meta-learning can aid in the efficient composition and adaptation of these hierarchical strategies. By mixing and adjusting learnt sub-policies, agents are able to deal with a wide range of events.
- Online meta-learning: Some real-world applications allow for the online emergence of tasks or for task changes that happen regularly. To help the agent adjust to new tasks as they come in, online meta-learning was created. The agent's excellent performance can be maintained when the environment changes thanks to this ongoing adaptability [17]. Because it gives agents the capacity to quickly adapt and generalize to changes in the environment or tasks, meta-learning combined with DRL is especially effective in dynamic real-world contexts. For intelligent agents to function successfully in challenging and constantly-evolving real-world scenarios, flexibility is essential.

1.5.1 Hybrid Approaches

Deep reinforcement learning (DRL) hybrid approaches relate to the integration of numerous techniques and methods from many areas to solve the difficulties and restrictions of implementing DRL in real-world scenarios. These situations frequently entail dynamic and complicated environments, where pure DRL techniques may be hindered by problems like sample inefficiency, security challenges, and high-dimensional state spaces. The goal of hybrid techniques is to improve performance, stability, and dependability by combining the benefits of DRL with those of other systems. Here is a summary of a few typical hybrid DRL methods for actual-world circumstances.

Model-based reinforcement learning: DRL techniques often fall under the heading of model-free learning, where the agent immediately picks up a policy or value function from interactions with the surroundings. The learning of an explicit model of the dynamics of the surroundings is a component of model-based techniques, on the other hand. By combining the two paradigms and simulating policy assessment and planning trajectories using learnt environment models, hybrid techniques can increase the sample reliability and effectiveness. Imitation learning, also known as behavioral cloning, is the process by which an agent learns from the examples of experts rather than just by doing it on their own [18]. By starting out by imitating expert actions and then fine-tuning through reinforcement learning, hybridizing DRL with imitation learning can aid the agent in bootstrapping its learning process. When there are professional demos available, this is really helpful.

Learning through transfer: Learning through transfer is the process of applying what you have learnt to one task to another that is closely related. A hybrid technique in the context of DRL might involve pre-training a neural network on a comparable task with plenty of data and then fine-tuning it on the target task with little data. As a result, learning will go more quickly, and performance in practical situations will be better. Real-world settings frequently feature an organizational framework where choices are made at many levels of abstraction. This is known as progressive reinforcement learning. In an ordered method, lower-level policies are in charge of fine-grained actions, and higher-level policies decide on high-level tactics. Taking up difficult activities in an orderly manner is made easier by doing this.

Multiple objectives that are in conflict frequently appear in real-world circumstances. To discover a compromise between these goals, multi-objective DRL is used [19]. To direct the RL agent toward a Pareto-optimal solution, combinations of techniques may use methods from multiple optimization objectives.

1.6 Online vs. Offline Reinforcement Learning

Online RL picks up knowledge by immediately engaging with the surroundings, whereas offline RL learns from a fixed dataset of previously gathered events.

- Meta reinforcement learning: In meta RL, agents discover ways to learn across a variety of tasks more efficiently. By combining DRL and meta RL, agents may become more adaptable to new tasks and more useful in dynamic real-world circumstances. The difficulties of using DRL in intricate, risky real-world circumstances are being addressed by these hybrid techniques. The particulars of the case, the data that are accessible, and the intended trade-offs between stability, performance, and sample efficiency all play a role in the decision of which hybrid technique to employ.

1.7 Human-in-the-Loop Systems

Deep reinforcement learning (DRL) systems that include human input in the learning process are known as "human-in-the-loop" systems. In order to solve the difficulties and restrictions of using DRL in real-world circumstances, these systems make use of the advantages of both automated learning and human decision-making. The following is a summary of DRL's human-in-the-loop systems for scenarios that occur in real life. A human-in-the-loop approach to interactive teaching and learning enhances education by combining human insight with technology for personalized instruction. When an agent is learning in DRL systems, humans may give them immediate feedback. A reward, a correction, or a ranking for certain activities may be included in this feedback. Accelerating learning and ensuring that the agent explores more promising areas of the action space are two benefits of interactive teaching.

- Expert demonstrations: Using expert presentations in human-in-the-loop DRL is a typical strategy [20]. With the guidance of experts, the agent can be shown the desired behaviors, which accelerates its learning process and prevents the agent from exploring ineffective options. To inform the agent's initial policy, these demonstrations might be employed as a method of imitation learning.
- Reward shaping: It might be difficult to design reward functions that precisely represent the goals of a real-world scenario. To improve learning outcomes, reward functions can be shaped or altered by human skill. People can convey their preferences and learning goals to the agent by leading it through rewards.
- Specification of the constraint: It is frequently necessary to abide by ethical and safety requirements in real-world situations. Humans can explicitly specify requirements or preferences that an AI agent should abide by in systems with a human in the loop. This guarantees that crucial constraints are respected during the learning process. In situations where the agent must deal with confusing or unknown circumstances, human assistance can be essential. When an agent's projections or behaviors are unknown, humans can give advice or make judgments, increasing the system's dependability. Exploration is a major challenge in RL, and adaptive exploration strategies are important [21]. Human guidance can be used in human-in-the-loop systems to point the agent's investigation in the direction of areas that are most likely to provide useful data. As a result, sample effectiveness in practical tasks may be greatly improved.
- Intervention: Human experts can still fine-tune an agent's behavior after it has gone through a learning phase. This is especially helpful when the agent encounters circumstances that were not covered in training or when the agent needs to modify their conduct as a result of evolving scenarios.
- Human feedback loops: Systems that include humans in the loop can establish iterative feedback loops in which an agent learns from user feedback, enhances performance, and then gets further feedback on the revised behavior. Continuous improvement is ensured by this circular process. Customization that is focused on the user: Human-in-the-loop systems can be created so that users can alter the behavior of AI agents to suit their tastes. AI systems may become more flexible and practical in a variety of real-world circumstances as a result of this personalization. The goal of incorporating humans in the loop of deep reinforcement learning (DRL) systems is to bridge the knowledge and decision-making gap between humans and AI agents.

1.8 Benchmarking and Standardization

Deep reinforcement learning (DRL) algorithm effectiveness in real-world scenarios can be evaluated, compared, and improved with the use of benchmarking and standards. These procedures aid in creating a standard framework for assessing various algorithms, encouraging reproducibility, and advancing the discipline. Here is a summary of DRL benchmarking and standards for actual situations.

a) Benchmarking in DRL: In order to compare various DRL algorithms, benchmarking entails creating a collection of standardized tasks or conditions. A fair and consistent method for evaluating and contrasting the effectiveness of various algorithms is what this project aims to achieve [22]. When comparing DRL algorithms in real-world circumstances, the selection of tasks entails locating a wide range of practical tasks that are representative of the difficulties and complexity of the intended applications. Designing environments for simulations that faithfully replicate tasks in the real world. These settings ought to have clearly defined state and action spaces, incentive systems, and dynamics.

- Performance measures: Creating suitable measures to assess the effectiveness of DRL techniques on the benchmark assignments. Depending on the particular goals of the real-world scenarios, these metrics may also include measures of task completion, efficiency, safety, resilience, and more.
- Baseline algorithms: Containing a number of baseline algorithms that are the cutting edge in DRL. For assessing the effectiveness of new algorithms and methodologies, these baselines serve as benchmarks.
- Reproducibility: Providing comprehensive documentation, code, and configurations to make sure that the benchmarking findings can be repeated by other researchers. To increase trust in the reported efficiency, repeatability is crucial.

b) Standardization in DRL: It is possible to improve the ability to compare and accuracy of DRL research by standardizing procedures, norms, and practices. Standardized initiatives can include things like the following:

- Code libraries and frameworks: creating and maintaining standardized libraries and frameworks that offer implementations of frequently utilized DRL algorithms. The likelihood of implementation errors is decreased and consistent experimentation is made easier [23]. Create and exchange standard datasets that reflect real-world scenarios. These datasets enable fair comparisons between various strategies by being utilized for the training, testing, and validation of DRL algorithms.
- Assessment protocols: Creating standards for how DRL methods should be assessed, including how data should be divided into training and test sets, how hyperparameters Encouragement for complete recording of techniques, hyperparameters, and experimental setups is part of the documenting and reporting requirements. Having consistent reporting formats can make it simpler to comprehend, reproduce, and contrast various findings. To develop and improve benchmarks, measurements, and best practices as a group, the research community should be encouraged to work together. Collaborative efforts can lead to stronger and more relevant standards, which may require adjustments and clear communication of outcomes.

c) Benchmarking and standardization have several advantages, namely:

- ➢ Fair comparison: Benchmarking and standardization ensure that various DRL algorithms are assessed on the same tasks and in the same environments, allowing for comparisons that are both accurate and objective. Developing new techniques to solve specific challenges in real-world scenarios: By highlighting the advantages and disadvantages of various algorithms, benchmarking can direct the creation of new methods that solve particular problems in real-world scenarios.
- ➢ Reproducibility: Standardization procedures improve research findings' ability to be replicated, enabling future investigators to validate and build on earlier findings.
- ➢ Resource allocation: Benchmarks can be used to direct research resources to areas that will have the greatest influence on practical applications, hence promoting faster development in such fields.
- ➢ Adoption in industry: By supplying precise evaluation criteria and proving the dependability of algorithms, standardization might make it easier for DRL techniques to be adopted in enterprises. For objective assessment, meaningful comparison, and long-term advancement in the field, benchmarking and standardization in DRL for actual scenarios are crucial. With the help of these procedures, DRL algorithms are created knowing exactly what they are capable of and what they cannot do to solve problems in the actual world.

1.9 Collaborative Multi-Agent Systems

Multi-agent collaboration is used in cooperative multi-agent systems (MARL) in deep reinforcement learning to address challenging problems in practical contexts. These systems use deep reinforcement learning methods to give agents the ability to learn and modify their behavior in either a cooperative or competitive way. There are numerous fields where collaborative MARL can be used, including robots, autonomous driving, traffic control, industrial automation, and more. The cooperative multi-agent systems in DRL for real-world scenarios are described here. Collaboration between numerous agents in a shared environment is referred to as multi-agent interaction (MARL). The effects of an agent's actions on their surroundings and other agents' behavior must be understood by the agent.

a) Competition and cooperation: Agents might engage in competitive competition for scarce resources or cooperate to accomplish shared goals. For successful collaboration, these factors must be balanced. Agents frequently need to coordinate their actions in order to get the best results in real-world circumstances. It is quite difficult to create coordinating techniques or communication protocols [24]. Multi-agent systems are capable of engaging in emergent behavior, in which the combined actions of agents produce results that are not immediately anticipated from their own actions.
b) Approaches and techniques: independent learning: Each agent develops its policy on its own, presuming that the policies of the other agents are predetermined. Although

straightforward, this method could not produce the best collaborative behaviors because agents do not consciously analyze how their actions would affect others. Agents are trained centrally while having access to the global state data during training, which is known as centralized training with decentralized execution (CTDE), but they use a decentralized decision-making process while carrying out their plans. Between coordination and scalability, this strategy achieves a balance.

- Multi-agent actor–critic (MAAC): MAAC expands the actor–critic framework to include multi-agent scenarios. It employs decentralized actors to choose actions while centralized critics are used to assess the value functions of the policies of the agents.
- Graph neural networks (GNNs): GNNs are used to represent interactions between agents in contexts where the interaction topology is important. They record inter-agent communication and interdependence.
- Communication networks: Some MARL methods include explicit channels for agent-to-agent communication. In order to communicate pertinent information for group decision-making, agents learn to exchange messages. With this method, several layers of agents with various levels of decision-making capabilities are introduced. Higher-level agents help lower-level agents cooperate effectively by giving them high-level direction.

c) Real-world applications:

- Robotics: Collaborative MARL can be used to control several robots to carry out operations like exploration, item manipulation, or area coverage.
- Traffic management: Drones, traffic signals, and autonomous vehicles can work together to improve safety, optimize traffic flow, and ease congestion.
- Inventory management: Order fulfilment and distribution can all be improved through the use of supply chain management by several entities.
- Energy systems can work together to balance the production, delivery, and consumption of energy through the use of smart grids.
- Multi-robot exploration: Agents are capable of cooperatively navigating unfamiliar surroundings while exchanging knowledge about places they find.

To investigate team dynamics and strategy, collaborative multi-agent systems in sports like basketball or soccer are simulated. DRL's collaborative multi-agent systems have the potential to address challenging real-world issues that call for the cooperation of several intelligent agents. To use these systems effectively across a range of areas, however, communication, coordination, and scalability issues still need to be resolved.

1.10 Transfer Learning and Domain Adaptation

In deep reinforcement learning (DRL), transfer learning and domain adaptation are potent methods for utilizing information acquired from one task or domain to enhance efficiency in a related task or domain. These methods offer methods to quicken learning and improve generalize in real-world situations where gathering data and training from scratch can be time- and resource-consuming. An overview of transfer learning and domain-specific adaptation in DRL for real-world scenarios is provided below.

a) Transfer learning: During transfer learning in DRL, an agent is trained on a source task (or domain) before applying what they have learnt to a target task (or domain). Transfer learning can be accomplished in DRL in numerous ways. Transfer of parameters: The target task's neural network is initialized with the parameters of a previously trained agent. Using information from the goal job, fine-tuning is then performed. Transferring a policy from one task to another is known as policy transfer [25]. A small network may need to be trained to map the policy of the source task to the policy of the target task in order to do this.
b) Transfer of value function: The target task's exploration and learning are influenced by the value function learned from the source task. Exploration efficiency may be improved by using this strategy. A neural network that has learned features for the source task extracts them and feeds them into the network for the target job. When activities share some underlying structure in common, this method can be quite helpful.
c) Domain adaptation: Adapting an agent's policy or value function from a source domain to work successfully in a different destination domain is the core goal of domain adaptation in DRL. When there are differences in the data or dynamic distributions between the source and target domains, this is very crucial.
d) Domain adaptation strategies: The agent is exposed to a variety of diverse domain situations during training by including random variations. This aids the agent in learning a more robust policy that is flexible enough to accommodate many real-world events. To motivate the agent to learn features that are domain-invariant, a domain discriminator is included in the domain adversarial training process. This lessens the impact of domain-specific variances on the learned policy.
e) Real-world applications: Data augmentation can assist close the gap between the source and target domains by generating fictitious data through the modification or transformation of actual data.

- ➢ Unsupervised domain adaptation: In order to adjust the agent's policy to the target domain, this technique uses labeled and unlabeled data from the source domain.
- ➢ Robotics: Transfer learning and domain adaptation allow for the speedy adaptation of robots to new settings and tasks by allowing them to draw on prior knowledge gained from completed tasks.
- ➢ Autonomous vehicles: In the real world, where data collecting is expensive and dangerous, agents can use what they learn about safe driving practices from simulations.
- ➢ Health: To account for changes in patient populations, agents can be pre-trained on data from one hospital and then fine-tuned on data from another hospital.
- ➢ Automation in the industrial sector: Agents can be simulated–trained and then adapted to actual manufacturing settings.

By utilizing previously taught language representations, transfer learning can aid language understanding models [26]. By allowing agents to draw on previous knowledge when confronted with new obstacles, transfer learning and domain adaptation both improve the effectiveness, resilience, and application of DRL algorithms in real-world contexts. But creating efficient transfer and adaptation techniques necessitates a thorough comprehension of the source and target tasks/domains and careful assessment of the transferability of learnt knowledge.

1.11 Hierarchical and Multimodal Learning

Deep reinforcement learning (DRL) uses cutting-edge methods like hierarchical and multimodal learning to solve problems that arise in real-world settings with complex surroundings and a variety of input sources. By structuring the learning process and integrating various types of input, these strategies seek to enhance the effectiveness, generalization, and performance of DRL agents. An outline of multimodal and hierarchical learning in DRL for real-world settings is provided below:

a) Understanding in hierarchy: By breaking a big activity down into smaller tasks or levels of abstraction, hierarchical learning is accomplished. In DRL, this is done by introducing numerous layers of agents, each in charge of a different facet of the issue. Multiple advantages of hierarchy learning include. Exploration that is conducted effectively uses high-level policies to choose subtasks, which narrows the exploration domain and speeds up learning. Hierarchies enable agents to learn actions at various timelines due to their ability to abstract time [27]. Long-term strategies are decided by high-level policies, while short-term activities are handled by low-level policies.

- ➢ Task decomposition: By breaking complex tasks down into simpler subtasks, we can improve policy optimization and learn more efficiently.
- ➢ Reusability: Shared pre-trained subtask policies encourage transfer learning and adaption across a variety of tasks.

Agents can successfully navigate complicated environments by combining high-level policies that choose subtasks with low-level policies that carry out actions.

b) Multimodal learning: Integration of data from several sources or modalities, including as vision, language, and sensor data, can help people make better decisions and perceive the world more clearly. This is possible in DRL by include many kinds of data in the learning process, including. To give an agent a more thorough picture of its surroundings, sensor fusion combines data from various sensors (such as cameras, lidars, and radars).
c) Text and language: Integrating natural language instructions or textual data can aid agents in understanding high-level objectives or human direction. Explicit monetary rewards, language criticism, or perceptual cues are just a few of the different reward modalities that agents can get.
d) Real-world applications:

- Robotic manipulation: Hierarchical learning can assist robots in breaking down difficult manipulation tasks into smaller tasks like gripping, lifting, and placing.
- Autonomous driving: By combining data from cameras, lidars, and maps, multimodal learning can help drivers make safer and more knowledgeable judgments.
- Healthcare: Patient information and medical imagery can be integrated with multimodal learning, while hierarchical learning can be used to break down a medical diagnosis into a series of choices.
- Gaming: Multimodal learning can combine visual and aural inputs, whereas hierarchical learning can help agents learn techniques at various game levels. Textual and visual context can both help with language understanding in multimodal learning, which can enhance language models. When it comes to managing the complexity and variety of real-world settings, both hierarchical and multimodal learning are effective approaches. DRL agents can improve their performance, flexibility, and robustness in complicated and dynamic contexts by organizing learning and combining many sources of information.

1.12 Imitation Learning and Human Feedback

When it comes to deep reinforcement learning (DRL) for real-world settings, imitation learning and human input are essential strategies. They entail utilizing human direction and experience to improve the learning process and the performance of DRL agents in challenging situations. Let us investigate these ideas in more detail.

- Imitation learning: It is possible to teach an AI agent by having it imitate the activities or behaviors of experts. This process is referred to as imitation learning, learning from demonstrations, or behavioral cloning. The agent picks up new skills by studying examples given by human professionals rather than just learning by doing. Several benefits come with this strategy.
- Learning that is effective: By using the knowledge of humans or pre-trained agents, imitation learning enables agents to swiftly learn useful policies.
- Data economy: Professional demos offer high-quality training data, obviating the requirement for in-depth investigation and sample gathering.
- Policy initialization: Imitation learning can help reinforce learning get off to a faster convergence and better performance.
- Safety and desired behavior: Professional examples make sure the agent starts off with safe and desirable activities, lowering the possibility of harmful exploration.
- Policy transferability: With little adaptation, practices learned through imitation may frequently be applied to similar activities or settings.
- Human feedback: To incorporate human input into an agent's learning process, human assessors' comments, suggestions, or corrections must be considered. Explicit corrections, rankings, comparisons, and prizes might all fall

under this category. Determining reward functions and optimizing learned policies are two issues that human feedback helps with.

- Reward shaping: By incorporating user feedback into the design process, reward functions that support desired objectives can be created, leading to improved policy optimization.
- Sparse incentives: When receiving frequent or instructive incentives is challenging, human input might fill the gap [28]. Exploration that is guided by human input directs an agent to areas of the state space that are most likely to produce advantageous results.
- Ethics and safety: Human input can enforce moral standards and safety restrictions, ensuring that learnt policies meet desirable criteria. In order to evaluate or compare various trajectories and optimize policies appropriately, agents might learn from human preferences.

a) Real-world applications:

- Robotics: Complex activities may be taught to robots via imitation learning and human input. This ensures that the robots operate safely and effectively in a variety of environments.
- Medical diagnosis or treatment plans can be learned from experts and modified by agents utilizing human feedback to ensure patient safety.
- Autonomous vehicles: Simulated human feedback and imitation learning allow self-driving cars learn how to properly negotiate challenging traffic situations.
- Language comprehension: Human input helps chatbots and virtual assistants better grasp and generate natural language.
- Gaming: By fusing the knowledge of humans and the power of artificial intelligence, imitation learning and human feedback help game-playing methods. Collaboration is a key component of AI development, as shown by the inclusion of imitation learning and human input in DRL.

1.13 Inverse Reinforcement Learning

Deep reinforcement learning (DRL) use the approach of inverse reinforcement learning (IRL) to discover the reward function underlying an observed behavior, enabling agents to comprehend and imitate expert behaviors in practical settings. By estimating rewards from observed behaviors, IRL overcomes the difficulty of designing appropriate reward functions for challenging activities. An overview of DRL's inverse reinforcement learning for real-world scenarios is provided below.

a) Reward inference: IRL is concerned with determining the underlying reward function that drove an expert's behavior. This is accomplished by studying the expert's actions and using them to determine the preferences or objectives that direct their decisions. Expert demonstrations are the series of states and actions that a person or an expert agent takes to complete a task in order to be used in IRL [29]. Learning how the reward function works is

based on these examples. An RL agent's learning process can be shaped by the reward function once it has been inferred. By tuning its policy in accordance with the implied rewards, the agent subsequently tries to imitate the expert's actions.
b) The reverse reinforcement learning process:

- Gather expert demonstrations: Compile expert demonstrations, which are the paths taken by the expert to execute the task in terms of states and actions.
- Reward function inference: The objective of IRL algorithms is to identify a reward function that optimizes the observed expert behavior. The reward function is often tweaked until it matches the observed behavior in this manner. The reward function is then utilized to direct the RL agent's learning process after it has been inferred. Agent behavior resembles that of an expert since it attempts to maximize expected inferred reward.

c) Benefits and applications:

- Learning from limited data: IRL enables agents to learn from a small number of expert demonstrations, making it particularly valuable when gathering data through RL exploration is difficult or costly.
- Imitating difficult to specify complex behaviors: IRL is successful at mimicking complex behaviors that are challenging to define using bespoke reward mechanisms.
- Transferring human expertise: By modeling expert behavior, IRL makes it possible for AI agents in a variety of fields, including robotics, driverless cars, and healthcare, to acquire the knowledge of humans [30]. IRL can assist agents in developing safe and ethical behaviors, making it appropriate for applications that have ethical considerations. A promising method for imparting knowledge from humans and teaching complex behaviors in real-world situations is inverse reinforcement learning.

1.14 Sim-to-Real Transfer

Deep reinforcement learning (DRL) uses a technique called "Sim-to-real transfer" in which an agent is trained in simulated settings before applying newly acquired behaviors or policies to actual situations. Using agents to train in actual environments can be costly, time-consuming, and unsafe, which is why this method is used in DRL to overcome these issues. Sim-to-real transfer strives to close the gap between simulated and actual domains, allowing DRL agents to function well in real, challenging contexts. A summary of Sim-to-real transfer in DRL for real-world settings is provided below. Simulated environments offer a managed and economical method of producing a lot of training data for DRL agents. In order to facilitate effective data collecting and experimentation, these environments resemble real-world situations.

Domain gap: The "domain gap" is the discrepancy between the simulation and the outside reality. Differences in perception, dynamics, sensor noise, lighting conditions, and other elements all have an impact on how well an agent performs while moving from a simulation to reality. Sim-to-real transfer use transfer learning approaches to modify simulation-taught policies so they work better in real-world settings. The agent's policy must be adjusted using data from the real world.

a) Process of Sim-to-real transfer: Using reinforcement learning methods, agents are trained in simulated situations. Using the simulated reward signals as a guide, they learn to optimize their policies.

- Domain adaptation: Using information gathered from the real-world environment, agents are adjusted to take into account the domain gap. By adjusting the policies, real-world dynamics and difficulties are better reflected in the laws.
- Testing for robustness: After domain adaptation, agents are put to the test in the actual world to gauge their effectiveness and robustness. Uncertainties, noise, and unmodeled dynamics may need to be dealt with during this procedure.

b) Benefits and applications:

- Data efficiency: When compared to real-world data collection, simulation environments enable agents to gather a significant volume of training data more quickly and affordably.
- Risk mitigation: Training through simulations lowers the likelihood of harming oneself or causing damage to tools in real-world learning situations, like those involving robotics or autonomous vehicles. Sim-to-real transfer enables quick creation and testing of DRL algorithms prior to their implementation in real-world contexts, which is known as fast prototyping.
- Scaling and generalization: Sim-to-real strategies can assist agents in applying their newly learnt behaviors to a variety of real-world settings and environments [31]. Applications that are essential to safety include teaching agents how to handle unusual and risky situations in a safe environment before they are exposed to them in the real world.

1.15 Conclusion

Although there are many obstacles in the way of using deep reinforcement learning in practical situations, each one also offers a chance for creativity and progress. The demand for intelligent decision-making in complex, unstructured situations is driving a quick evolution of the area. The insights provided here are not only informative but also represent a call to action for researchers, practitioners, and organizations to harness the full potential of DRL, ultimately reshaping industries, improving efficiency, and contributing to the betterment of society. This paper serves as a cornerstone for those who seek to explore the frontiers of DRL and participate in the ongoing transformation of our technological landscape.

Researchers and practitioners can realize the full potential of DRL and develop solutions that help society across a wide range of areas by addressing the issues of sample efficiency, safety, generalization, and ethics.

References

1. Srinath, H., Sharma, A.K.V., Akhil, M.R., Reinforcement Learning in Real-World Scenarios: Challenges, Applications, and Future Directions. *Int. J. Res. Eng. Sci. Manage.*, 6, 7, 40–45, 2023.
2. Li, Y., Reinforcement learning in practice: Opportunities and challenges, p. 2, 2022, arXiv preprint arXiv:2202.11296.
3. AlMahamid, F. and Grolinger, K., Reinforcement learning algorithms: An overview and classification, in: *2021 IEEE Canadian Conference on Electrical and Computer Engineering (CCECE)*, IEEE, pp. 1–7, 2021, September.
4. Sivamayil, K., Rajasekar, E., Aljafari, B., Nikolovski, S., Vairavasundaram, S., Vairavasundaram, I., A systematic study on reinforcement learning based applications. *Energies*, 16, 3, 1512, 2023.
5. Chen, X., Yao, L., McAuley, J., Zhou, G., Wang, X., Deep reinforcement learning in recommender systems: A survey and new perspectives. *Knowl.-Based Syst.*, 264, 110335, 2023.
6. Lin, Y., Liu, Y., Lin, F., Zou, L., Wu, P., Zeng, W., Miao, C., A survey on reinforcement learning for recommender systems. *IEEE Trans. Neural Netw. Learn. Syst.*, 2023.
7. Kiran, B.R., Sobh, I., Talpaert, V., Mannion, P., Al Sallab, A.A., Yogamani, S., Pérez, P., Deep reinforcement learning for autonomous driving: A survey. *IEEE Trans. Intell. Transp. Syst.*, 23, 6, 4909–4926, 2021.
8. Torrado, R.R., Bontrager, P., Togelius, J., Liu, J., Perez-Liebana, D., Deep reinforcement learning for general video game ai, in: *2018 IEEE Conference on Computational Intelligence and Games (CIG)*, IEEE, pp. 1–8, 2018, August.
9. Dulac-Arnold, G., Levine, N., Mankowitz, D.J., Li, J., Paduraru, C., Gowal, S., Hester, T., Challenges of real-world reinforcement learning: definitions, benchmarks and analysis. *Mach. Learn.*, 110, 9, 2419–2468, 2021.
10. Ding, D., Wei, C.Y., Zhang, K., Ribeiro, A., Last-Iterate Convergent Policy Gradient Primal-Dual Methods for Constrained MDPs, p. 7, 2023, arXiv preprint arXiv:2306.11700.
11. Amin, S., Gomrokchi, M., Satija, H., van Hoof, H., Precup, D., A survey of exploration methods in reinforcement learning, p. 7, 2021, arXiv preprint arXiv:2109.00157.
12. Dabney, W., Rowland, M., Bellemare, M., Munos, R., Distributional reinforcement learning with quantile regression, in: *Proceedings of the AAAI Conference on Artificial Intelligence*, vol. 32, no. 1, 2018, April.
13. Belogolovsky, S., Korsunsky, P., Mannor, S., Tessler, C., Zahavy, T., Inverse reinforcement learning in contextual MDPs. *Mach. Learn.*, 110, 9, 2295–2334, 2021.
14. Bengio, Y., Lecun, Y., Hinton, G., Deep learning for AI. *Commun. ACM*, 64, 7, 58–65, 2021.
15. Bertsekas, D., *Reinforcement learning and optimal control*, Athena Scientific, U.S.A, 2019.
16. Ie, E., Jain, V., Wang, J., Narvekar, S., Agarwal, R., Wu, R., Boutilier, C., Reinforcement learning for slate-based recommender systems: A tractable decomposition and practical methodology, 2019, arXiv preprint arXiv:1905.12767.
17. Chen, X., Qu, G., Tang, Y., Low, S., Li, N., Reinforcement learning for decision-making and control in power systems: Tutorial, review, and vision, 2021, arXiv.
18. Co-Reyes, J.D., Miao, Y., Peng, D., Real, E., Levine, S., Le, Q.V., Faust, A., Evolving reinforcement learning algorithms, 2021, arXiv preprint arXiv:2101.03958.

19. Abbas, A.N., Chasparis, G.C., and Kelleher, J.D. Hierarchical framework for interpretable and specialized deep reinforcement learning-based predictive maintenance. *Data Knowl. Eng.*, 149, 102240, 2024.
20. Garcıa, J. and Fernández, F., A comprehensive survey on safe reinforcement learning. *J. Mach. Learn. Res.*, 16, 1, 1437–1480, 2015.
21. Gottesman, O., Johansson, F., Komorowski, M., Faisal, A., Sontag, D., Doshi-Velez, F., Celi, L.A., Guidelines for reinforcement learning in healthcare. *Nat. Med.*, 25, 1, 16–18, 2019.
22. Haydari, A. and Yılmaz, Y., Deep reinforcement learning for intelligent transportation systems: A survey. *IEEE Trans. Intell. Transp. Syst.*, 23, 1, 11–32, 2020.
23. Henderson, P., Islam, R., Bachman, P., Pineau, J., Precup, D., Meger, D., Deep reinforcement learning that matters, in: *Proceedings of the AAAI conference on artificial intelligence*, vol. 32, no.1, 2018, April.
24. Henry, R. and Ernst, D., Gym-ANM: Reinforcement learning environments for active network management tasks in electricity distribution systems. *Energy AI*, 5, 100092, 2021.
25. Heuillet, A., Couthouis, F., Díaz-Rodríguez, N., Explainability in deep reinforcement learning. *Knowl.-Based Syst.*, 214, 106685, 2021.
26. Jaderberg, M., Mnih, V., Czarnecki, W.M., Schaul, T., Leibo, J.Z., Silver, D., Kavukcuoglu, K., Reinforcement learning with unsupervised auxiliary tasks, 2016, arXiv preprint arXiv:1611.05397.
27. Jiang, N. and Li, L., Doubly robust off-policy value evaluation for reinforcement learning, in: *International Conference on Machine Learning*, PMLR, pp. 652–661, 2016, June.
28. Kirk, R., Zhang, A., Grefenstette, E., Rocktäschel, T., A survey of generalisation in deep reinforcement learning, 2021, arXiv e-prints, arXiv-2111.
29. Kober, J., Bagnell, J.A., Peters, J., Reinforcement learning in robotics: A survey. *Int. J. Rob. Res.*, 32, 11, 1238–1274, 2013.
30. Xu, H., Ho, C.Y., Abdelmoniem, A.M., Dutta, A., Bergou, E.H., Karatsenidis, K., Kalnis, P., GRACE: A compressed communication framework for distributed machine learning, in: *2021 IEEE 41st international conference on distributed computing systems (ICDCS)*, IEEE, pp. 561–572, 2021, July.
31. Oymak, S., Honors and Awards, University of California, Berkeley, 2015, 2014.

2

Deep Reinforcement Learning: A Key to Unlocking the Potential of Robotics and Autonomous Systems

Saksham and Chhavi Rana*

Department of Computer Science and Engineering, University Institute of Engineering and Technology, Maharshi Dayanand University, Rohtak, Haryana, India

Abstract

Deep reinforcement learning (DRL) has emerged as a transformative paradigm in the field of robotics and autonomous systems, enabling machines to learn complex tasks through interaction with their environments. This chapter provides a comprehensive overview of the current state of research at the intersection of DRL and robotics. We systematically analyze and synthesize the literature to highlight key developments, methodologies, applications, and challenges within this dynamic field. This chapter is organized into thematic sections, covering fundamental concepts of DRL and its applications in robot control, navigation, object manipulation, and autonomous vehicles. It explores various general methods for investigation, evaluating their effectiveness in addressing the unique challenges posed by real-world robotics applications. Through the lens of case studies and real-world applications, the remarkable capabilities of DRL in enabling robots to perform tasks with unprecedented autonomy and adaptability is shown. The intricacies of hardware and software configurations employed in these applications is discussed, shedding light on the practical considerations that impact DRL's deployment in the field. Furthermore, this chapter also covers the present challenges and restrictions, such as sample efficiency, safety concerns, and scalability, that continue to challenge the widespread adoption of DRL in robotics. While delving into these challenges, this chapter also outline potential future research directions and the evolving landscape of evaluation metrics for assessing DRL algorithms in robotic contexts. In conclusion, this chapter serves as a valuable resource for researchers, practitioners, and enthusiasts interested in the fusion of deep reinforcement learning and robotics. It synthesizes the current knowledge, highlights the impressive progress achieved, and outlines the exciting avenues for further exploration, ultimately contributing to the advancement of robotics and autonomous systems in an era defined by machine learning and artificial intelligence.

Keywords: Deep reinforcement learning (DRL), robotics, navigation, object manipulation, safety concerns, scalability, research directions, evaluation metrics

**Corresponding author*: chhavi.uiet@mdurohtak.ac.in

Shubham Mahajan, Pethuru Raj and Amit Kant Pandit (eds.) Deep Reinforcement Learning and Its Industrial Use Cases: AI for Real-World Applications, (29–50) © 2024 Scrivener Publishing LLC

2.1 Introduction

The fusion of deep reinforcement learning (DRL) and robotics represents a groundbreaking synergy at the forefront of modern technological advancement. As articulated by Sutton and Barto (2018), reinforcement learning constitutes a computational approach to understanding and automating the acquisition of intelligent behaviors in autonomous agents. In the realm of robotics and autonomous systems, this approach has materialized into an exciting convergence that promises to redefine the boundaries of what machines can achieve.

Reinforcement learning, as a subfield of machine learning, emphasizes the development of agents capable of learning optimal behaviors through interaction with their environments. It mirrors the process of human learning, where actions are reinforced by rewards or penalties, enabling the agent to discern favorable strategies. In traditional robotics, this has translated into the meticulous design and programming of explicit rules and models to govern robotic behavior. However, the advent of DRL introduces a profound departure from this conventional approach.

DRL introduces the power of neural networks and deep learning into the realm of reinforcement learning. This transformational leap in computational capacity, as seen in the works of Mnih *et al.* (2015) and Silver *et al.* (2016), enables robots and autonomous systems to grapple with far more complex and dynamic environments. The essence of DRL lies in its ability to process high-dimensional sensory inputs, learn abstract representations, and discover optimal actions through iterative trials and errors. This self-improving mechanism has proven instrumental in training robots to perform a myriad of tasks, from autonomous navigation and obstacle avoidance to fine-grained object manipulation.

In the context of robotics and autonomous systems, DRL heralds a new era of adaptability and autonomy. Robots no longer rely solely on pre-programmed routines; they are now capable of learning from their experiences, adapting to uncertainties, and evolving their strategies to tackle novel challenges. This paradigm shift holds immense promise across industries, from revolutionizing manufacturing processes and healthcare assistance to redefining the landscape of self-driving vehicles.

This chapter embarks on a comprehensive journey to unravel the evolving landscape of DRL in robotics and autonomous systems. Through a rigorous analysis of existing literature, it aims to elucidate the key developments, general methods for investigation, applications, and challenges that characterize this dynamic intersection. By doing this, it hopes to give researchers, professionals, and fans a comprehensive viewpoint as they work to harness DRL's transformative potential and influence the development of robotics and autonomous systems in the future.

2.1.1 Significance of DRL Field

The importance and relevance of the topic of "deep reinforcement learning (DRL) in robotics and autonomous systems" are multifaceted and extend across various domains of science, technology, and industry. Figure 2.1 shows the importance and relevance of the topic of "deep reinforcement learning (DRL) in robotics and autonomous systems" across various domains of science, technology, and industry. Understanding why this topic is significant is essential for appreciating its potential impact. Here are key points explaining its importance and relevance.

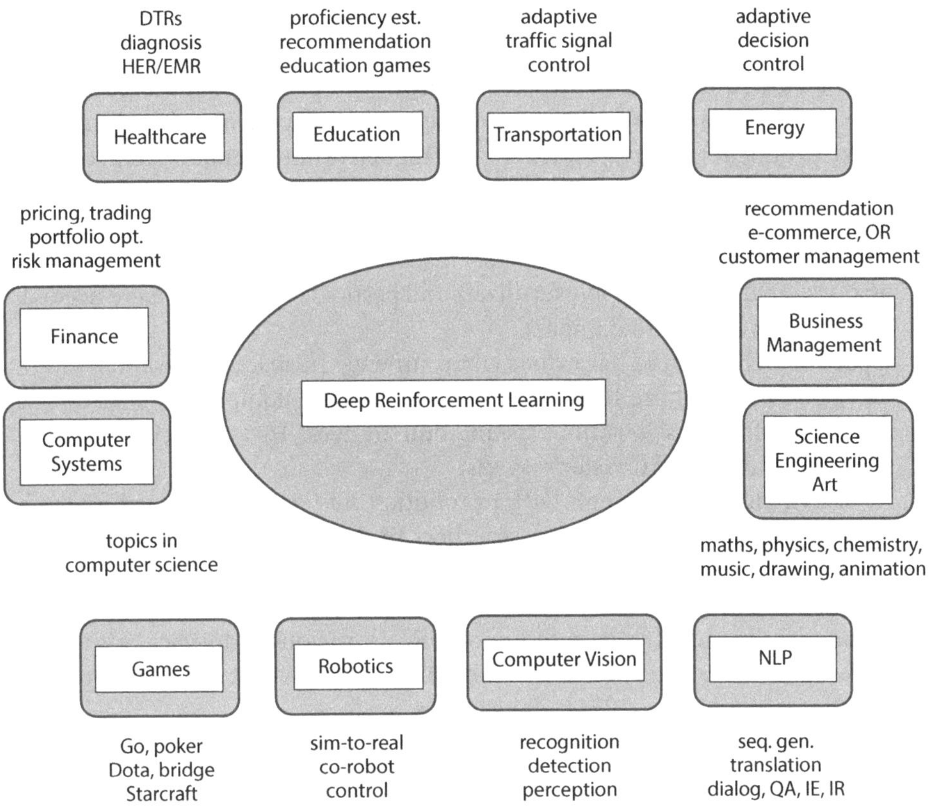

Figure 2.1 Yuxi Li, deep reinforcement learning, arXiv, 2018.

1. Advancing robotics technology: DRL offers the potential to significantly advance the capabilities of robots and autonomous systems. Robots that can learn and adapt in real-time can perform tasks more efficiently, effectively, and with greater autonomy. This is vital for industries such as manufacturing, healthcare, agriculture, and logistics, where robots are increasingly used.
2. Autonomous vehicles: DRL plays a pivotal role in the development of autonomous vehicles, including self-driving cars, drones, and unmanned aerial vehicles. These technologies have the potential to revolutionize transportation, reducing accidents and improving traffic management, thus leading to safer and more efficient mobility.
3. Reducing human intervention: By enabling robots and autonomous systems to learn and make decisions independently, DRL reduces the need for constant human supervision and manual programming. This is particularly relevant in situations where it may be impractical or hazardous for humans to be directly involved.
4. Complex and dynamic environments: Many real-world environments are highly complex and dynamic, making it challenging to pre-program robots with fixed instructions. DRL equips robots with the ability to adapt to

changing conditions and unforeseen obstacles, making them more versatile and capable.

5. Precision and efficiency: DRL can improve the precision and efficiency of tasks performed by robots—for example, in manufacturing, robots can optimize production processes, leading to higher product quality and reduced waste.
6. Resource efficiency: DRL can help conserve resources by optimizing resource usage in various applications—for instance, in agriculture, autonomous drones can use DRL to apply fertilizers and pesticides precisely where needed, reducing environmental impact.
7. Space exploration and hazardous environments: In space exploration and in environments that are hazardous to humans, DRL-equipped robots can be sent to perform tasks without risking human lives. This is crucial for scientific exploration and disaster response.
8. Research and development: DRL in robotics is a fertile ground for research and innovation. It challenges researchers to develop new algorithms, hardware, and software solutions, leading to advancements not only in robotics but also in artificial intelligence and machine learning.
9. Economic impact: The integration of DRL in various industries can have a substantial economic impact by improving productivity, reducing costs, and creating new opportunities for businesses.
10. Sustainability: DRL can contribute to sustainability efforts by optimizing resource usage, reducing energy consumption, and improving the efficiency of systems like smart grids and energy management.

2.1.2 Transformative Advantages of DRL Field

Deep reinforcement learning (DRL) is a paradigm-shifting technology in the realm of robotics and autonomous systems, with numerous specific advantages. DRL-enabled robots succeed in real-world scenarios by learning from their interactions with their surroundings, which eliminates the need for explicit programming, which is especially beneficial in complex and dynamic contexts.

DRL's capacity to manage non-stationary conditions, which are common in real-world circumstances, is one of its unique features. DRL agents are extremely adaptable and versatile, altering their behavior to changing environmental conditions. In disaster response scenarios, for example, a robot can prioritize safety at first and then seamlessly adjust its focus when survivors are located, demonstrating the crucial significance of dynamic goal management in effective autonomous decision-making.

DRL systems often work in complex and dynamic environments where balancing several goals is necessary. Consider a self-driving automobile as an example. The AI agent has to optimize various aspects, like speedy arrival at the destination, passenger safety, and adherence to traffic laws. There may be moments when these goals clash since aggressive driving can result in faster arrival times at the expense of higher safety concerns.

In conclusion, DRL is revolutionizing the robotics and autonomous systems sectors with its ability to manage several goals at once and adjust to changing priorities. It increases these systems' efficiency and adaptability in a wide range of applications by enabling them to handle goal conflicts in real-world circumstances with effectiveness.

2.2 Fields of Investigation

Deep reinforcement learning in robotics and autonomous systems should clearly outline the goals and aims of research. These fields of investigation guide the entire review process and help structure it to study. Here are some sample fields of investigation for this chapter:

1. To provide an overview: To systematically review the existing literature on the integration of deep reinforcement learning (DRL) techniques in the field of robotics and autonomous systems, offering a comprehensive overview of the current state of research.
2. To analyze methodologies: To assess and analyze the various methodologies, algorithms, and techniques employed in DRL for robotics, identifying common practices and innovative approaches.
3. To examine applications: To investigate the diverse applications of DRL in robotics, including but not limited to robot control, navigation, object manipulation, and autonomous vehicles, and to summarize the key findings and challenges within each application domain.
4. To evaluate case studies: To scrutinize real-world case studies and projects that showcase the practical implementation and effectiveness of DRL in robotic systems, highlighting successful deployments and lessons learned.
5. To address challenges: To identify and discuss the challenges and limitations of DRL in robotics and autonomous systems, including issues related to sample efficiency, safety, scalability, and real-world deployment.
6. To suggest future directions: To propose potential future research directions and emerging trends in the field, emphasizing areas where further investigation and innovation are needed.
7. To explore evaluation metrics: To explore the evaluation metrics and benchmarks commonly used to assess the performance of DRL algorithms in robotic applications, discussing their relevance and limitations.
8. To offer insights: To provide insights and recommendations for researchers, practitioners, and stakeholders interested in leveraging DRL for enhancing the capabilities of robotic systems.
9. To contribute to knowledge: To contribute to the academic and practical understanding of the role of DRL in advancing robotics and autonomous systems, consolidating existing knowledge, and identifying gaps for future research.
10. To facilitate decision-making: To offer a resource that aids decision-makers in industries considering the adoption of DRL-based solutions in their robotic systems by providing a comprehensive summary of the state-of-the-art and best practices.

2.2.1 General Methods for Investigation

1. Data sources
   - Key finding: A wide range of data sources enhanced our literature search, including scholarly journals and conferences in robotics and machine learning, as well as well-known repositories like PubMed, IEEE Xplore, and Google Scholar.
   - For instance, the abundance of research findings available on IEEE Xplore was a valuable resource for the investigation of current developments in deep reinforcement learning (DRL) in robotics and autonomous systems, with a particular emphasis on the previous 5 years. Access to recent articles on DRL's breakthroughs and uses in autonomous systems was made possible *via* this repository.
2. Search strategy
   - Key finding: Precise and pertinent results were obtained by carefully crafting the search strategy. It involves the deliberate use of certain keywords, such as "robotics," "autonomous systems," and "deep reinforcement learning," in addition to the deliberate application of Boolean operators to refine search searches.
   - Example: By including both "deep reinforcement learning" and "robotics" in the search terms, it was possible to directly target academic articles that discussed the use of DRL to robotics.
3. Inclusion and exclusion criteria
   - Key finding: Strict inclusion and exclusion standards were set in order to preserve the calibre and applicability of the study. The aforementioned criteria comprised elements such as the reliability of the published sources, emphasizing peer-reviewed journals and respectable conference proceedings, and the incorporation of recent articles from the previous 5 years.
   - For instance, papers that were not accepted for publication in conferences or peer-reviewed journals were purposefully left out in order to guarantee the research's intellectual rigor.
4. Paper selection
   - Key finding: Titles, abstracts, and full-text publications were thoroughly examined as the first step in the paper selection process, which followed the methodical standards of PRISMA (Preferred Reporting Items for methodical Reviews and Meta-Analyses), which places an emphasis on transparency and rigor in research synthesis. A fraction of the papers that were included were the consequence of a very selective screening procedure.
   - Example: The selection of a subset of the pool of initially identified papers highlights the process of careful and deliberate paper selection, which is in line with PRISMA's features for systematic reviews and meta-analyses. This methodology guarantees the transparency and resilience of the research synthesis.

5. Data extraction
   - Key finding: Every important detail from every chosen paper was extracted using a precise approach. These included information about titles, authors, publishing years, abstract content, methodology, main conclusions, and pertinent statistical data, among other things. Ref management software was used to efficiently organize the gathered data.
   - Example: A thorough analysis and synthesis was made possible by the extraction of important insights from each work, including publishing specifics and noteworthy discoveries. This process was further aided by effective data management with the use of reference management software.
6. Quality assessment
   - Key finding: The approach included a thorough quality evaluation procedure that examined the robustness of the research methodology and its alignment with the study objectives.
   - Example: High-reliability studies were identified using quality evaluation techniques; studies that did not have appropriate experimental controls were given less weight throughout the analysis.
7. Data synthesis
   - Key finding: In order to facilitate the detection of emerging trends and patterns, the articles were systematically categorized into topic divisions as part of the data synthesis process. To obtain deep insights, both qualitative and quantitative analyses including content analysis were carried out.
   - Example: Papers were arranged systematically by application fields, including manufacturing, healthcare, and autonomous cars, thanks to thematic categorization. This paradigm facilitated thorough analysis and offered clarity.
8. Review protocol
   - Key finding: The chapter closely adhered to accepted standards like PRISMA, Cochrane as detailed in the study. The particular circumstances, therefore, did not call for the official registration of a review procedure.
9. Ethical considerations
   - Key finding: Strict attention to ethical norms, including acquiring authorization for the use of proprietary data and adhering to copyright restrictions during data collection and analysis, was important due to the pivotal role that ethical concerns played in the research process.
   - Example: To ensure ethical research practices, in cases involving proprietary data, stringent measures were taken to get the required permits and steadfast adherence to data usage agreements was maintained.
10. Limitations
    - Key finding: A crucial component of the research was acknowledging potential limitations. These restrictions included the potential to leave out pertinent but unindexed material as well as the subjectivity that comes with the information selection process itself.

- Example: Despite the research's extensive use of search sources, it may have been limited by the omission of unpublished studies or gray literature, which could have affected the review's thoroughness.

2.3 Background

The convergence of deep reinforcement learning (DRL) and robotics represents a significant milestone in the evolution of autonomous systems. DRL, as a subfield of machine learning, has its roots in the fundamental principles of reinforcement learning, a concept that traces its origins to the foundational work of Sutton and Barto (1998). Reinforcement learning, in essence, is an approach that enables agents to learn by interacting with their environments, akin to how humans acquire knowledge through trial and error.

Sutton and Barto's reinforcement learning framework hinges on the concept of an agent that interacts with an environment, taking actions to maximize cumulative rewards while navigating through states. The agent learns optimal policies through a process of exploration and exploitation. This paradigm has been influential in various domains, including game-playing agents and recommendation systems, but its integration into robotics and autonomous systems heralds an era of adaptive, intelligent machines.

The hallmark of DRL is the infusion of deep neural networks into the reinforcement learning framework. This integration significantly amplifies the capacity of autonomous systems to handle high-dimensional sensory data and learn abstract representations. The pivotal work by Mnih *et al.* (2015) demonstrated the remarkable potential of deep Q-networks (DQNs) in mastering complex tasks, particularly in the realm of Atari game-playing, through the assimilation of raw pixel inputs.

The application of DRL in robotics and autonomous systems has, since then, burgeoned into a fertile ground for exploration and innovation. Robots equipped with DRL algorithms have showcased remarkable capabilities, such as autonomous navigation through dynamic environments, agile object manipulation, and even participation in competitions like the DARPA Robotics Challenge (DRC). These advances underline the adaptability and versatility that DRL brings to the field, empowering robots to learn and adapt to diverse tasks and unforeseen challenges.

While the promise of DRL in robotics is undeniable, it is not without its challenges. Scalability, sample efficiency, and safety remain critical issues in deploying DRL in real-world applications, as observed by researchers like Hessel *et al.* (2018). Figure 2.2 shows procedures for various learning approaches in deep reinforcement learning (DRL). Consequently, the field is continually evolving, with researchers exploring novel algorithms, benchmark environments, and evaluation metrics to overcome these impediments and maximize the potential of DRL in robotics.

This chapter endeavours to encapsulate the transformative journey of DRL in robotics and autonomous systems. Through a comprehensive analysis of existing literature, this chapter aim to shed light on the key developments, general methods for investigation, applications, and challenges within this dynamic intersection, contributing to the collective understanding of how DRL is reshaping the landscape of autonomous systems.

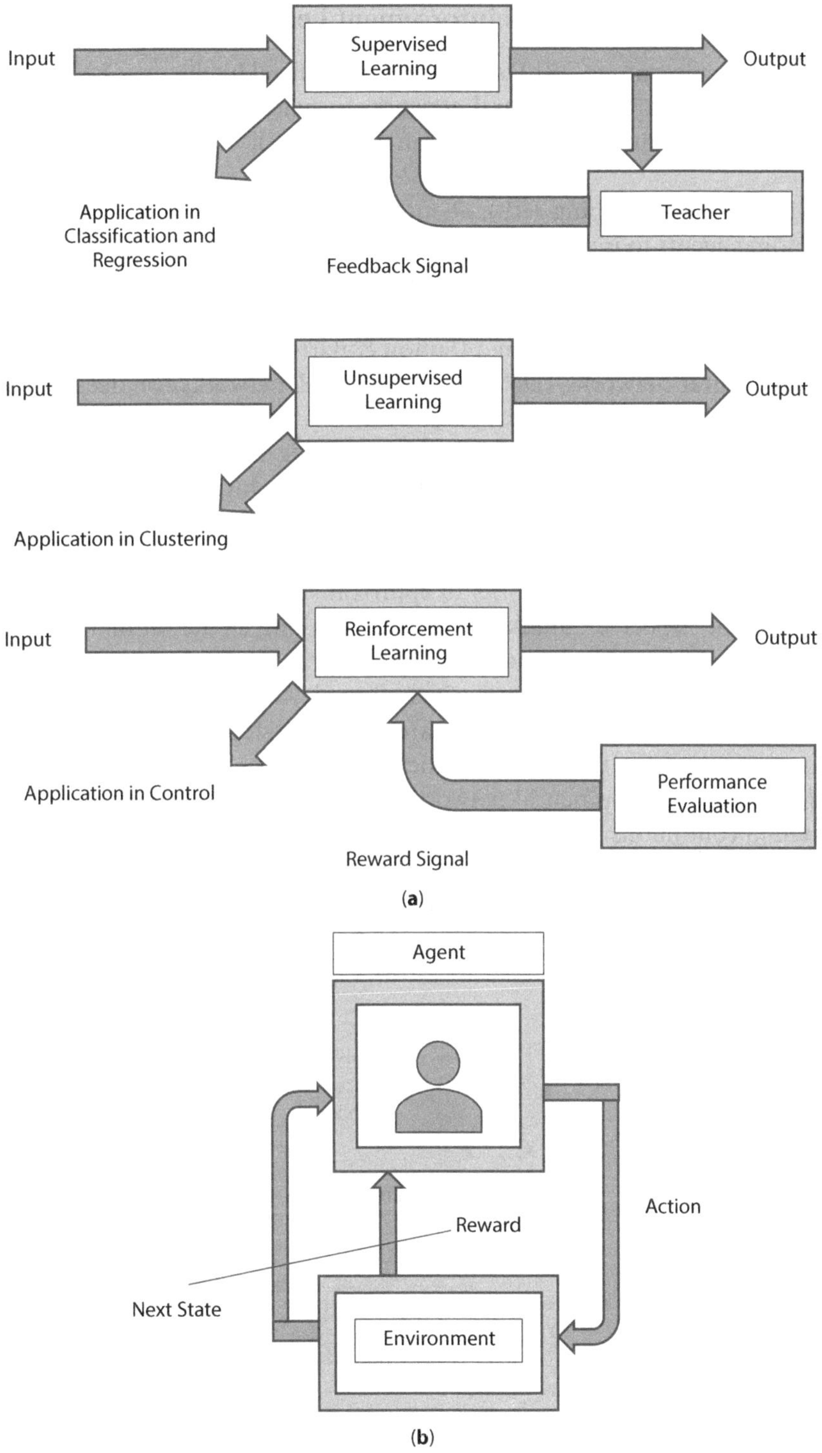

Figure 2.2 Procedures for various learning approaches.

2.3.1 Fundamentals of Deep Reinforcement Learning (DRL)

Deep reinforcement learning (DRL) is a subfield of artificial intelligence (AI) that combines reinforcement learning and deep neural networks to enable agents to learn and make decisions in complex environments (Arulkumaran *et al.*, 2017).

1. Reinforcement learning (RL): Reinforcement learning is a framework for training agents to interact with an environment and learn optimal behaviors through trial and error. It is based on the concept of reward maximization, where agents aim to maximize a cumulative reward signal by taking actions that lead to desirable outcomes (Sutton & Barto, 2018).
2. Deep neural networks (DNNs): Deep neural networks, or deep learning, involve the use of artificial neural networks with multiple hidden layers. DNNs are capable of automatically extracting hierarchical features from raw data, making them well-suited for processing high-dimensional sensory inputs like images and sensor data (Goodfellow *et al.*, 2016).
3. Deep Q-networks (DQNs): Deep Q-networks are a class of DRL algorithms that use deep neural networks to approximate the Q-function, which represents the expected cumulative reward of taking an action in a given state and following a particular policy (Mnih *et al.*, 2015). DQNs have been instrumental in solving complex tasks, including playing video games.
4. Policy networks: Policy networks are another category of DRL algorithms that directly learn the optimal policy, which maps states to actions. These algorithms use deep neural networks to approximate the policy and can be used in continuous action spaces (Silver *et al.*, 2014).
5. Value iteration: Value iteration is a core concept in DRL that involves iteratively estimating the value of each state in the environment. This process helps the agent assess the desirability of different states and guides its decision-making (Sutton & Barto, 2018). Figure 2.3 shows the process action

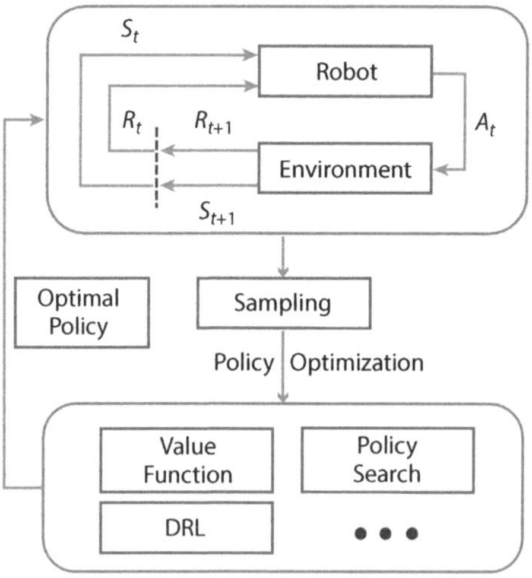

Figure 2.3 Policy networks in DRL algorithms.

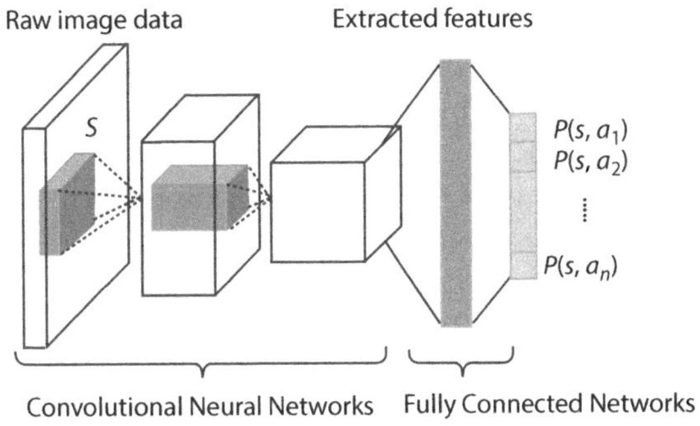

Figure 2.4 Convolutional neural network for image based DRL.

of optimal policy and involves iteratively estimating the value of each state in the environment. Figure 2.4 shows convolutional neural network (CNN) which is a type of artificial neural network used primarily for image recognition and processing, due to its ability to recognize patterns in images.

6. Exploration vs. exploitation: Exploration and exploitation are essential trade-offs in DRL. Agents must explore new actions and states to discover better policies while also exploiting their current knowledge to maximize rewards (Sutton & Barto, 2018).
7. Experience replay: Experience replay is a technique used in DRL to improve data efficiency and stabilize training. It involves storing past experiences in a replay buffer and randomly sampling mini-batches for training to break temporal correlations in the data (Mnih *et al.*, 2015).
8. Policy gradients: Policy gradient methods are used to directly optimize the policy of an agent by adjusting its parameters through gradient ascent (Sutton *et al.*, 2000). These methods are suitable for problems with continuous action spaces.
9. Actor–critic architectures: Actor–critic architectures combine value estimation (the critic) with policy optimization (the actor) in a single framework. The critic evaluates actions, and the actor adjusts the policy based on the critic's feedback (Konda & Tsitsiklis, 2000).
10. Reward shaping: Reward shaping is a technique used to design custom reward functions that guide the learning process. It can help accelerate training by providing informative feedback to the agent (Ng *et al.*, 1999).

2.4 Deep Reinforcement Learning (DRL) in Robot Control

Deep reinforcement learning (DRL) has revolutionized robot control by enabling machines to learn complex control policies from high-dimensional sensor data and make decisions in real-time. This section explores how DRL is applied to robot control across various domains.

1. Robotic arm control: DRL has significantly advanced robotic arm control. Robots equipped with DRL algorithms can learn precise control strategies for tasks like pick-and-place, assembly, and manipulation (Andrychowicz *et al.*, 2020). These algorithms enable robots to adapt to varying object shapes and sizes, making them versatile for manufacturing and logistics.

2. Autonomous navigation: DRL is pivotal in autonomous navigation for mobile robots and drones. Self-driving cars use DRL to interpret sensor data, plan trajectories, and navigate complex urban environments (Zhang *et al.*, 2020). Aerial robots employ DRL to fly through cluttered environments, adapting to obstacles and dynamic scenarios (Kasemsuppakorn *et al.*, 2020). This technology is crucial for applications like delivery, search and rescue, and exploration.
3. Legged robot locomotion: DRL plays a key role in legged robot locomotion, enabling robots to traverse challenging terrains. Quadrupeds and bipeds can learn to walk, run, and jump through DRL (Hwangbo *et al.*, 2019). These robots adapt their gaits to different environments and maintain stability on uneven surfaces, making them suitable for tasks in agriculture, disaster response, and planetary exploration.
4. Humanoid robots: Humanoid robots leverage DRL for natural and agile movements. DRL algorithms teach robots tasks like dancing, acrobatics, and sports (Cruz *et al.*, 2018). This technology has applications in entertainment, rehabilitation, and human–robot interaction.
5. Swarm robotics: DRL is also applied in swarm robotics, where multiple robots cooperate to accomplish tasks. DRL algorithms enable swarm robots to coordinate their actions, such as exploring unknown environments (Pereira *et al.*, 2020). This technology has implications for environmental monitoring and surveillance.
6. Underwater robotics: Underwater robots use DRL for navigation, localization, and object manipulation in challenging underwater environments (Pereira *et al.*, 2020). They can explore deep-sea regions, inspect underwater infrastructure, and assist in marine research.
7. Exoskeletons and prosthetics: DRL empowers wearable robotics like exoskeletons and prosthetic limbs to provide natural and adaptive assistance to individuals with mobility impairments (Gopinath *et al.*, 2020). These devices adapt to the user's movements and intentions, offering improved mobility and quality of life.
8. Industrial automation: In industrial automation, robots with DRL capabilities optimize manufacturing processes, reduce downtime, and enhance production efficiency (Zhang *et al.*, 2018). These robots can adapt to changing production demands and execute tasks with precision.

2.4.1 Navigation and Localization

Navigation and localization are fundamental challenges in the field of robotics and autonomous systems, and deep reinforcement learning (DRL) has emerged as a powerful approach for addressing these challenges (Arulkumaran *et al.*, 2017).

Navigation

DRL-based navigation enables robots and autonomous vehicles to plan and execute complex trajectories in dynamic environments. Deep Q-networks (DQNs), proximal policy optimization (PPO), and other DRL algorithms have been applied to autonomous navigation tasks.

For instance, self-driving cars employ DRL to navigate urban streets safely and efficiently. Companies like Waymo and Tesla use DRL techniques to interpret sensor data, make real-time decisions, and control the vehicle's movements (Zhang *et al.*, 2020). These systems can handle complex scenarios, such as merging onto highways and navigating through dense traffic.

Aerial robots, including drones, utilize DRL to autonomously fly through cluttered and GPS-denied environments. Researchers have demonstrated DRL-based navigation in scenarios like forest exploration and search and rescue missions (Kasemsuppakorn *et al.*, 2020). DRL enables these robots to adapt to dynamic obstacles and environmental changes.

Localization

DRL also plays a vital role in localization, allowing robots to estimate their position and orientation accurately. Simultaneous localization and mapping (SLAM) is a common problem in robotics, and DRL has been applied to enhance SLAM algorithms.

Deep neural networks, including convolutional neural networks (CNNs), are used to process sensor data such as images and LiDAR scans for robust localization. DRL agents can integrate sensory information with historical data to estimate the robot's pose accurately (Gupta *et al.*, 2017).

In underwater robotics, autonomous submarines use DRL to navigate and localize themselves in challenging GPS-denied environments. DRL techniques enable these vehicles to create maps of underwater terrain and accurately determine their position relative to

Table 2.1 Comparison of navigation and localization.

Aspect	Navigation	Localization
Techniques	Robots can plan and carry out complex paths in dynamic environments while adjusting to changing circumstances thanks to DRL approaches.	Accurate position and orientation estimation is aided by DRL.
DRL algorithms	Examples include proximal policy optimization (PPO), Deep Q-networks (DQNs), and others.	For reliable localization, deep neural networks (such as CNNs) handle sensor data.
Applications	Autonomous aerial robots (such as drones) employ DRL for flying in crowded and GPS-denied situations, such as forest exploration and search and rescue operations. Self-driving automobiles use DRL for safe urban navigation, including highway merging.	DRL is used by autonomous submarines to map underwater terrain and for landmark-based location. Mobile robots in indoor environments such as warehouses and hospitals employ DRL for precise indoor localization using optical and sensor data.

known landmarks (Pereira *et al.*, 2020). Table 2.1 shows the comparison of Navigation and Localization based on various aspects.

Additionally, DRL has applications in indoor localization scenarios. Mobile robots equipped with DRL capabilities can navigate indoor spaces, such as warehouses or hospitals, by leveraging visual and sensor data for precise localization (Dosovitskiy *et al.*, 2017).

In summary, DRL has transformed the fields of navigation and localization in robotics and autonomous systems. It enables robots and vehicles to navigate complex environments autonomously, adapt to changing conditions, and accurately estimate their positions, making it a critical technology for a wide range of applications.

2.4.2 Object Manipulation

Object manipulation is a fundamental task in robotics, and deep reinforcement learning (DRL) has emerged as a powerful approach for enhancing the dexterity and precision of robotic manipulation systems (Andrychowicz *et al.*, 2020).

DRL-powered robots have demonstrated remarkable capabilities in grasping and manipulating objects in various applications—for instance, in a study by OpenAI, the robotic hand system called Dactyl used DRL to achieve human-level dexterity in manipulating a Rubik's Cube (OpenAI, 2018). By training in a simulated environment and then fine-tuning in the real world, the robot learned complex manipulation skills that involve both perception and control. Table 2.2 shows few applications of DRL with their exampled and respective references.

Table 2.2 Few applications of DRL.

Description	Example	Reference
DRL is essential for improving the dexterity and accuracy of robotic object manipulation systems.	Using a Rubik's cube, Dactyl demonstrates dexterity comparable to that of a human.	Andrychowicz *et al.*, 2020
Robots may now carry out industrial automation tasks including pick-and-place operations thanks to DRL algorithms.	Robots can change how objects are shaped, sized, and posed.	Zhang *et al.*, 2018
Surgical robots with DRL-based precision improve minimally invasive surgery.	Using the da Vinci Surgical System, surgeons have better control.	Marescaux *et al.*, 2001
Exoskeletons driven by DRL and prosthetic limbs help people with limited movement manipulate objects.	Wearable technology responds to user intentions and movements	Gopinath *et al.*, 2020
Autonomous vehicles with DRLs experiment with object manipulation for last-mile logistics operations.	Vehicles move objects to make delivery more effectively.	Pan *et al.*, 2020

DRL algorithms have also been employed in industrial automation. Robots equipped with DRL are capable of performing tasks such as pick-and-place operations with high accuracy and efficiency (Zhang *et al.*, 2018). These robots can adapt to varying object shapes, sizes, and poses, making them suitable for versatile manufacturing processes.

In healthcare, DRL-based surgical robots are revolutionizing minimally invasive surgery. The da Vinci Surgical System, which integrates DRL techniques, provides surgeons with enhanced control and precision during procedures (Marescaux *et al.*, 2001). Surgeons can perform intricate tasks with reduced hand tremors and improved accuracy, leading to better patient outcomes.

Furthermore, DRL-powered exoskeletons and prosthetic limbs offer significant advancements in assisting individuals with impaired mobility (Gopinath *et al.*, 2020). These wearable devices leverage DRL to adapt to the user's movements and intentions, allowing for more natural and intuitive control.

DRL's impact on object manipulation extends beyond traditional robotic platforms. Autonomous vehicles equipped with robotic arms and DRL capabilities are being explored for package delivery and other last-mile logistics tasks (Pan *et al.*, 2020). These vehicles can manipulate objects, such as packages, to complete deliveries efficiently.

In summary, DRL has substantially improved the capabilities of robotic systems in the domain of object manipulation. From fine-grained tasks like solving Rubik's cubes to critical applications in surgery and logistics, DRL-powered robots are showcasing their ability to adapt and excel in complex manipulation scenarios.

2.5 Applications and Case Studies

1. Autonomous vehicles: Autonomous vehicles represent one of the most prominent applications of DRL—for example, Waymo, a subsidiary of Alphabet, has harnessed DRL techniques to develop self-driving cars capable of navigating complex urban environments (Zhang *et al.*, 2020). DRL algorithms enable these vehicles to make real-time decisions based on sensory input, improving safety and efficiency on the road.
2. Aerial robotics: DRL has empowered drones to navigate challenging environments. In a case study, drones have been trained to fly through dense forests for search and rescue missions (Kasemsuppakorn *et al.*, 2020). By leveraging DRL, these drones can autonomously navigate through obstructed terrain to locate missing persons.
3. Robotic arm manipulation: In industrial automation, DRL has found applications in robotic arm manipulation—for instance, researchers have demonstrated the use of DRL to optimize the grasping and manipulation of objects by robotic arms (Andrychowicz *et al.*, 2020). Such advancements enhance manufacturing efficiency and adaptability.
4. Healthcare: DRL is making strides in healthcare robotics, particularly in surgical applications. The da Vinci Surgical System, for example, employs DRL for precise instrument control during minimally invasive surgeries (Marescaux *et al.*, 2001). DRL algorithms assist surgeons in achieving steady and controlled movements.

5. Warehouse automation: In the realm of logistics, DRL is revolutionizing warehouse automation. Companies like Amazon have implemented DRL-based systems to optimize the operations of autonomous mobile robots (Xu *et al.*, 2018). These robots navigate warehouses, picking and transporting items efficiently, thus contributing to faster order fulfilment.
6. Game playing: DRL has achieved remarkable success in game playing—for instance, DeepMind's AlphaGo algorithm defeated world-class Go players (Silver *et al.*, 2016). This case study highlights the ability of DRL algorithms to learn complex strategies through self-play and has led to advancements in game AI.
7. Humanoid robotics: DRL is also contributing to advancements in humanoid robotics. Researchers have employed DRL to teach humanoid robots tasks that involve complex movements, such as dancing (Cruz *et al.*, 2018). These robots can adapt and improve their performance over time, opening up possibilities in entertainment and human-robot interaction.
8. Agriculture: Precision agriculture benefits from DRL as well. Autonomous drones equipped with DRL algorithms monitor and manage crops more effectively (Pinto *et al.*, 2017). By optimizing the application of fertilizers, pesticides, and water, these drones improve crop yields and resource efficiency.
9. Environmental monitoring: In environmental science, underwater robots equipped with DRL capabilities are making a significant impact. These robots autonomously navigate oceans to monitor environmental conditions and study marine life (Pereira *et al.*, 2020). They can explore underwater caves, map uncharted regions, and contribute to oceanographic research.
10. Disaster response: DRL-equipped robots have applications in disaster response scenarios. These robots can navigate through debris, locate survivors, and deliver supplies to affected areas (Hwangbo *et al.*, 2019). DRL's ability to handle complex and dynamic environments enhances the efficiency and safety of rescue missions.

These applications and case studies illustrate the versatility and transformative potential of DRL in robotics and autonomous systems. From improving transportation safety to revolutionizing industries like healthcare, logistics, and environmental science, DRL continues to push the boundaries of what autonomous systems can achieve.

2.6 Challenges and Future Directions

Challenges

One of the primary challenges in the application of DRL in robotics and autonomous systems is the issue of sample efficiency (Schulman *et al.*, 2017). DRL algorithms often require a large number of interactions with the environment to learn effective policies. This high sample complexity can be impractical or costly in real-world scenarios. Improving sample efficiency remains a critical research challenge in the field (Hessel *et al.*, 2018).

Another critical challenge is ensuring the safety of DRL-controlled systems (Chen *et al.*, 2020). Safety concerns are paramount, especially in applications like autonomous vehicles and healthcare robotics. Ensuring robustness to uncertainties, sensor noise, and unforeseen events is an ongoing challenge that requires careful algorithmic design (Gupta *et al.*, 2017).

Real-world deployment poses its own set of challenges (Hussein *et al.*, 2017). Transitioning DRL models from simulation to real-world environments can be challenging due to domain gaps and discrepancies between simulated and real conditions. Addressing these domain adaptation challenges is crucial to the practical deployment of DRL-powered systems (Tobin *et al.*, 2017).

Generalization is another persistent challenge (Zhang *et al.*, 2020). DRL models often struggle to generalize from the training environment to unseen conditions or variations in tasks. Overfitting to specific conditions can limit the applicability of learned policies. Developing techniques to improve generalization capabilities is a key research direction (Finn *et al.*, 2017).

In robotics DRL, sample efficiency is still a major obstacle. It alludes to the significant amount of data that DRL algorithms need in order to function well. With practical applications, acquiring adequate data can be a major challenge, particularly with autonomous systems. In the case of autonomous driving, a DRL-based system needs considerable real-world data to learn safe and effective driving behaviors across a variety of scenarios and conditions. Collecting data for training frequently requires time-consuming and expensive operations. Large-scale and diversified datasets are required, which present practical difficulties for data processing and acquisition.

In robotics and autonomous systems, safety must always come first. DRL agents may experiment with less-than-ideal or even dangerous acts as they learn since they pick up new skills through decision-making and interaction with their surroundings. Safety issues are brought up by this, especially for uses like autonomous surgery and medical robotics. A robotic surgical assistant that has received DRL training may experiment with activities that, although necessary for learning, may have ramifications for safety. During this process of learning, ensuring patient safety is a major concern that calls for creative solutions.

Scalability is a recurring issue in the design of autonomous and robotic systems. The rising complexity and diversity of occupations and situations necessitates the evolution of DRL algorithms. These methods' scalability may also be compromised, as this adaptation typically necessitates significant computing resources. In industrial automation, task scope can vary widely, ranging from simple assembly procedures to complicated manufacturing jobs. More study is required to handle the challenge of guaranteeing that DRL-based robots can efficiently scale from one task to another, regardless of task complexity.

Future Directions

One promising future direction is the development of improved DRL algorithms (Silver *et al.*, 2017). Continual advancements in algorithms, including better exploration strategies, more efficient policy optimization techniques, and enhanced neural network architectures, will contribute to more effective and practical DRL solutions (Haarnoja *et al.*, 2018).

Transfer learning techniques will play a crucial role in the future of DRL (Fang *et al.*, 2017). Enabling DRL models to adapt quickly to new tasks or environments with minimal data will be essential for their versatility and efficiency (Rusu *et al.*, 2016).

Multi-modal sensing is an exciting direction for enhancing perception and decision-making capabilities (Xu *et al.*, 2019). Integrating various sensor modalities, such as vision, lidar, and audio, will enable DRL-based systems to gather more comprehensive information about their environment.

Human–AI collaboration is another promising area (Knox *et al.*, 2013). Exploring ways in which DRL-powered robots and autonomous systems can collaborate effectively with humans, particularly in shared control scenarios or human-in-the-loop learning, holds great potential for real-world applications.

Ethical considerations will continue to gain importance (Amodei *et al.*, 2016). Research on ethical AI practices, including addressing issues of fairness, transparency, and accountability, will be essential to ensure the responsible deployment of DRL in robotics and autonomous systems.

In summary, addressing challenges related to sample efficiency, safety, real-world deployment, and generalization, while pursuing future directions such as improved algorithms, transfer learning, multi-modal sensing, human–AI collaboration, and ethical AI practices, will shape the trajectory of DRL in robotics and autonomous systems. These efforts will lead to more adaptable, efficient, and capable autonomous systems across various domains.

Improved generalization capabilities of DRL agents to adapt to new and unknown contexts can be the topic of research. This would allow robots and autonomous systems to respond to novel and unexpected events successfully without requiring substantial retraining. Consider the usage of drones for environmental monitoring. The goal of research can be to improve their ability to adapt quickly and effectively to previously unknown terrains or weather conditions, allowing for more diverse and adaptive operations.

Developing ways for analyzing DRL agents' decision-making processes is crucial, especially in safety-sensitive applications. This research topic entails developing tools and strategies to help operators and users understand how DRL agents make decisions. In the case of autonomous inspection robots, for example, having an interpretability framework in place allows operators to fully comprehend why the robot makes specific choices during key tasks. This not only improves safety but also fosters trust in technology.

2.7 Evaluation and Metrics

Evaluating the performance of DRL algorithms in the context of robotics and autonomous systems is a critical aspect of research and development. Rigorous evaluation provides insights into the capabilities and limitations of DRL models in various applications. This section discusses key evaluation considerations and metrics, drawing from recent studies in the field.

1. Sample efficiency and learning curves: One fundamental metric in DRL evaluation is sample efficiency. It refers to the number of interactions with the environment required for the algorithm to achieve a certain level of performance. Sample efficiency is often assessed by analyzing learning curves, which depict the algorithm's progress in terms of cumulative rewards over time (Mnih *et al.*, 2015).

2. Task-specific metrics: Metrics must align with the specific task and application—for instance, in autonomous driving, success can be measured by collision rates, adherence to traffic rules, and navigation time (Zhang *et al.*, 2020). In object manipulation, success rates in grasping or manipulation tasks are vital metrics (Levine *et al.*, 2016).
3. Generalization and domain transfer: Evaluating how well DRL models generalize to new conditions is crucial. Metrics should assess the model's performance when transferred from a simulated environment to the real world (Tobin *et al.*, 2017). Domain adaptation metrics, such as domain discrepancy measures, help quantify the model's adaptability (Ganin *et al.*, 2016).
4. Safety metrics: Ensuring safety is paramount in autonomous systems. Metrics include collision rates, safety violations, and the assessment of worst-case scenarios to verify that DRL-controlled systems operate safely (Chen *et al.*, 2020).
5. Real-time performance: Real-time performance metrics assess the computational efficiency of DRL algorithms, which is critical for applications like self-driving cars or drones (Choi *et al.*, 2019). Low-latency execution is essential for timely decision-making.
6. User-centric metrics: In human–robot interaction scenarios, user satisfaction, comfort, and trust are essential considerations. Metrics for assessing the user experience can include subjective surveys or human-rated performance metrics (Driggs-Campbell *et al.*, 2017).
7. Comparison with baselines: Comparative metrics are valuable for benchmarking DRL approaches against traditional control methods or alternative machine learning techniques. These metrics highlight the advantages and shortcomings of DRL (Pan *et al.*, 2020).
8. Ethical and responsible AI metrics: As ethical considerations become increasingly important, metrics that evaluate fairness, transparency, and accountability of DRL models should be incorporated (Russell *et al.*, 2015).

The advancement of deep reinforcement learning (DRL) integration in robotics and autonomous systems depends critically on effective evaluation and the application of the right metrics. It is the responsibility of researchers and practitioners to carefully design thorough evaluation protocols that are both reliable and adaptable, ensuring that DRL models not only meet but also exceed the desired performance standards and objectives across a wide range of real-world scenarios and contexts.

2.8 Summary

In summary, deep reinforcement learning (DRL) has emerged as a transformative force in the field of robotics and autonomous systems. Its ability to enable machines to learn, adapt, and make decisions in complex, dynamic environments has opened up new horizons across a wide range of applications. Throughout this chapter, it explored the fundamental concepts, applications, challenges, and future directions of DRL in this domain.

Robotics and autonomous systems have entered a transformative era thanks to the fundamental ideas of deep reinforcement learning (DRL), which are firmly founded in the fusion of reinforcement learning and deep neural networks. This ground-breaking paradigm has fundamentally altered how robots and other autonomous entities perceive, pick up information from, and interact with their surroundings. With the addition of DRL, these systems have gained a previously unheard-of capacity to set out on autonomous journeys, manipulate objects with an unmatched level of precision, make crucial decisions in real time, and work effectively across domains that were previously thought to be insurmountably difficult.

The chapter has highlighted applications in autonomous vehicles, aerial robotics, healthcare, warehouse automation, and beyond. Case studies have illustrated how DRL has already begun to revolutionize industries, improve safety, and enhance efficiency.

However, there are a number of difficult obstacles on the way to fully realizing deep reinforcement learning's (DRL) limitless potential in the field of robotics. The need for ongoing commitment in terms of research and development is still strong as challenges ranging from sample efficiency, safety considerations, and real-world deployment limits to the crucial aspect of generalization still loom big. The development of the field is facilitated by addressing these complex problems.

Moreover, as we traverse this evolving landscape, it becomes increasingly evident that fostering interdisciplinary collaboration, as well as adhering to stringent ethical norms and the principles of responsible AI, is an imperative of utmost significance. In this interconnected and dynamic arena, the synergy of diverse expertise and the integration of ethical considerations are key not only to the advancement of DRL in robotics but also to shaping the way forward with consciousness and responsibility.

The future of DRL in robotics holds exciting prospects. Improved algorithms, transfer learning techniques, multi-modal sensing, and human–AI collaboration are among the many avenues of exploration. These directions will enable DRL-powered systems to become more adaptable, efficient, and capable, opening up new possibilities for innovation and automation.

References

1. Arulkumaran, K., Deisenroth, M.P., Brundage, M., Bharath, A.A., Deep reinforcement learning: A brief survey. *IEEE Signal Process Mag.*, *34*, 6, 26–38, 2017.
2. Sutton, R.S. and Barto, A.G., *Reinforcement learning: An introduction*, MIT Press Cambridge, Massachusetts, London, England, 2018.
3. Goodfellow, I., Bengio, Y., Courville, A., Bengio, Y., *Deep learning*, MIT Press Cambridge, London, England, 2016.
4. Mnih, V., Kavukcuoglu, K., Silver, D., Rusu, A.A., Veness, J., Bellemare, M.G., Hassabis, D., Human-level control through deep reinforcement learning. *Nature*, *518*, 7540, 529–533, 2015.
5. Silver, D., Huang, A., Maddison, C.J., Guez, A., Sifre, L., Van Den Driessche, G., Dieleman, S., Mastering chess and shogi by self-play with a general reinforcement learning algorithm, 2016, arXiv preprint arXiv:1712.01815.
6. Silver, D., Lever, G., Heess, N., Degris, T., Wierstra, D., Riedmiller, M.A., Deterministic policy gradient algorithms. *Proceedings of the 31st International Conference on International Conference on Machine Learning*, vol. *32*, 2014.

7. Konda, V.R. and Tsitsiklis, J.N., Actor-critic algorithms, in: *Advances in Neural Information Processing Systems*, pp. 1008–1014, 2000.
8. Ng, A.Y., Harada, D., Russell, S., Policy invariance under reward transformations: Theory and application to reward shaping. *Proceedings of the Sixteenth International Conference on Machine Learning*, 1999.
9. Dong, Y., Zhao, X., Wang, S., Huang, X., Reachability Verification Based Reliability Assessment for Deep Reinforcement Learning Controlled Robotics and Autonomous Systems, 2022, arXiv preprint arXiv:2210.14991.
10. Dong, Y., Zhao, X., Huang, X., Dependability analysis of deep reinforcement learning based robotics and autonomous systems, 2021.
11. Dong, Y., Zhao, X., Huang, X., Dependability analysis of deep reinforcement learning based robotics and autonomous systems through probabilistic model checking, in: *2022 IEEE/RSJ International Conference on Intelligent Robots and Systems (IROS)*, IEEE, pp. 5171–5178, 2022, October.
12. Rayhan, A., Artificial intelligence in robotics: From automation to autonomous systems, 2023.
13. Carlucho, I., De Paula, M., Wang, S., Petillot, Y., Acosta, G.G., Adaptive low-level control of autonomous underwater vehicles using deep reinforcement learning. *Rob. Auton. Syst.*, *107*, 71–86, 2018.
14. Yan, C., Xiang, X., Wang, C., Fixed-Wing UAVs flocking in continuous spaces: A deep reinforcement learning approach. *Rob. Auton. Syst.*, *131*, 103594, 2020.
15. Shibata, K., Jimbo, T., Matsubara, T., Deep reinforcement learning of event-triggered communication and consensus-based control for distributed cooperative transport. *Rob. Auton. Syst.*, *159*, 104307, 2023.
16. Xianjia, Y., Queralta, J.P., Heikkonen, J., Westerlund, T., Federated learning in robotic and autonomous systems. *Proc. Comput. Sci.*, *191*, 135–142, 2021.
17. Tsurumine, Y., Cui, Y., Uchibe, E., Matsubara, T., Deep reinforcement learning with smooth policy update: Application to robotic cloth manipulation. *Rob. Auton. Syst.*, *112*, 72–83, 2019.
18. Kouppas, C., Saada, M., Meng, Q., King, M., Majoe, D., Hybrid autonomous controller for bipedal robot balance with deep reinforcement learning and pattern generators. *Rob. Auton. Syst.*, *146*, 103891, 2021.

3

Deep Reinforcement Learning Algorithms: A Comprehensive Overview

Shweta V. Bondre[1]*, Bhakti Thakre[2], Uma Yadav[3] and Vipin D. Bondre[4]

[1]Information Technology Department, Shri Ramdeobaba College of Engineering and Management, Nagpur, India
[2]Cyber Security Department, St. Vincent Pallotti College of Engineering and Technology, Nagpur, India
[3]Computer Science and Engineering Department, Shri Ramdeobaba College of Engineering and Management, Nagpur, India
[4]Department of Electronics and Tele-Communication, Yeshwantrao Chavan College of Engineering, Nagpur, India

Abstract
Deep reinforcement learning (DRL) algorithms have become a key intersection of deep learning and reinforcement learning, providing answers to challenging decision-making problems in a variety of fields. The abstract presents the basic idea of reinforcement learning, which is the process by which an agent learns to interact with its environment to maximize a cumulative reward signal. The abstract then moves on to discuss the use of deep learning strategies, emphasizing how neural networks make it possible to represent and approximate intricate state–action mappings. A notable DRL approach is Q-learning, which calculates action values, and its extension, DQN, which approximates values using deep neural networks. The abstract outlines significant algorithmic developments, including deep Q-networks (DQN), trust region policy optimization (TRPO), proximal policy optimization (PPO), and others, outlining their unique workings, advantages, and drawbacks. In the context of multi-agent scenarios, DRL is extended to address both cooperative and antagonistic interactions between agents. Additionally, the abstract discusses reward structuring, transfer learning, and exploration-exploitation trade-offs in DRL situations. Proximal policy optimization (PPO) and trust region policy optimization (TRPO) are two examples of policy gradient approaches that directly learn action policies. A2C and A3C are two well-known examples of actor-critical techniques that combine value function estimates and policy optimization. Examples of cross-domain applications include robots, gaming, self-driving cars, finance, and healthcare. The abstract explains how DRL algorithms have accomplished ground-breaking feats, outperforming human performance in challenging situations. The abstract also highlights ongoing issues such as sample inefficiency, stability, and generalization, which spurs ongoing study. Deep reinforcement learning (DRL) has indeed achieved remarkable feats across various domains such as game playing, robotics, healthcare, autonomous vehicles, language understanding, etc.

**Corresponding author*: shwetakharat1510@gmail.com

Shubham Mahajan, Pethuru Raj and Amit Kant Pandit (eds.) Deep Reinforcement Learning and Its Industrial Use Cases: AI for Real-World Applications, (51–74) © 2024 Scrivener Publishing LLC

Keywords: Deep reinforcement learning, deep learning, value-based methods, policy-based methods

3.1 Introduction

A machine learning method called reinforcement learning concentrates on the appropriate actions that software agents need to do in a certain setting. This neural network learning method shows how to maximize a certain parameter over several phases or accomplish a challenging objective. Reinforcement learning (RL) is a subfield of both machine learning and artificial intelligence (AI). The learning approach comprises interactions with an environment and learning from them in order to maximize a cumulative reward signal. A fundamental aspect of reinforcement learning is the concept of trial and error. A reinforcement learning agent (RL) performs a sequence of tasks in a dynamic environment and receives feedback in the form of rewards and penalties, represented by a reward function, in order to learn from experience.With the experience gathered, the AI agent should be able to maximize the several goals presented in the form of cumulative rewards. Finding the optimal path, or a mapping between states and actions that maximizes predicted cumulative reward, is the agent's main objective.

One of the three machine learning paradigms is reinforcement learning (RL), which goes beyond supervised and unsupervised learning. It uses agents that assume the role of human domain experts to carry out tasks. RL learns from experiences by interacting with the world, observing what happens, and responding to the results, as opposed to needing tagged data. Figure 3.1 illustrates how "Markov decision process (MDP)" can be used to explain reinforcement learning. Each habitat is represented by a state that captures the current conditions there. The reinforcement learning agent acts in the environment, generating a new state and altering the current one, and is rewarded based on the results.The agent receives positive feedback for good action and negative feedback for bad action in order to evaluate the completed activity in a particular state and learn from experiences. One of the most widely used RL applications has been video games, and they have also served as the primary testing and evaluation platforms for RL algorithms. However, RL also has additional uses and may be used in a variety of fields, including autonomous robots, natural language processing (NLP), self-driving automobiles, and many more. RL algorithms come in a wide variety and have many distinct iteration. Therefore, it is crucial to comprehend

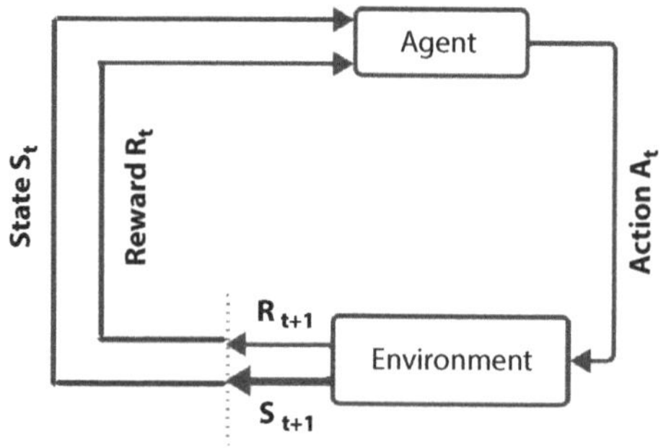

Figure 3.1 Generic reinforcement learning model's action–reward feedback loop.

the variations between RL algorithms and choose the one that is best suited to the work at hand and the environment type.

3.1.1 How Reinforcement Learning Works?

Suppose you wanted to teach your dog a new skill. Since the dog cannot understand human languages, including English, we are unable to instruct him. We instead use a different approach [1].

The dog makes many attempts to react in the circumstances we imitate. We will feed the dog meat if his answer is what we want.As a result, the dog now acts in a similar manner with even greater passion in the hopes of earning more food rewards whenever it comes into the same situation. Similar to how dogs are learning "what to do" *via* rewarding experiences [2]. The dog also gains knowledge about what to avoid when presented with unpleasant situations.

3.2 Reinforcement Learning Algorithms

A machine learning paradigm called reinforcement learning (RL) enables an agent to interact with its environment and make decisions in a sequential fashion to maximize a cumulative reward signal. Depending on the issue, its complexity, and the resources at hand, there are several ways to implement RL algorithms. A reinforcement learning algorithm may be implemented using one of three methods.

3.2.1 Value-Based Algorithms

Algorithms for value-based reinforcement learning concentrate on determining the relative importance of various states or state–action pairings in a given context. The primary goal of these algorithms is to identify a value function that measures the benefits of an agent staying in a particular state or taking a particular action in that state. This value function is then used by the agent to guide decisions that will maximize its anticipated cumulative reward.

3.2.1.1 Q-Learning

Depending on the agent's current state, Q-learning, a model-free, value-based, off-policy technique, will determine the best plan of action. Quality is represented by the letter "Q". The significance of the activity for optimizing the potential benefits is indicated by its quality. The model-based algorithms assess the best policy and build the model using transition and reward functions. Model-free algorithms, on the other hand, learn *via* experience without transition and reward functions about the effects of their actions. The value-based approach teaches the value function to identify the more valuable state and take appropriate action. Policy-based solutions, on the other hand, instruct the policy explicitly on what to do in a certain condition.

For example, to reach its goal, a robot has to make its way through a maze. There are mines, so the robot can only move one tile at a time. Robots that step on mines are destroyed. The robot has to arrive at the location in the shortest possible time as shown in Figure 3.2.

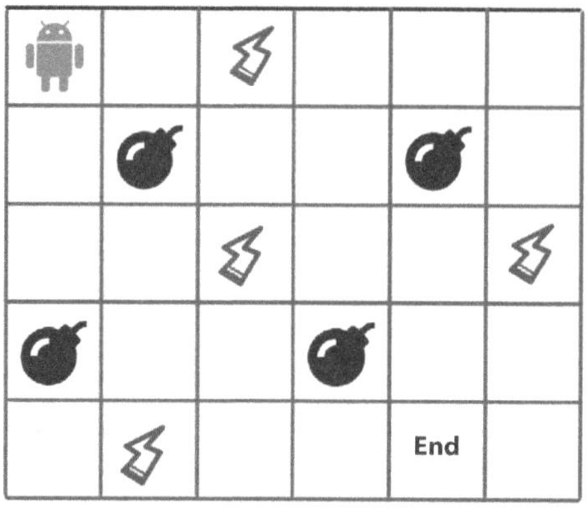

Figure 3.2 Robot travels the shortest route and reaches its destination.

The mechanism for rewards and score is as follows: The robot loses one point for each move. The robot will travel the shortest route and reach its destination as soon as it is feasible by doing this. If the robot steps on a mine and loses 100 points, the game is lost. If the robot gains power, it earns 1 point. If the robot makes it to the finish line, it gets 100 points.

a. Q-table

A Q-table, which is a straightforward lookup table, is used to calculate the maximum projected future rewards for acting in each condition.In simple terms; this chart will show us what to do in each state.

Figure 3.3 depicts four different numbers of actions will be present at each non-edge tile. A robot in a particular condition can move left, right, up, or down. The states and actions are represented by the rows and columns of the Q-table, respectively as shown in Figure 3.4.

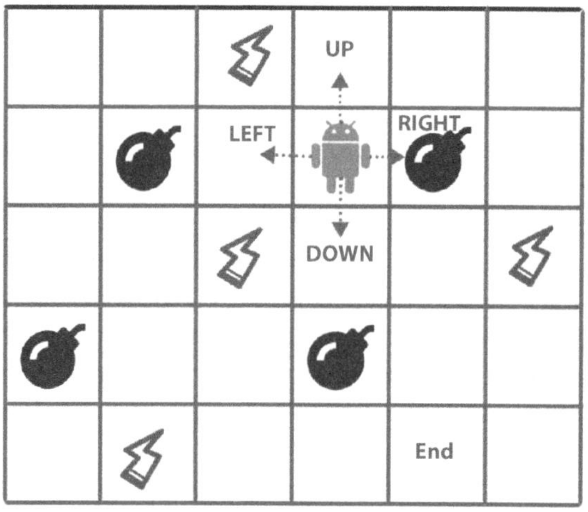

Figure 3.3 States and actions of robot.

Actions:	↑	→	↓	←
Start				
Nothing / Blank				
Power				
Mines				
END				

Figure 3.4 Numbers of actions.

The Q-table score represents the maximum projected future reward for any action the robot would take at that point. This is an iterative process because we must improve the Q-table at each iteration.

b. Q-function

State (s) and action (a) are the two inputs needed by the Q-function, which applies the Bellman equation shown in Figure 3.5.

We may obtain the Q values for the table's cells using the previously mentioned method. The Q-table has zero values at the beginning. The values are updated in an iterative process. The "Q-function" constantly modify the "Q-values" in the table continuously to give us approximations that are progressively more accurate. Now let us examine the updating process.

c. Q-learning algorithm process

First, a Q-table will be constructed. There are n columns, where n represents the number of activities. There are m rows where m is the number of states. Initially, all values will be set to zero as shown in Figure 3.6.

We have four actions (a=4) and five states (s=5) in our robot example. Therefore, a table with four columns and five rows will be constructed.

Steps 2 and 3 are to select and carry out an activity.

It is not stated how long this series of events will last. This means that until we halt the training or the training loop ends as indicated by the code, this phase will keep running.

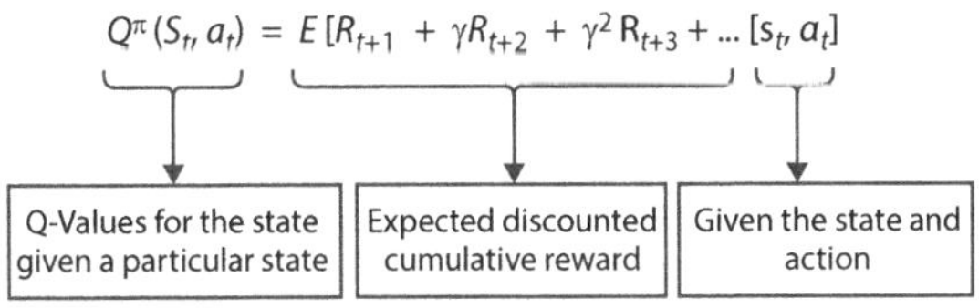

Figure 3.5 Q-function [3].

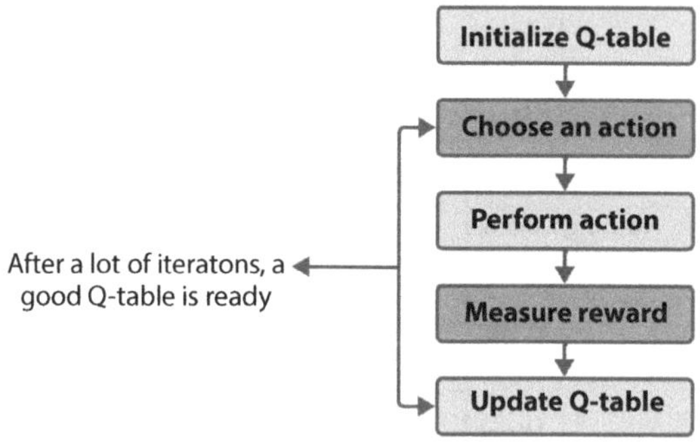

Figure 3.6 Initialize the Q-table in step one [3].

Actions:	↑	→	↓	←
Start	0	0	0	0
Nothing / Blank	0	0	0	0
Power	0	0	0	0
Mines	0	0	0	0
END	0	0	0	0

Figure 3.7 Q-table.

The action (a) in the state (s) will be chosen based on the Q-table. But each Q-value is zero at the start of the episode, as was previously stated.

Steps 4 and 5: evaluate.

We have now carried out a task and seen the outcome and benefit. The Figure 3.8 depicts new Q(s,a) function has to be updated.

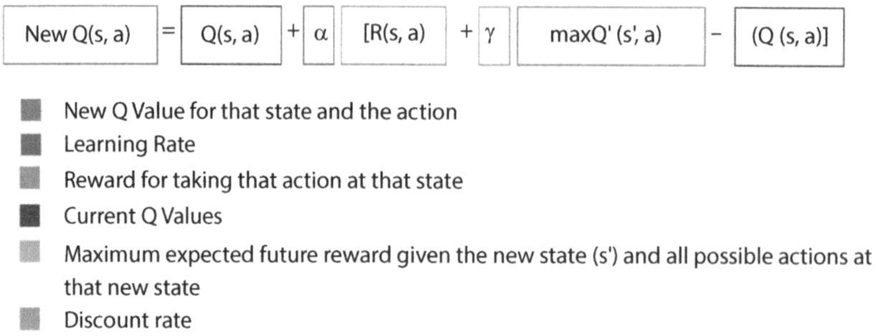

Figure 3.8 New Q(s, a) function.

3.2.1.2 Deep Q-Networks (DQN)

Instead of a two-dimensional array, this employs a neural network. Because they are unable to estimate and update values for states they are unaware of, Q-learning agents and algorithms cannot do so.In other words, regular Q-learning won't be able to predict the value if there is a completely new and unknown condition. Additionally, Q-learning uses 2-D arrays in a dynamic programming technique.Therefore, for improved value calculation in DQN, neural networks are used in place of the array. Convolution neural networks (CNNs) are composed of layers, and each layer has a variety of kernels (functions) that cover the picture in a variety of ways. An input image that is 3 × N × N dimensional (also known as an input picture with three-color channels, each of which has "N × N pixels") may produce numerous convolution maps when a convolution layer is applied, each of which has the same kernel weight. When working with images, a CNN performs better than any "DNN (MLP-based deep neural network)" equivalent, producing comparable accuracy since only one vector needs to be optimized because each layer's kernel weights stay constant, but a CNN's output is a multi-dimensional tensor that cannot be entered into a regression (value estimate) or classification model later on. In order to feed data into a "Soft-Max activation layer" for classification or a "linear activation layer" for regression, the last convolution layer of a CNN is connected to one or more flat layers. Selecting the output class with the highest class-probability (argmax), which is produced by the "Soft-Max activation layer" for each class that needs to be classified, is the optimal course of action.

The Figure 3.9 depicts action–value function, which connects environmental conditions to the expected return (i.e., the total of future rewards) for each conceivable action, is approximated by DQNs using a neural network. The DQN's objective is to discover the best path of action, or the policy, which will maximize each state's expected return. In order to train the DQN, the agent interacts with the surroundings by doing actions, earning rewards, and gaining feedback. These experiences are kept in a memory buffer by the agent, who uses them to update the DQN on a regular basis. Experience replay is a technique used to update the DQN that involves randomly sampling a group of experiences from the memory buffer. This procedure can stabilize the learning process and allows the agent to draw lessons from a larger range of events.

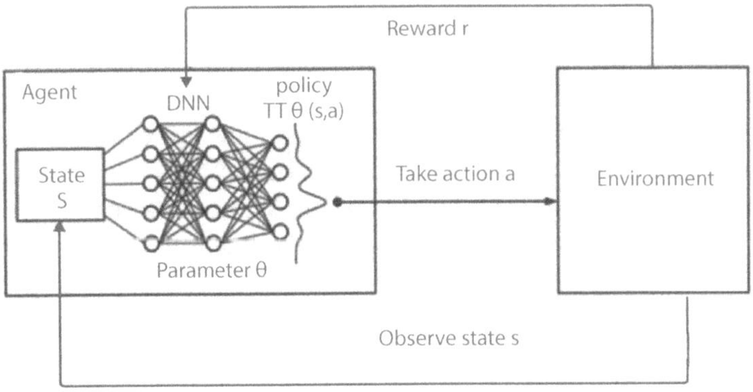

Figure 3.9 Deep Q-networks [3].

3.2.1.3 *Double DQN*

The "DQN" approach has a flaw in that the Q-values overestimate the agent's actual return, hence overestimating the true rewards. A simple solution to this issue is to divide action selection and assessment, as suggested by the authors of the double DQN algorithm [3]. They modify the Bellman equation as follows, instead of utilizing the same one as in the DQN method:

$$Q(s,a;\theta) = r + \gamma Q(s', argmax_{a'} Q(s',a';\theta);\theta'$$

After the main neural network chooses the best feasible path of action, the target neural network evaluates the chosen path of action to determine its Q-value. Reducing overestimations using this simple strategy has been shown to improve final policies.

3.2.1.4 *Dueling DQN*

The advantage function A(s, a) and the value function V(s) are divided into two pieces by this method, which also splits the Q-values. The amount of compensation we will receive from state s can be found using the value function V(s). Furthermore, the advantage function A(s, a) reveals the relative superiority of one action over the others. We may determine the Q-values by summing the value V and the benefit A for each activity.

$$Q(s,a) = V(s) + A(s,a)$$

The last layer of the comparable neural network is divided into two parts: one portion is used to estimate the advantage function (A(s, a)) for each action a and the other part is used to estimate the state value function for state s (V(s)). In the end, the dueling DQN algorithm [4] merges both components into a single output to estimate the Q-values. This adjustment is advantageous because, in certain cases, understanding the state–value function could be sufficient instead of understanding the precise value of each action. Nevertheless, the neural network cannot be trained by just adding the value and benefit functions. We are unable to discover the values of V and A in the equation Q=V+A because they are "unidentifiable" given the function Q. An equivalent situation would be if I informed you that Q equals 20 and asked you to determine which two integers (20 = V + A) add up to 20. There are countless ways to respond to this. The study's method sets the maximum Q-value at V, which causes the advantage function's maximum value to be zero and all other values to be negative [5]. This will tell us the precise value for V, allowing us to compute all of the benefits and find a solution. Although adding the value and benefit functions on their own cannot train a neural network, this is how we would train it. We are unable to ascertain the values of V and A in the equation Q=V+A since they are "unidentifiable" given the function Q. In a similar situation, I might tell you that Q is worth 20 and ask you to determine which two integers (20 = V + A) add up to 20. There are an infinite number of possible responses. The approach presented in the study sets the maximum Q-value to V, which results in the maximum value in the advantage function being zero and all other values being negative.

This will provide us with the exact value for V, from which we can calculate all of the advantages and resolve the issue. This is how we would train it:

$$Q(s,a) = V(s) + (A(s,a) - max_{a' \in |A|} A(s,a))$$

It proposes that we calculate the mean instead of the max; therefore we will do that (see the paper for more information). We will train our network in this manner:

$$Q(s,a) = V(s) + \left(A(s,a) - \frac{1}{|A|} \sum_{a'} A(s,a) \right)$$

3.3 Policy-Based

The main objective of "policy-based reinforcement learning algorithms" is to automatically determine which policy governs an agent's behavior in various scenarios. These algorithms do not evaluate the value of states or state–action combinations; instead, they aim to find the optimal policy that maximizes the expected cumulative reward.

3.3.1 Policy Gradient Methods

A group of "reinforcement learning techniques" known as "policy gradient methods" are used to learn the best policies in challenging situations. They have been extensively employed in fields like robotics, gaming, and autonomous control since they were specifically created for issues with high-dimensional or continuous action spaces. The policy, or the method by which an agent chooses actions in an environment, is directly parameterized and optimized *via* policy gradient methods. The main ideas and elements of policy gradient methods are as follows:

a. Policy: A policy is a chart that links states to actions. In a particular scenario, it describes what an agent should do. The goal of "policy gradient approaches" is to maximize the expected cumulative reward by optimizing these parameters, which are commonly set as a neural network for the policy [6].
b. Policy gradients: These techniques calculate the gradient of the objective function J(θ) with regard to the policy parameters in order to optimize the policy. This gradient explains how to adjust the policy's settings to boost predicted return.
c. Exploration: Exploration techniques are frequently used in policy gradient methodologies. They naturally stimulate exploration by selecting actions from the distribution of the policy since they optimize stochastic policies. Finding better policies will depend on this.
d. Sampling: Policy gradient approaches engage in interactions with the environment to gather sequences of states, actions, and rewards, which are then

used to compute an estimate of the gradient. The REINFORCE algorithm is the most well-known technique for doing this.

e. Baseline: Policy gradient methods can include a baseline, which is a value function that calculates the expected return for each state, to lower the variance of gradient estimates and speed up learning. The variation of the gradient estimations can be decreased by deducting the baseline from the returns.
f. Convergence and stability: It is critical to make sure that policy gradient approaches are stable and convergent. To prevent policy collapse and divergence problems, significant consideration must be given to the design of algorithms, exploration tactics, and parameter settings.

3.3.2 REINFORCE (Monte Carlo Policy Gradient)

An agent is used in the model-free Monte Carlo (MC) approach to learn directly from the experience data of episodic tasks. A collection of experience samples (x, u, r, and x′) produced in a hypothetical or actual setting are needed for the MC technique [7]. To create the step reward and the next probable state of the agent in a simulated learning environment, a model is necessary; however, policies and actions are not optimized with respect to the model's knowledge. The MC technique only averages the sample results produced by agent environment interactions to solve the RL difficulties. After an episode concludes, the value and policy estimates in the MC approach are revised. In MC policy evaluation approaches, the agent tries to estimate the value function V(x) for a given policy by averaging sample returns from the environment after every visit to the state x. The state values converge to the predicted state values after a sufficient number of visits are conducted. An episode may consist of multiple visits to any given state x. When the all visit MC method is utilized instead of the first visit MC method, all journeys to a state are taken into account, not only the first visit during an episode.

In the context of "Markov decision processes (MDPs)" and other related environments, the Monte Carlo policy gradient algorithm, also known as "REINFORCE (REward Increment = Nonnegative Factor × Offset Reinforcement × Characteristic Eligibility)", is a reinforcement learning algorithm for training policy-based models. An example of a policy gradient approach is this algorithm. A detailed explanation of the Monte Carlo policy gradient algorithm is provided below:

1. Initialization:
   Set the policy parameters to θ.
2. Data gathering
   - For each episode:
     - Initialize the state s.
     - Collect a trajectory (list of state–action pairs)
     - Choose your actions based on the existing policy: $a \sim \pi(a|s, \theta)$
     - Execute the chosen actions to interact with the surroundings.
     - Record all of the episode's states, actions, and prizes.
3. Calculate returns:
   - Calculate the return G_t for each time step t in every episode. The return is the total of the incentives earned up until time step t in the episode.

4. Update on policy gradient:
   - For each time step t in each episode:
     - Compute the policy gradient:
       - $\nabla\theta J(\theta) = \Sigma(G\_t * \nabla\theta \log \pi(a\_t|s\_t, \theta))$
     - Using the gradient estimate, update the policy parameters:
       - $\theta <- \theta + \alpha * \nabla\theta J(\theta)$
     - Here, $\nabla\theta J(\theta)$ is the gradient of the expected return $J(\theta)$ with respect to the policy parameters θ, and α is the learning rate.
5. Repeat:
     - Until convergence or a predetermined number of episodes, repeat steps 2 through 4.
6. Policy improvement (optional):
     - You can assess the effectiveness of the updated policy on a regular basis or after a predetermined number of episodes, and you can modify the policy's parameters as needed.

By changing the policy parameters in a way that raises expected return, Monte Carlo Policy Gradient seeks to directly optimize the expected return. The policy gradient with regard to the expected return is used to perform stochastic gradient ascent on the policy parameters.

One of the main problems with this method is that the policy gradient estimates can have a lot of variance. In response, several techniques have been devised to improve training's stability and efficacy. These include baselines, which lower variance, and more complex policy gradient techniques like "proximal policy optimization (PPO)" and "trust region policy optimization (TRPO)".

3.3.3 Actor–Critic Methods

Actor-critical techniques are a class of "reinforcement learning algorithms" that combine features of policy-based and value-based approaches to solve Markov decision processes (MDPs) and train agents to make decisions in sequential decision-making tasks [8]. Due to their capacity to balance the advantages of both strategies, these methods are frequently utilized in reinforcement learning. The following is a summary of actor–critical techniques:

a. Actor (policy network): The actor is a representation of the policy, which is a plan or a mapping from situations to actions. In order to estimate the policy, it often uses a parameterized function, often a neural network. Depending on the issue, the actor network output is either a probability distribution of actions or a deterministic action. The actor wants to discover a policy that maximizes predicted cumulative benefits over time.

b. Critic (value network): The critic makes an approximation of the "value function", which calculates the potential cumulative predictable reward of adopting a particular course of actions. By assessing how good actions or states are, the value function sends a feedback signal to the actor. The expected return associated with various actions or states is estimated by the critic network, which aids the actor in making more informed judgments.

c. Actor–critic interaction: The critic assesses the state–action pairs or state values while the actor chooses actions based on the policy during training. The "temporal difference (TD) error", which is the difference between the estimated value and the actual return, is used to update the value network's (critic) parameters. The actor adjusts its policy settings based on the critic's evaluation. For this update, policy gradient approaches or advantage-based approaches are frequently used; the advantage is the variation between the estimated value and the value of the present state.

d. Training: Iterative training is used to prepare actors and critics. The actor modifies the policy after receiving feedback from the critic, who uses TD learning techniques to determine the value of state–action combinations. The training procedure goes on until convergence, which is normally determined by some stopping condition. Actor–critical approaches have a number of benefits over pure policy-based approaches, including increased sampling efficiency and stability. Since it can be difficult to directly optimize the policy for continuous action space problems, they are especially well suited for such. The fact that they can manage both "on-policy and off-policy" learning offers flexibility in a variety of reinforcement learning scenarios.

3.3.4 Natural Policy Gradient Methods

A class of RL algorithms known as "natural policy gradient (NPG)" methods aims to enhance the stability and convergence characteristics of "policy gradient methods". NPG techniques use the natural gradient, which takes into consideration the geometry of the parameter space, as opposed to conventional policy gradients. This can result in learning that is more effective and stable, particularly when working with complicated policy parameterizations. The key concepts and features of natural policy gradient methods are as follows:

Policy optimization: In a reinforcement learning situation, NPG approaches concentrate on optimizing a parameterized policy $\pi(\theta)$. The policy is frequently expressed as a parameterized function that links states to actions, like a neural network.Like other reinforcement learning techniques, the objective of NPG is to maximize the projected cumulative return. "Natural policy gradient" is based on the "minorize–maximization algorithm (MM)", which maximizes discounted rewards for a policy.

In Figure 3.10,
θ - parameterizes our policy model,
η - is the discounted reward function and
θ^* - is the optimal policy we want to find.

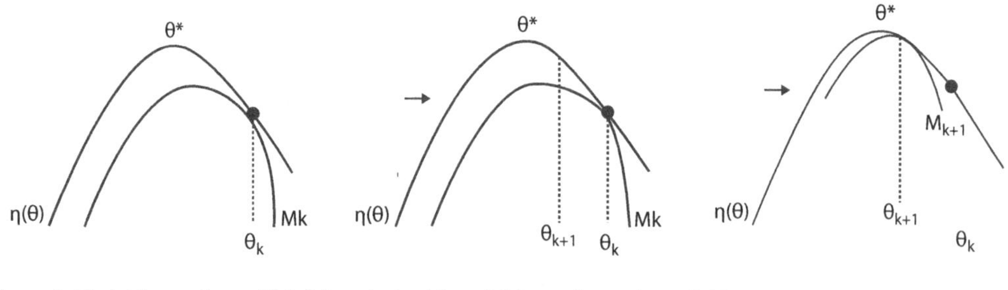

Figure 3.10 (a) Lower bound M, (b) optimize M, and (c) new lower bound, M.

First, we start with a random or educated policy as the current policy.

1. (Left figure a) We determine a lower bound M, represented in blue, for the discounted reward function η, and it approximates η locally at the current policy (the black dot).
2. (Middle b) After that, we optimize M and set the optimized policy as the one that applies right now.
3. (Right c) We optimize the current policy and re-establish a new lower bound, M.

Iterate through these steps repeatedly, optimizing a lower bound function each time, until the current policy converges. We call this optimization of the lower bound function a trust region optimization.

$$maximize L_{\pi_{\theta'}}(\pi_\theta)$$
$$subject\ to\ ||\Delta\theta||_F < \delta$$

Specifically, we are only looking for an ideal location for L within a trust region (for $\Delta\theta$ within δ). Because we cannot trust the accuracy of the improvement, even though points beyond the trust region may have higher computed rewards, we will ignore them.

3.4 Model-Based Reinforcement Learning

The goal of "model-based reinforcement learning (MBRL) algorithms" is to develop a model of the dynamics of the environment, which is then used to plan and make decisions. With the use of these algorithms, the environment's reactions to various actions made by the agent are to be accurately represented. Developing a model of the environment that can be utilized for planning and decision-making in a reinforcement learning setting is the aim of "model-based reinforcement learning (MBRL) algorithms" [9]. Which of the numerous MBRL algorithms should be applied will depend on the particular issue at hand as well as the characteristics of the surrounding environment [10]. These are some prominent MBRL algorithms:

3.4.1 Probabilistic Ensembles with Trajectory Sampling (PETS)

Probabilistic Ensembles with Trajectory Sampling (PETS) is a state-of-the-art "model-based reinforcement learning (MBRL) algorithm" designed to address complex and high-dimensional control problems. PETS leverages an ensemble of probabilistic models to capture the dynamics of the environment and uses trajectory sampling to plan and optimize policies efficiently. The algorithm has been successful in a wide range of challenging control tasks and robotics applications.

Key components and features of PETS include the following:

1. **Probabilistic models**: PETS employs an ensemble of neural network models, each representing a distinct hypothesis of the dynamics of the environment. With respect to the current state and action, these models are trained to predict the subsequent state and reward. Importantly, the models output not just a single prediction but a distribution, capturing the uncertainty in the model's predictions.
2. **Planning and optimization**: PETS uses a combination of random shooting and cross-entropy methods to generate diverse and informative trajectories in the model's learned dynamics. The trained model does trajectory optimization to identify an order of actions that maximizes the expected cumulative reward.
3. **Ensemble diversity**: The ensemble of models helps address the problem of model uncertainty. By considering multiple hypotheses, PETS can mitigate the effects of inaccurate models and provide more robust predictions and planning.
4. **Sample efficiency**: PETS is known for its sample efficiency. By simulating trajectories within the model, it reduces the total number of real-world interactions required to optimize a policy. This is particularly advantageous in scenarios where obtaining real data can be costly or risky.
5. **Exploration strategies**: PETS employs various exploration strategies, including action noise and trajectory sampling, to explore different areas of the state space. This allows the algorithm to discover optimal policies and adapt to different situations.
6. **Evaluation and control**: PETS not only uses the models for policy optimization but also for policy evaluation. By simulating trajectories and evaluating the expected return of candidate policies, PETS can make informed decisions about which policies to explore further.
7. **Transfer learning**: The ensemble approach in PETS facilitates transfer learning, as knowledge gained in one environment can be useful in a related one. The ensemble models can be adapted for new tasks with minimal real-world data collection.
8. **Real-world applications**: PETS has demonstrated impressive results in real-world applications, including robotic control, locomotion tasks, and manipulation tasks. Its ability to handle high-dimensional action spaces and complex dynamics makes it a powerful tool in the area of "deep reinforcement learning".

3.4.2 Probabilistic Inference for Learning Control (PILCO)

"Probabilistic inference for learning control (PILCO)" is a "model-based reinforcement learning (MBRL)" algorithm that focuses on learning and planning under uncertainty. PILCO is designed to optimize control policies while taking into account the inherent uncertainty in the dynamics of the environment [11]. It was created by Carl Edward Rasmussen and Marc Deisenroth and has been used for many control applications, especially in robotics.

The key features and concepts of PILCO include the following:

1. **Gaussian process models**: "PILCO" uses "Gaussian process (GP)" models to represent the dynamics of the environment. GPs are a powerful tool for modeling the uncertainty associated with complex and high-dimensional state spaces.
2. **Probabilistic models**: PILCO models the transition dynamics as a probabilistic process, taking into account the uncertainty in state transitions. Instead of predicting a single next state, PILCO's GP models predict a distribution over possible next states.
3. **Policy optimization with uncertainty**: PILCO employs analytical techniques to optimize control policies under uncertainty. The policy optimization process accounts for the uncertainty in the model, aiming to minimize risk while maximizing expected return.
4. **Information theoretic model predictive control (iMPC)**: PILCO uses an approach called information theoretic model predictive control to optimize policies. iMPC considers both the expected return and the uncertainty in the model to make control decisions that balance exploration and exploitation effectively.
5. **Sample efficiency**: PILCO is known for its sample efficiency. By using probabilistic models to capture uncertainty, the algorithm reduces the need for large amounts of training data and allows for more efficient learning and optimization.
6. **Data efficiency**: PILCO can make use of limited data to learn and optimize control policies. This makes it suitable for tasks where collecting real-world data is costly or time-consuming.
7. **Transfer learning**: PILCO facilitates transfer learning by adapting learned models to new tasks or environments with minimal data. The knowledge acquired in one scenario can be applied to related scenarios.
8. **Stochastic policies**: PILCO can handle stochastic policies, which means the algorithm can optimize policies that generate a distribution over actions. This can be particularly useful in situations where risk-sensitive control is required.

3.4.3 Model Predictive Control (MPC)

"Model predictive control (MPC)" is a widely used control strategy that is employed in a variety of fields, including control systems, robotics, process control, and autonomous vehicles. MPC is a model-based control strategy that solves an optimization problem at each time step in order to optimize a control action over a finite time horizon. This approach allows MPC to handle complex systems with constraints and uncertainty effectively.

The key features and concepts of model predictive control (MPC) include the following:

1. **Model of the system**: MPC relies on a dynamic model of the system it aims to control. This model represents the system's behavior and how it evolves over time in response to control inputs.

2. **Prediction horizon**: MPC looks into the future by considering a prediction horizon, a finite number of future time steps. This horizon is typically shorter than the entire trajectory, focusing on near-term planning.
3. **Control horizon**: Within the prediction horizon, MPC defines a control horizon, which represents the sequence of control actions to be applied in a receding horizon manner. The first control action in the sequence is executed, and the process repeats at the next time step.
4. **Optimization problem**: At each time step, MPC solves an optimization problem to determine the best sequence of control inputs over the control horizon. The objective function is defined to optimize a certain performance criterion, such as minimizing cost, maximizing a reward, or achieving a specific goal.
5. **Constraints**: MPC can handle constraints on both the system's state variables and control inputs. These constraints ensure that the system operates within safe and acceptable limits.
6. **Recursiveness**: MPC is a receding horizon control strategy. After each time step, it recalculates the optimal control sequence using updated information and a shifted prediction horizon. This allows MPC to react to changes in the system or disturbances in real-time.
7. **Feedback control**: MPC inherently provides feedback control, as it recomputes the control action at each time step based on the most recent system measurements. This allows the controller to adapt to changing conditions and uncertainties.
8. **Applications**: MPC is widely used in various applications, including autonomous driving, industrial process control, robotics, and energy management. It is suitable for systems with complex dynamics, nonlinearities, and constraints.
9. **Linear and nonlinear MPC**: MPC can be implemented with linear or nonlinear models, depending on the system's characteristics. Linear MPC is computationally efficient and well-suited for systems with linear dynamics, while nonlinear MPC can handle more complex systems.
10. **Real-time implementation**: MPC requires solving an optimization problem at each time step, which can be computationally intensive. Efficient numerical solvers and parallel computing techniques are often used to ensure real-time implementation.

3.4.4 Model-Agnostic Meta-Learning (MAML)

"Model-agnostic meta-learning (MAML)" is a machine learning technique that focuses on the problem of few-shot learning. "Few-shot learning" refers to the ability of a model to quickly adapt to new tasks or environments with only a little amount of data. MAML is a "meta-learning algorithm" that aims to learn an initialization of a model's parameters that can be fine-tuned with minimal data to perform well on a new task. It has been used in a number of fields, such as computer vision, natural language processing, and reinforcement learning.

The key features and concepts of model-agnostic meta-learning (MAML) include the following:

1. **Meta-learning**: "MAML" seeks to understand the learning process itself by functioning in a meta-learning environment. It gains the ability to quickly adjust a model's parameters to a new task by learning its initialization.
2. **Parameter initialization**: MAML starts with an initial set of model parameters that are shared across tasks. These parameters are then fine-tuned using fewer amounts of data from the specific task.
3. **Task-specific gradients**: MAML uses a two-step optimization process. In the inner loop, it performs gradient descent on the task-specific data to revise the model parameters. In the outer-loop, it adjusts the initial parameters based on the performance of the fine-tuned model.
4. **Few-shot learning**: When there are few examples provided for every new task, MAML works notably effectively in few-shot learning circumstances. It allows the model to adapt quickly, making it useful for tasks with limited data.
5. **Adaptation and generalization**: MAML's key idea is to balance adaptation (learning to perform well on a specific task) and generalization (the ability to perform well on a wide range of tasks) through its two-step optimization process.
6. **Applications**: MAML has been applied to a wide range of tasks, including object detection, image classification, natural language processing and reinforcement learning. It is particularly valuable in settings where the model needs to be versatile and adaptable.
7. **Meta-training**: MAML requires a meta-training phase, where it is exposed to a variety of tasks during training. It learns a good initialization of model parameters through this process, which allows it to adapt to new tasks more effectively.
8. **Transfer learning**: MAML facilitates transfer learning by leveraging the knowledge gained from the meta-training phase to quickly adapt to new, related tasks. It minimizes the need for training from scratch.

3.4.5 Soft Actor–Critic with Model Ensemble

Soft actor–critic with model ensemble (SAC-ME) is an advanced model-based reinforcement learning algorithm designed to address complex control tasks and enhance sample efficiency. SAC-ME combines elements of model-free reinforcement learning (soft actor–critic) with a model-based ensemble approach to improve the agent's performance.

The key features and characteristics of SAC-ME include the following:

1. **Model ensemble**: SAC-ME employs an ensemble of learned models of the environment. These models predict the next-state and reward based on the current state and action. Using multiple models helps mitigate model uncertainty and provides a range of predictions.

2. **Soft actor–critic base**: SAC-ME builds upon the "soft actor–critic (SAC) algorithm", which is a state-of-the-art model-free reinforcement learning algorithm known for its robustness and performance. SAC provides the foundation for SAC-ME's policy optimization.
3. **Sample efficiency**: The model ensemble in SAC-ME contributes to enhanced sample efficiency. By simulating trajectories in the learned models, the algorithm reduces the need for a large number of real-world interactions.
4. **Planning and optimization**: SAC-ME uses the ensemble of learned models for planning and action optimization. The agent generates a range of possible trajectories and evaluates their expected return, allowing it to make informed decisions.
5. **Model uncertainty handling**: The ensemble approach helps SAC-ME account for model uncertainty. The agent considers the range of possible outcomes when planning and making decisions, which can lead to more robust performance.
6. **Regularization and exploration**: SAC-ME may incorporate entropy regularization and exploration strategies to encourage effective exploration while maintaining a balance between exploration and exploitation.
7. **Multiple model predictions**: SAC-ME leverages the ensemble's predictions for generating multiple possible action sequences, which are then evaluated to select the best course of action.
8. **Real-world applications**: SAC-ME has demonstrated success in real-world applications, particularly in control and robotics tasks. Its ability to handle model uncertainty and reduce the number of real-world interactions makes it attractive in challenging scenarios.
9. **Ensemble size and diversity**: The performance of SAC-ME can be influenced by the size and diversity of the model ensemble. Researchers may experiment with various ensemble configurations to optimize results.

3.4.6 Deep Deterministic Policy Gradients with Model (DDPG with Model)

Deep deterministic policy gradients with model (DDPG with model) appears to be an extension or variation of the deep deterministic policy gradients (DDPG) algorithm that incorporates a learned model of the environment. DDPG is a popular RL algorithm for solving continuous action space problems, and adding a model of the environment can enhance its performance, especially in terms of sample efficiency.

Here is an overview of DDPG with model and its key components:

1. **DDPG background**: DDPG is an off-policy actor–critic algorithm that combine elements of deep Q-learning and deterministic policy gradients. It is designed to handle continuous action spaces and is known for its stability and performance.
2. **Model of the environment**: A neural network is commonly used to represent the environment in DDPG with Model, where the agent learns a model of it.

This paradigm is similar to model-based reinforcement learning in that it makes predictions about the next state and reward based on the present state and action.

3. **Actor–critic architecture**: DDPG consists of two neural networks, an actor–network that determines the policy (action) and a critic network that estimates the "action–value function (Q-function)". The model of the environment can be used to augment the actor–critic architecture.
4. **Planning and simulation**: The learned model of the environment allows the agent to simulate and plan in a virtual environment. This can reduce the number of interactions with the real environment and enhance sample efficiency.
5. **Exploration strategies**: DDPG with model may use the model to explore uncertain or unknown parts of the state space, guiding the agent's actions in the simulated environment.
6. **Model uncertainty**: Dealing with model uncertainty is crucial in DDPG with model. Techniques for handling model uncertainty may be employed, such as using ensemble models, model ensembles, or incorporating Bayesian neural networks.
7. **Hybrid approach**: DDPG with model is considered a hybrid approach, combining elements of "model-free and model-based reinforcement learning". It leverages the benefits of both paradigms.
8. **Stabilization and optimization**: Like standard DDPG, various techniques can be applied to stabilize and optimize the training process, including experience replay and target networks.

3.5 Characteristics of Reinforcement Learning

Reinforcement learning (RL) is a machine learning paradigm that aims to train agents to maximise a cumulative reward signal by making decisions in a particular order in the environment. It has several key characteristics, which distinguish it from other types of machine learning:

1. **Sequential decision-making**: RL deals with problems where an agent interacts with an environment over a sequence of discrete time steps. The agent takes actions, and the environment responds with new states and associated rewards. The agent's goal is to study a policy that determines which actions to take in each state to maximize long-term rewards.
2. **Trial and error**: RL agents learn through trial and error. They explore different actions and observe the consequences (rewards and state transitions) of those actions. Over time, they adapt their behaviour based on these observations.
3. **Delayed rewards**: In RL, the consequences of an action may not be immediately evident. Rewards can be delayed, and the agent must learn to make decisions that optimize cumulative rewards over time.

4. **Exploration and exploitation**: Balancing exploration (trying new actions to learn more about the environment) and exploitation (choosing actions that are believed to be the best based on current knowledge) is a fundamental challenge in RL.
5. **Markov decision process (MDP)**: Markov decision processes are a common formulation for RL problems, wherein it is assumed that the dynamics of state transitions meet the Markov property. Thus, the reward and state in the future are solely dependent on the deed and situation in the present.
6. **Policy and value functions**: RL uses concepts like policies, which define the agent's strategy, and value functions, which estimate the expected cumulative reward under a given policy. Value functions are used for guiding the agent's decision-making process.
7. **Model-free and model-based**: RL can be categorized into model-free (directly learning the policy without modeling the environment) and model-based (learning an explicit model of the environment and using it for decision-making) approaches.
8. **Function approximation**: Value functions and policies are approximated in various reinforcement learning contexts using function approximation approaches like neural networks. This enables RL to handle high-dimensional state and action spaces.
9. **Reward signal**: The agent learns from a reward signal, which is a scalar value provided by the environment after each action. The reward signal provides feedback on the desirability of actions and is used to shape the agent's learning.
10. **Credit assignment problem**: RL faces the challenge of credit assignment, determining which actions or decisions led to specific outcomes. The agent needs to attribute rewards or penalties to the right actions to improve its behavior effectively.
11. **Partially observable environments**: RL can handle environments where the agent has partial or noisy observations of the state. Techniques like partially observable Markov decision processes (POMDPs) are used to address such scenarios.
12. **Exploration strategies**: To explore the environment effectively, RL agents use various exploration strategies, such as epsilon-greedy, Thompson sampling, and exploration bonuses.
13. **Off-policy and on-policy learning**: RL algorithms can be categorized as on-policy (learning from the current policy) or off-policy (learning from different policies). Each has its advantages and disadvantages.
14. **Convergence and exploration trade-off**: RL algorithms need to strike a balance between exploration for learning and exploiting the best-known strategies. This trade-off affects learning efficiency and the quality of the final policy.

3.6 DRL Algorithms and Their Advantages and Drawbacks

Advantages of DRL:

1. **Autonomous learning:**
   - **Advantage:** DRL enables systems to learn and adapt autonomously through trial and error. This autonomy is especially beneficial in scenarios where explicit programming is challenging or impractical.
2. **Versatility:**
   - **Advantage:** DRL algorithms are versatile and can be applied to a broad range of tasks across different domains, from playing games and robotics to healthcare and finance.
3. **Complex decision-making:**
   - **Advantage:** DRL excels in situations with high-dimensional and complex decision spaces. This makes it suitable for problems where traditional rule-based systems may struggle.
4. **Generalization:**
   - **Advantage:** DRL models can generalize their learned knowledge to perform well in unseen situations. This ability to transfer learning is crucial in real-world applications where conditions may change.
5. **Real-time adaptability:**
   - **Advantage:** DRL systems can adapt to real-time changes in the environment, making them suitable for dynamic and evolving scenarios, such as robotics and autonomous vehicles.
6. **Reduced need for feature engineering:**
   - **Advantage:** DRL algorithms can automatically learn relevant features from raw data, reducing the need for extensive manual feature engineering in certain applications.

Drawbacks of DRL:

1. **Sample inefficiency:**
   - **Drawback:** DRL algorithms often require a large number of samples (interactions with the environment) to learn effectively. This can be impractical or expensive in scenarios where each interaction is resource-intensive or time-consuming.
2. **Lack of interpretability:**
   - **Drawback:** DRL models, especially deep neural networks, can be complex and lack interpretability. Understanding why a model makes a specific decision can be challenging, limiting their applicability in critical domains where interpretability is crucial.
3. **Reward specification:**
   - **Drawback:** Designing a suitable reward function is crucial in reinforcement learning. Poorly designed reward functions may lead to suboptimal

or unintended behaviors. Specifying rewards that align with the desired behavior is a non-trivial task.

4. **Training stability:**
   - **Drawback:** Training deep networks, which are often used in DRL, can be challenging and unstable. Issues such as vanishing gradients or diverging training may hinder the learning process.
5. **Exploration–exploitation trade-off:**
   - **Drawback:** Balancing exploration (trying new actions to discover their effects) and exploitation (choosing known good actions) is a fundamental challenge. DRL algorithms must find the right balance to avoid getting stuck in suboptimal solutions.
6. **Transfer learning limitations:**
   - **Drawback:** While DRL can generalize learning to some extent, transferring knowledge from one domain to another may still be a challenge, particularly in vastly different environments.
7. **Computational resources:**
   - **Drawback:** Training sophisticated DRL models, especially deep neural networks, often requires substantial computational resources, including powerful GPUs or TPUs. This can be a barrier for small research groups or companies with limited resources.

3.7 Conclusion

Reinforcement learning is a powerful frame-work for training agents to make decision in complex, dynamic environments, and its characteristics make it suitable for a wide range of application, including game playing, robotics, recommendation systems, and autonomous systems. Value-based reinforcement learning is effective for solving a broad range of problems; including recommendation systems, control tasks, and game playing, particularly when the agents needs to learn the best policy by estimating state or state–action values. These methods have contributed to significant advancements in the area of reinforcement learning. Policy-based reinforcement learning is widely used in a variety of domains, including natural language processing, autonomous control, robotics, and game playing. Its focus on learning a policy can make it more suitable for challenging tasks where the action space is complex, high-dimensional, or continuous model-based RL is a promising method that combines data-driven learning with simulation-based planning to achieve efficient and effective decision-making in a wide range of domains.

References

1. Naeem, M., Rizvi, S.T.H., Coronato, A., A Gentle Introduction to Reinforcement Learning and its Application in Different Fields. *IEEE Access*, 8, 209320–209344, 2020, doi: 10.1109/ACCESS.2020.3038605.

2. Nian, R., Liu, J., Huang, B., A review On reinforcement learning: Introduction and applications in industrial process control. *Comput. Chem. Eng.*, 139, 106886, Aug. 2020, doi: 10.1016/j.compchemeng.2020.106886.
3. Sewak, M., Deep Q Network (DQN), Double DQN, and Dueling DQN, in: *Deep Reinforcement Learning*, pp. 95–108, Springer Singapore, Singapore, 2019, doi: 10.1007/978-981-13-8285-7_8.
4. Qiu, H. and Liu, F., A State Representation Dueling Network for Deep Reinforcement Learning, in: *2020 IEEE 32nd International Conference on Tools with Artificial Intelligence (ICTAI)*, IEEE, pp. 669–674, Nov. 2020, doi: 10.1109/ICTAI50040.2020.00107.
5. Mnih, V. *et al.*, Human-level control through deep reinforcement learning. *Nature*, 518, 7540, 529–533, Feb. 2015, doi: 10.1038/nature14236.
6. Peters, J., Policy gradient methods. *Scholarpedia*, 5, 11, 3698, 2010, doi: 10.4249/scholarpedia.3698.
7. Williams, R.J., Simple statistical gradient-following algorithms for connectionist reinforcement learning. *Mach. Learn.*, 8, 3–4, 229–256, May 1992, doi: 10.1007/BF00992696.
8. Konda, V.R. and Tsitsiklis, J.N., Actor-Critic Algorithms. *SIAM J. Control Optim.*, 42, 4, 1143–1166, Jan. 2003, doi: 10.1137/S0363012901385691.
9. Moerland, T.M., Broekens, J., Plaat, A., Jonker, C.M., Model-based Reinforcement Learning: A Survey. *Found. Trends® Mach. Learn.*, 16, 1, 1–118, 2023, doi: 10.1561/2200000086.
10. Fürnkranz, J. *et al.*, Model-Based Reinforcement Learning, in: *Encyclopedia of Machine Learning*, pp. 690–693, Springer US, Boston, MA, 2011, doi: 10.1007/978-0-387-30164-8_556.
11. Rose, D.C., Mair, J.F., Garrahan, J.P., A reinforcement learning approach to rare trajectory sampling. *New J. Phys.*, 23, 1, 013013, Jan. 2021, doi: 10.1088/1367-2630/abd7bd.

4

Deep Reinforcement Learning in Healthcare and Biomedical Applications

Balakrishnan D.[1]*, Aarthy C.[2], Nandhagopal Subramani[3], Venkatesan R.[4] and Logesh T. R.[5]

[1]*Department of Computer Science and Business Systems, KGiSL Institute of Technology, Coimbatore, Tamil Nadu, India*
[2]*Department of Computer Science & Engineering, DhanalakshmiSrinivasan University, Tiruchirappalli, Tamil Nadu, India*
[3]*Department of Computer Science Engineering (Artificial Intelligence & Machine Learning), KPR Institute of Engineering and Technology, Coimbatore, Tamil Nadu, India*
[4]*Department of Information Technology, Bannari Amman Institute of Technology, Erode, Tamil Nadu, India*
[5]*Department of Information Technology, KSR Institute for Engineering and Technology, Namakkal, Tamil Nadu, India*

Abstract

In contrast to conventional supervised learning, which frequently employs one-shot, exhaustive, and supervised reward signals, deep reinforcement learning addresses sequential decision-making challenges with sampling, evaluative, and delayed feedbacks at the same time. Deep learning approaches are a good candidate to generate strong solutions in a wide range of healthcare domains where diagnosing choices or treatment regimens are frequently characterized by an extended time frame with delayed feedbacks due to this particular feature. This chapter examines deep learning applications in a wide range of healthcare domains, including dynamic treatment regimes in chronic diseases and critical care, automated medical diagnosis, and numerous other control or scheduling issues that have permeated every aspect of the healthcare system. This necessitates an in-depth review of the theoretical foundations as well as significant approaches in deep reinforcement learning research. Furthermore, the obstacles and unaddressed problems in the present investigation and highlight a few possible fixes and research possibilities for the future. Through this chapter it looks on the possibilities of deep reinforcement learning (DRL) in biological applications and healthcare. It highlights the utilization of the double deep Q-network (DDQN) and its prospective uses in medical imaging, pharmacological research, robot-assisted surgery, and predictive methods of analysis. The chapter additionally highlights the necessity of data potency, ethical concerns, protection, easy access, and clarity in ensuring DRL's sustainability for subsequent investigations as well as healthcare.

***Keywords*:** Deep reinforcement learning (DRL), healthcare, medical diagnosis, robotic surgery, machine learning, deep learning, biomedical, deep Q-network

**Corresponding author*: bks.28.01@gmail.com

Shubham Mahajan, Pethuru Raj and Amit Kant Pandit (eds.) Deep Reinforcement Learning and Its Industrial Use Cases: AI for Real-World Applications, (75–90) © 2024 Scrivener Publishing LLC

4.1 Introduction

A formidable subset of artificial intelligence named deep reinforcement learning (DRL) combines deep learning and reinforcement learning methodologies. Due to its capacity to make wise decisions in challenging, dynamic circumstances, it bears unique potential for applications in the biomedical and healthcare domains.

DRL can be utilized for healthcare in order to enhance personalized therapy, dosages of medications, and treatment plans. In an effort to suggest therapies that maximize the intended effects while reducing challenges, it may be gained from patient data. DRL can also be used in medical imaging to perform tasks, including the process of segmentation detection and categorization, increasing the accuracy of diagnostics.

Pharmacological studies and advancements are included in biomedical uses. DRL may assist with trial strategy, therapeutic molecule maximizing efficiency, and treatment efficacy prediction. It can also be used to control medical supplies, such as robotic surgery systems, strengthening functioning safety and precision.

Overall, the application of deep reinforcement learning to the biomedical and medical services sectors has immense potential for reshaping medical practices, medication development, and patient care. However, while employing these cutting-edge techniques in fragile fields like healthcare, it is imperative to ensure moral considerations, solitude, and compliance with regulations.

4.2 Related Works

Yang, J. *et al.* (2023): The model was explicitly assessed to predict COVID-19 for patients consulting hospital urgent care centers, with the intention of reducing any site (hospital)-specific and ethnicity-based biases in the information that was under consideration, demonstrating that the breakthroughs offer clinically efficient screening performances while greatly enhancing result fairness when compared to contemporary benchmarks and state-of-the-art machine learning technologies by applying an entirely new incentive function and training strategy. The methodology also underwent assessment on a patient intensive therapy unit evacuation position challenge, indicating model applicability [17].

Wang, C. *et al.* (2023): Progress in artificial intelligence have resulted in intricate algorithms and data analysis and processing tools, enabling intelligent health monitoring as well as high-precision forecasts and assessments. A closed-loop system with real-time monitoring, the gathering of data, online investigation, assessment, and remedy recommendations may be produced by amalgamating the Internet of Things, artificial intelligence, and health surveillance equipment [16].

Qiu, J. *et al.* (2023): After being pertained, large AI models perform well in a number of downstream tasks. ChatGPT is a notable example, whose capabilities have ignited people's curiosity in the far-reaching effects that massive AI models may have, as well as their potential for altering various facets of our lives. Massive AI models have been treated as introduced into health informatics, culminating in new perspectives for methodology development. The value of multi-modal data in the biomedical and health industry has perpetually risen, especially since the community adopted the machine learning (ML) age.

This shows out the foundations for enormous AI models to be built, assessed, and apprised as a way to advance in medical areas [13].

Baucum, M. *et al.* (2020): Healthcare reinforcement learning is exacerbated by the fact that, for evident reasons of morality, medical procedures cannot always be grasped in real time from patients who exist. There are two main strategies for removing this restriction. The initial strategy is to learn policies by examining existing data sets. Off-policy reinforcement learning permits an agent to use information obtained from a different policy in order to transform patient states into the most effective actions. Another approach is to build an "environment model" using prior patient data that gives response to a reinforcement learning agent on any state/action pair. This enables innovative on-policy learning, which has been shown to outperform off-policy learning [3].

Jayatilake, S. M. D. A. C., & Ganegoda, G. U. (2021): In health care, machine learning approaches are being used for computational decision making when a critical analysis of data on biological data is necessary to find hidden links or anomalies that are not visible to humans. Implementing algorithms to perform such tasks is difficult in and of itself, but enhancing the accuracy of the algorithm while cutting the time needed for the program to operate makes it much more difficult. Processing large amounts of medical data was a big difficulty in the early days, which led to the acceptance of AI in the medical arena [9].

Castiglioni, I. *et al.* (2021): Deep learning models which are multi-layered artificial/convolutional neural networks that allow us to instantly assess imagery, envision signage, envision annotations, metadata convergence, collective learning, and adaptation for deviations in image processing methods that frequently establish interference in non-AI imaging investigations are all included [5].

Azghadi, M. R. *et al.* (2021): Deep learning in the biological processing of signals and healthcare presents a lot of significant promise for professionals and the individuals they treat. Deep neural networks are used in order to enhance the excellence of life of patients with chronic illnesses by allowing ubiquitous screening for anomalies, reducing the pressure on medical resources. This presented an even understanding of how memristors could result in optimum equipment computation on deep Q-networks and simulated neural networks, as well as the hurdles that must be surmounted before they can be thoroughly set up. While this lesson and review emphasizes the hardware implementation of many different algorithms used in deep learning, the reader ought to become aware that hardware development is an essential but insufficient condition for effective medical-AI convergence [2].

Qayyum, A. *et al.* (2020): Despite ML/DL's impressive performance, there remain worries over its resilience in medical settings (which has historically been seen as difficult due to the multitude of privacy and security concerns involved), especially given the recent finding that ML/DL are susceptible to malicious attempts [12].

Han, J. *et al.* (2018): Digital segmentation of objects is a vital but challenging undertaking in the field of computer vision. Using a deep learning reinforcement framework, the challenge has been investigated as a Markov decision process, and models undergo training to segregate object regions. Learning agents for segmentation are difficult to flesh out because segmentation is a practically continuous decision-making process involving a large number of agents and action stages that span from the seed pixels to the entirety object mask. To address this issue, the learning of a fragmentation agent shrinks to that of a cutting-agent, which has just a handful of action subunits and can come together in a few steps of action [7].

Zhang, Y. D. *et al.* (2021): The increasing capabilities of deep learning in recognizing diseases and understanding medical data will empower physicians and expedite taking decisions in clinical settings. The use of contemporary healthcare equipment and the computerization of medical treatment have led to vast amounts of biological data in recent years. These new issues, needs, as well as possibilities for novel AI approaches and computational models for quick data processing, analysis, and simulation with created data, which is vital for medicinal motives as well as grasping the fundamental biological process, are outlined [19].

Alaniz, S. (2018): The possibility of constructing a model-based learning through a reinforcement agent that is capable of vying with model-free approaches, particularly DQN, has been investigated. An amalgamation of acquiring an intermediate foundation with a network of deep neural networks and a version of Monte Carlo tree search was utilized to bolster such an agent. Experiments demonstrate that learning a meaningful transition model requires far less instructional information than training Q-values of equal deep Q-network scoring on a Mine Craft block-placing assignment. Because the Monte Carlo tree searching agent uses a tree search to determine the most effective behavior, it takes longer than the deep Q-network agent for it to carry out just one action. As a result, the methodology is fascinating in cases when obtaining samples for training the environment is prohibitively costly to purchase [1].

Wang, S. *et al.* (2018): Deep learning, a technology based on artificial neural networks, is now recognized as an effective means for machine learning in the past few years, with the possibility to completely alter the future of artificial intelligence. Recent advances in processing power, quick retrieval of information, and the use of parallelization in addition to the technology's predictive ability and capacity to derive automatically optimum fundamental properties and semantic understanding from input data further contributed to its swift adoption [15].

Li, Z., & Xia, Y. (2020): When trying to access disease progression and alternatives to therapy statistically, reliable and automatic lymphatic node labeling is crucial. Lymphatic node identification is a difficult endeavor due to the immense variation in the lymphatic node physique and obstacles of acquiring voxel-wise annotations that are manual. since the Response Evaluation Criteria in Solid Tumors (RECIST) annotation, which reveals precisely the position, dimension, and dimension of a lymph node, is readily available in the healthcare facility repository, it can be utilized in order to weakly supervise learning on the aforementioned classification work [10].

Dash, S. *et al.* (2022): The goal involves offering statistical strategies for accumulation, revisions, and influencing facts within artificially intelligent systems, alongside methods of learning that help us infer knowledge from data. When direct algorithmic solutions are unavailable, formal models are lacking, or knowledge of the application the area is insufficiently pointed out, it is valuable. The Web of Things has the ability to influence the ways we live and work throughout the decades that follow. These computational methods are also crucial for engineering design and optimization. With the growing popularity and impact brought about by the Internet of Things notion, there is more appetite for AI (artificial intelligence) addressed greater than ever previously [6].

Papandreou, G. *et al.* (2015): Deep convolutional neural networks (DCNNs) trained on an enormous amount of snaps with strong pixel-level annotations have dramatically improved the state-of-the-art in semantic image segmentation. Investigated is the more difficult endeavor of learning DCNNs for semantic image segmentation utilizing either weakly annotated training data such as box boundaries or pixel-level labels, possibly a mixture of a couple of well-labeled photos and many weakly annotated images from any number of datasets. This work investigates the application of EM algorithms for training semantic segmentation models, particularly effective when dealing with limited labeled data situations [11].

Zhou, S. K. *et al.* (2021): Deep reinforcement learning (DRL) leverages the computational power of deep networks of neurons with a reinforcement training framework to learn a set of responses that maximize the intended reward. DRL applications for healthcare imaging are broken down into three types: (i) parametric medical image analysis tasks such as point recognition, object/lesion recognition, authority, and view plane localization; (ii) optimization-related obligations such as tuning the hyper parameter optimization, augmentation strategy decision-making, and neural framework search; and (iii) an assortment of services such as destructive signal differentiated instruction [20].

Jabbar, M. A. *et al.* (2021): Healthcare and artificial intelligence (AI), as well as how AI has influenced the medical field, include an in-depth examination of works applying deep learning in health informatics in the domains of medical imaging, electronic medical records, genomics, and detection, as well as emphasizing numerous obstacles connected to utilizing profound learning in the medical field [8].

Caballero, M. *et al.* (2019): The majority of medical inspections encompass image-based techniques, and European healthcare organizations generate enormous amounts of biomedical information, notably imagery. Considering the majority untapped their value originates from expert interpretations, these information sources continue to evolve and provide a large untapped repository of experience. In most of that point, this procedure is done manually, and global data interchange is cumbersome. Health research discovery and innovation will occur to move quickly under the so-called fourth paradigm of science, which is based on automating and expediting the examination of health information and procedures drawing together formerly distinct and diverse efficient computation and big data analysis tools [4].

Zemouri, R. *et al.* (2019): Deep neural networks are the key machine learning techniques used in all uses in medicine, including omics, bio and healthcare imaging, the body and brain machine user experience, and health care management. CNNs are the basic deep brain architecture, according to the literature, and have shown to be highly effective in conveying information between seemingly disparate tasks such as categorization via weight the act of transference. The overwhelming majority of transfer learning pertains with biological imaging applications. The most prominent emergent designs used in biomedical applications are generative adversarial networks. These delightful neural networks provide a data augmentation tackle to boost the deep representations in their dearth of substantial annotated training stuff. The majority of GANs serve a purpose in biological imaging applications [18].

Rayan, R. A. *et al.* (2022): Machine learning (ML) is an effective tool to recognize concealed information in internet of things (IoT) data. The intelligent composite fields enhance decision making in a range of fields, notably training, safety, commerce, and the field of biomedical sciences. For more accurate forecasting and systems for suggestion, ML improves the Internet of Things in explaining concealed patterns in massive quantities of data. Because of IoT, deep learning (DL) and machine learning (ML) were both previously integrated in medicine, so autonomous computers can now create medical data, assess illness, and, perhaps most importantly, examine patients in a continuous fashion [14].

4.3 Deep Reinforcement Learning Framework

Deep reinforcement learning (DRL) approaches have the subsequent basic components:

Environment: The agent's environment is the extrinsic entity with which it interfaces. This might involve a replicated receptive, a medical imagery dataset, or a drug synthesis simulation in healthcare.
State (S): The present circumstances or composition of the environment is symbolized by the state (S). In the case of a patient, it could constitute their vital statistics, medical history, and so forth.
Action (A): The decisions or manipulations that the agent can make in each state. Recommendation of a treatment, revision of medicine dosage, or choice of a region that is intriguing in medical imaging is a few such serves in healthcare.
Reward (R): This is constructive criticism signal the fact that the agent obtains after accomplishing an activity. It assesses the instantaneous reward or expenditure of the undertaking. In healthcare, it could be a measure of patient health advancement, affordability, or diagnostic precision.
Policy (π): The methodology or compilation of rules whereby the agent use when choosing actions in various circumstances. In DRL, this is frequently depicted by a neural network, which is a network that takes a state as input and produces an average distribution of probabilities among behaviors.
Value function (V or Q): It weighs the expected cumulative beneficial effect that comes from being in a specific state or performing a certain action. This enables the agent to gauge the desirability of different circumstances or actions.
Exploration vs. exploitation: Balancing among trying new actions to uncover possibly superior tactics (exploration) and adopting the most prevalent possibility based on current regulations (exploitation).
Learning algorithm: This is the algorithm's recommendation that influences the policy and/or value function in accordance with the experiences (state, action, reward) obtained via interactions with the environment. In DRL, this often entails back propagation throughout time.

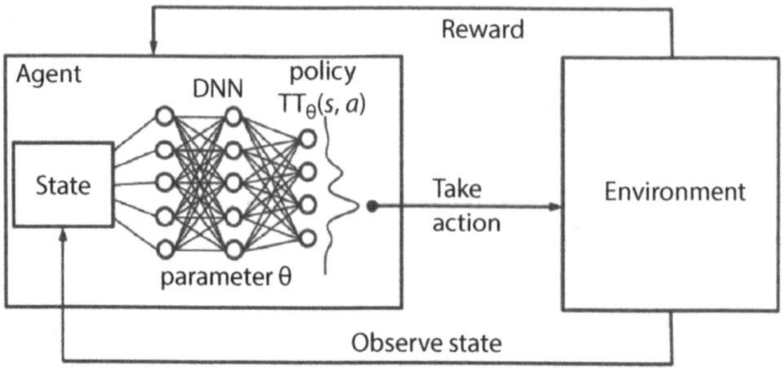

Figure 4.1 Framework of deep reinforcement learning.

The agents in Figure 4.1 have the option to learn directly from high-dimensional, raw visual inputs. Generally speaking, DRL relies on training deep neural networks to select the best values and optimal policies π^*.

4.4 Deep Reinforcement Learning Applications in Healthcare and Biomedicine

DRL (deep reinforcement learning) has tremendous upside in healthcare and biomedical applications. Here are a few examples of the key domains where DRL makes significant contributions:

Personalized treatment planning: Using their own distinctive histories of illness and responses to interventions, DRL can optimize treatment plans for individual patients. It is able to modify treatment tactics over time in order to guarantee the greatest efficacy.

Drug discovery and development: DRL may assist with speeds up the process of identifying future research medication prospects and optimizing molecular frameworks. It can predict a compound's biological activity while conducting trials to validate its safety and effectiveness.

Medical imaging and diagnosis: DRL can help with tasks including the process of segmentation detection, and classification regarding medical images. It influences the accuracy and efficiency of radiological being diagnosed.

Robot-assisted surgery: DRL enables surgical machines to function with dexterity and adapt to ever-shifting surgical instances. It aids in reducing the number of surgical complications and the enhancement of patient outcomes.

Predictive analytical techniques and persistent tracking: DRL models can assess uninterrupted patient data streams, making instantaneous forecasts regarding patient states and recommending actions that are appropriate.

Clinical trial enhancement: DRL can plan and enhance research investigations by allocating resources efficiently, selecting groups of patients, and changing trial procedures based on evolving facts.

Physical therapy treatment and rehabilitation: DRL can decide on adaptable rehabilitation regimens for patients recuperating from injuries or surgery. It tweaks exercises to individual ratings and tracks progress.

Chronic disease maintenance: DRL can help with management of chronic illnesses by offering customized treatment plans which have been amended based on consumer input and health position.

Genomic medicine: DRL can analyze data from genomics to find genetic characteristics that are associated with illnesses, optimize approaches to therapy depending on genetic characteristics, and are expecting disease risks.

Healthcare operations and resource management: DRL can optimize distribution of resources in health care organizations, from bed management to workforce scheduling, which can improve efficiency and safeguard costs.

Chatbots and virtual health assistants: Based on patient input, DRL-powered virtual assistants is able to provide personalized medical advice, reminders, and even exploratory receiving a diagnosis.

DRL can aid in the onset of therapies for psychological problems through modifications to treatment processes in accordance with client advancement and behaviors. While DRL has the potential aimed at transforming healthcare and biomedical applications, it also offers shortcomings in terms of moral quandaries, legal compliance, and privacy concerns. As a result, the implementation must be undertaken with prudential behavior and professionalism.

4.5 Deep Reinforcement Learning Employs Efficient Algorithms

4.5.1 Deep Q-Networks

It is helpful to design tailored treatment strategies for chronic disease patients by selecting relevant interventions over time. DQN can account for the ever-shifting nature of patient instances and alter therapies accordingly and also learns rapidly from sequential clinical information, dipped what is needed for big databases.

4.5.2 Policy Differentiation Techniques

Determine the best medication dosage for each patient based on their distinct features and treatment responses. Appropriate for concerns with ongoing action spaces, making dosage shifts conceivable. PPO plays a role in policy stability during training, which is crucial in healthcare settings.

4.5.3 Hindsight Experience Replay (HER)

Rehabilitation exercises ought to be tailored for patients recuperating from illnesses or a surgical procedure. HER addressed a cause of sparse reward signals, a phenomenon prevalent during activities related to rehabilitation and promotes learning by recovering experiences with different desired outcomes.

4.5.4 Curiosity-Driven Exploration

Establish rehabilitation programs that are tailored to the distinctive requirements and abilities of every client. By encouraging an assortment of exercises, curiosity-driven exploration fosters more thorough recuperation for patients. Robust learning aids in the formation of robust policies interact with fluctuating patient conditions.

4.5.5 Long Short-Term Memory Networks and Recurring Neural Network Designs

Monitor information regarding patients in the present moment to detect fluctuations in their wellness and anticipate illness economic growth. RNNs and LSTMs are well-suited for jobs involving chronological connections in data. LSTMs can recognize long-lasting patterns in patient data, allowing for greater precision predictions.

4.5.6 Multi-Agent DRL

During an operation, enhance the coordination among multiple robotic surgical instruments. Facilitates agents to come up with collaborative tactics which contribute to superior surgical outcomes. Real-world addresses the intricate details of real-world surgical circumstances.

4.6 Semi-Autonomous Control Based on Deep Reinforcement Learning for Robotic Surgery

Automation has been made extensively employed in surgical assistive technology in recent years in order to assist surgeons as they carry out complicated procedures despite dropping labor during the operation. The great majority of feasible treatments rely on demonstration learning, which involves the practice of small-incision surgical procedures from a skilled surgeon. However, gathering such a dataset would previously regress on a preset trajectory, hard and arduous tasks can place significant demands on one's knowledge specialists.

4.6.1 Double Deep Q-Network (DDQN)

The double deep Q-network (DDQN) is a deep reinforcement learning technology used in robotic surgery. It is a version of the deep Q-network (DQN) approach, which is a type of reinforcement learning system that uses a neural network to approximate the Q-value function. The Q-value function is used to determine the expected reward of a certain action in a particular condition. DDQN outperforms DQN by utilizing two neural networks rather than one. One network determines the best course of action, while the other evaluates its Q-value. This helps to reduce overestimation of the Q-value in DQN when the neural network's accuracy is insufficient.

In robotic surgery, double deep Q-networks (DDQN) might have various real-time applications:

Autonomous surgical services: DDQN may be used to independently teach a robotic agent to execute certain surgical procedures. It might help in suturing, cutting, or manipulating tissues.

Optimization of instrument pose: DDQN may be used to learn ideal surgical tool stances and motions for certain operations. This can assist enhance surgical precision and efficiency.

Navigation and path planning: The system can design and navigate ideal pathways for robotic surgical tools while accounting for anatomical limits and safety concerns.

Control and feedback using adaptive control: DDQN can be used to dynamically alter the control strategy of robotic instruments in response to changes in the surgical environment, such as patient movement or tissue property variations.

Transferring and training skills: By learning from expert demonstrations, DDQN may be used to teach surgical robots to do specific operations. This has the potential to speed up the training process for new surgical methods.

Detection and correction of errors: During a surgical operation, the DDQN model may be programmed to detect and fix problems in real time. It might, for example, change the direction of a surgical tool if it deviates from the desired course.
Improved teleoperation: To help surgeons perform complex tasks with greater precision and stability, DDQN can be integrated into teleoperated robotic surgical systems.
Dynamic tissue variability response: Based on feedback from sensors that monitor tissue properties, the algorithm can adjust the surgical strategy in real time. This is especially useful when dealing with anatomical variations or unexpected changes in tissue properties.
Surgeon fatigue reduction: DDQN can help reduce the cognitive load on surgeons by automating certain aspects of the surgical procedure, allowing them to focus on critical decision-making and more complex tasks.
Error prevention and safety: DDQN can be used to implement safety mechanisms that actively prevent the robotic system from performing potentially hazardous or unintended movements during surgery.

It should be noted that the use of such technologies in real-world surgical settings would necessitate significant validation, regulatory approval, and careful adherence to safety standards. Furthermore, as a fail-safe safeguard, human monitoring and intervention should always be available. The use of DDQN in robotic surgery is a promising route for enhancing surgical results and expanding robotic surgical system capabilities.

4.6.2 Materials and Methods

As depicted in Figure 4.2(a), the customizable emulator is predicated on the Acoustic Multi-Body Frame (AMBF). The intention is to achieve the goal of making the peg transference activity semi-autonomous. The job has been divided into two sections: automatic broad control and manual precise control override. Regulating the mechanism that allows the gripper to get closer to the end of the peg and shift how it is positioned to a permitted grasp, the position serves as a component of the coarse control.

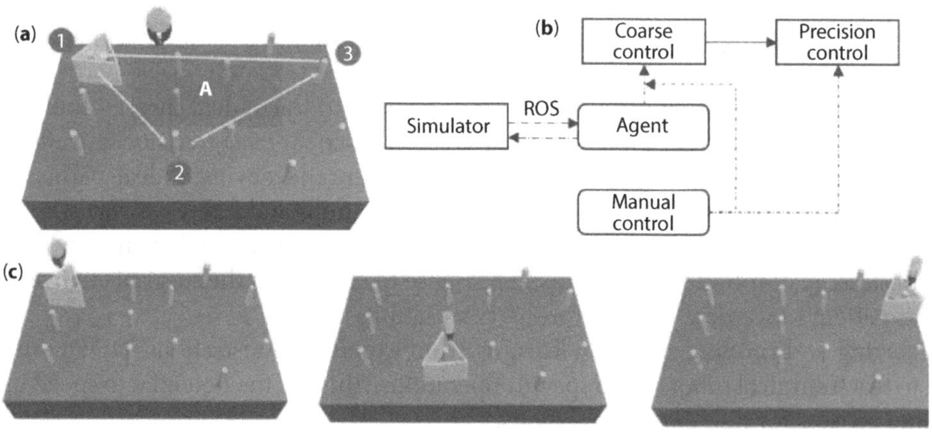

Figure 4.2 Displays the assessment task (a), the supervision process diagram (b), and the concluded structure, which consists of a succession of initializations with evolving initializations (c).

Precision controlling entails fine-tuning how to adjust the position of the gripper, grasping and shifting the peg to the desired position. Figure 4.2(b) exhibits the control process chart. We established an environment using robot operating system (ROS) that allows an agent to perform its tasks in a real-world situation using deep reinforcement learning techniques. The interface permits for the surroundings to serve up feedback on reward, apparent frame to show and information revealing whether or not the transition received.

The double deep Q-network (DDQN) was utilized to enhance the agent's performance as a tool in coarse control automation. Similarly, a hand-held interface that allowed the user to change the device's settings yet enabling precise administration was required. Control on a broad scale: We viewed coarse control as a Markov decision process encapsulated in the quintet S, A, T, and R, which stand for state of the environment, action and response space, likelihood of transition, incentive activity, and reduction ratio. Because this was an auditory trial, the interpreter was only provided with one image framework. After going on an excursion without knowing the current situation, perception of reality provides the agent with relevant knowledge.

The possibility of deducing the changing objective value. In this instance, the study's results reduced the entire frame to reduce the area of purpose to decrease determining load and then superimposed four frames in succession to feed the deep acquisition input neural network, therefore allowing it to infer at its current state. It would like to keep the end-effector at a suitable level so that it does not collide with other objects. The action vector was denoted by dx, dy, d, and it reflected the position displacement along the x and y axes in Cartesian coordinates as well as the roll inclination associated with the end-effector in Euler region.

The movement of the region has been compartmentalized with 6 mm, 8 mm, and 10 rad precision with ranges of [6 mm, 6 mm], [8 mm, 8 mm], and [10 rads, 10 rads]. Concentrating the action region through the discretization process can result in faster convergence while also saving training time and computation. The incentives operation, as shown in the initial equation, was designed to motivate the agent to come closer to the object being targeted and change its perspective when the distance between them, d, is smaller than the minimal threshold measurement, dt, of 10mm.

Where dt and t designate the length of the path to the intended target and how far it deviates from the necessitate orientations angle orthogonal to the direction of the target's that is closest side during time step t, correspondingly. The discount rate has been preset at 0.95 percentage.

$$r_{t+1} = \begin{cases} (d_t - d_{t+1})\,|\,d_t - d_{t+1}\,|, & \text{if } dt+1 > d_{\text{threshold}} \\ (\Delta\Theta_t - \Delta\Theta_{t+1})|\,\Delta\Theta_t - \Delta\Theta_{t+1}|, & \text{otherwise} \end{cases} \tag{4.1}$$

$$Q(S_t,a_t|\Theta)=r(S_t,a_t,S_{t+1})+Q(S_{t+1},a^*_{t+1}|\Theta`) \tag{4.2}$$

The uncoupling of the ideal selection of actions and action estimate of value for the next condition could mitigate underestimation and consequently calm the education procedure. Additionally, employing the use of objective network may aid in stabilizing the learning process. As stated in Figure 4.3(a), we envisioned and built a portable controller for accurate discretionary bypass operation. Using the OpenCV libraries, a comprehensiveness camera was utilized to track the 3-D establish of the tip of the 3-D printed handheld gadget controller, and an IMU detector was affixed at the end of the controller to track its 3-D their orientation. The position was then assigned to the simulator's gripper for manual control. Another instrument utilized to control the gripper's clutch was a foot pedal.

4.6.3 Results

The agent's retraining necessitates roughly 150 instances to accomplish integration as demonstrated in Figure 4.3(c). Moreover, following the integration the phases crucial to complete every instance merged, as shown in Figure 4.3(d). Figure 4.2(c) exhibits the end frames of instances featuring different initialized target positions. The agents are capable of managing the gripper's movement when approaching the target and shift the gripper's orientation to an optimal configuration for gripping in all three target positions. In accordance with manual overridden control, we qualitatively investigated the relation that existed between the traced gripping itinerary and the controller's progression, as shown in Figure 4.3(b).

It signifies that the grasp direction may be analogous to the controller's track. To validate the suggested solution methodology, we conducted an individual assessment. Figure 4.2(a) exhibits how to assess task. First, the gripper needs to comprehend the component in the initial position and push it towards position 2. The device that grasps is then reset towards an orientation within the area known as A. The method of estimation follows suit in order to shift the object of interest from the second spot to spot 3 and then return it to spot 1. The participants were given instructions to go through the procedure nine times in succession. For assessments, the standard actuator journey path M and time taken to finish the task T were determined. Table 4.1 summarizes what was an assessed outcome. It was uncovered

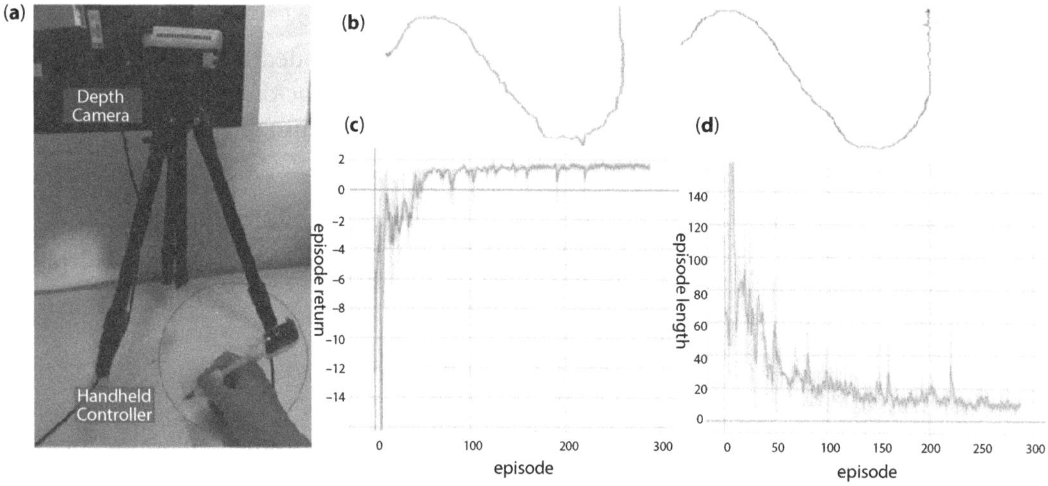

Figure 4.3 The configuration for manually overridden controller (a), the results of the qualitative analysis (b)—grasping track (red) and controlling track (blue)—series vs. series exchange (c), and series against series frequency (d).

Table 4.1 Evaluation results.

	Manual	Semi-autonomous
M	329 mm	136 mm
T	94 s	76 s

that the suggested framework shortened time spent travelling by roughly 58.7% and the duration of completion by almost 19.1%.

4.6.4 Discussion

Semi-autonomous control based on deep reinforcement learning for robotic surgery leverages the DDQN for establishing automated coarse supervision, freeing the person using it solely focused on precise control and making decisions at critical moments. Pursuant to the user research, the technique may considerably decrease the controller travel dimension while also accelerating the time required for completion. This demonstrates the recommended method's likelihood of streamlining routine procedures and lowering cognitive demands on medical professionals in MIS interventions. However, the time required for completion lowering margin was not quite as significant as anticipated because, when starting the fine-tuning stage immediately following the expansive control stage, the user was frequently forced to recognize the relative stance of the end-effector to the destination by tilting the transmitter's position significantly. As a result, future research will focus on providing tactile or force feedback to facilitate smooth interactive manipulation. Furthermore, an additional effort will be done on integrating what is currently learned interface to the da Vinci operating automation setup.

4.7 Conclusion

In a nutshell, the application of deep reinforcement learning (DRL) possesses tremendous promise for revolutionizing healthcare and biomedical applications. The capacity it has to learn convoluted making decisions processes from data has the potential to improve regimens for treatment, streamline health care processes, and personalized the provision of healthcare. Nevertheless, different obstacles must be surmounted beforehand it can achieve its maximum potential. The greater significance of data effectiveness while moral considerations and security considerations should not be emphasized. It has to be vital bringing about that DRL simulation undergo training on diverse, high-quality data while shielding patient confidentiality and preserving standards of ethics. Accessibility and comprehensibility are also crucial in obtaining the faith of healthcare providers and patients. Extending DRL models to shifted clinical contexts, demographics of patients, and patient heterogeneity seem to be an important concern. Transferable learning techniques and strategies for tweaking must be used. In healthcare tasks, the formulation of promotional processes and the establishment of pertinent indicators of performance are indispensable.

Balancing plenty aims and taking into consideration consequences over time are challenging issues that must be carefully taken into account. It is also critical for integrating with existing medical procedures and systems for electronic health records. DRL systems ought to blend easily into the routines of healthcare professionals, boosting productivity and promoting the outcomes for patients. Despite such obstacles, the prospective positive effects of DRL in healthcare and biological applications are immense if machine learning scientists, healthcare practitioners, ethicists, and regulatory agencies cooperate collaboratively. Resolving the aforementioned obstacles will pave an avenue for DRL to possess an invaluable part in promoting medical care and scientific inquiry in the decades to come. Overcoming these obstacles will pave the path for DRL to play a transformative role in advancing health services and scientific inquiry in future generations.

References

1. Alaniz, Stephan (2018) Deep reinforcement learning with model learning and montecarlo tree search in minecraft. The 3rd Multidisciplinary Conference on Reinforcement Learning and Decision Making (RLDM) 2017. DOI: 10.48550/arXiv.1803.08456.
2. Azghadi, M.R., Lammie, C., Eshraghian, J.K., Payvand, M., Donati, E., Linares-Barranco, B., Indiveri, G., Hardware implementation of deep network accelerators towards healthcare and biomedical applications. *IEEE Trans. Biomed. Circuits Syst.*, 14, 6, 1138–1159, 2020.
3. Baucum, M., Khojandi, A., Vasudevan, R., Improving deep reinforcement learning with transitional variational autoencoders: A healthcare application. *IEEE J. Biomed. Health Inf.*, 25, 6, 2273–2280, 2020.
4. Caballero, M., Gómez, J.A., Bantouna, A., Deep-learning and hpc to boost biomedical applications for health (deephealth), in: *2019 IEEE 32nd International Symposium on Computer-Based Medical Systems (CBMS)*, IEEE, pp. 150–155, 2019, June.
5. Castiglioni, I., Rundo, L., Codari, M., Di Leo, G., Salvatore, C., Interlenghi, M., Sardanelli, F., AI applications to medical images: From machine learning to deep learning. *Phys. Med.*, 83, 9–24, 2021.
6. Dash, S., Kumar Pani, S., Rodrigues, J.J.P.C., & Majhi, B. (Eds.). (2022). Deep Learning, Machine Learning and IoT in Biomedical and Health Informatics: Techniques and Applications (1st ed.). CRC Press, Boca Raton, Florida, USA, https://doi.org/10.1201/9780367548445.
7. Han, J., Yang, L., Zhang, D., Chang, X., Liang, X., Reinforcement cutting-agent learning for video object segmentation, in: *Proceedings of the IEEE conference on computer vision and pattern recognition*, pp. 9080–9089, 2018.
8. Jabbar, M.A., Abraham, A., Dogan, O., Madureira, A.M., & Tiwari, S. (Eds.). (2021). Deep Learning in Biomedical and Health Informatics: Current Applications and Possibilities (1st ed.). CRC Press, Boca Raton, https://doi.org/10.1201/9781003161233.
9. Jayatilake SMDAC, Ganegoda GU. Involvement of Machine Learning Tools in Healthcare Decision Making. *J Healthc Eng.* 2021 Jan 27;2021:6679512, 20 pages. doi: 10.1155/2021/6679512. PMID: 33575021; PMCID: PMC7857908.
10. Li, Z. and Xia, Y., Deep reinforcement learning for weakly-supervised lymph node segmentation in CT images. *IEEE J. Biomed. Health Inf.*, 25, 3, 774–783, 2020.
11. Papandreou, G., Chen, L.C., Murphy, K.P., Yuille, A.L., Weakly-and semi-supervised learning of a deep convolutional network for semantic image segmentation, in: *Proceedings of the IEEE international conference on computer vision*, pp. 1742–1750, 2015.

12. Qayyum, A., Qadir, J., Bilal, M., Al-Fuqaha, A., Secure and robust machine learning for healthcare: A survey. *IEEE Rev. Biomed. Eng.*, 14, 156–180, 2020.
13. J. Qiu *et al.*, "Large AI Models in Health Informatics: Applications, Challenges, and the Future," in *IEEE Journal of Biomedical and Health Informatics*, vol. 27, no. 12, pp. 6074–6087, Dec. 2023.
14. Rayan, R.A., Zafar, I., Rajab, H., Zubair, M.A.M., Maqbool, M., Hussain, S., Impact of IoT in Biomedical Applications Using Machine and Deep Learning, in: *Machine Learning Algorithms for Signal and Image Processing*, pp. 339–360, 2022.
15. Wang, S., Fu, L., Yao, J., Li, Y., The application of deep learning in biomedical informatics, in: *2018 International Conference on Robots & Intelligent System (ICRIS)*, IEEE, pp. 391–394, 2018, May.
16. Wang, C., He, T., Zhou, H., Zhang, Z., Lee, C., Artificial intelligence enhanced sensors-enabling technologies to next-generation healthcare and biomedical platform. *Bioelectron. Med.*, 9, 1, 17, 2023.
17. J., Soltan, A.A.S., Eyre, D.W. *et al.* Algorithmic fairness and bias mitigation for clinical machine learning with deep reinforcement learning. *Nat. Mach. Intell.* 5, 884–894 (2023), https://doi.org/10.1038/s42256-023-00697-3.
18. Zemouri, R., Zerhouni, N., Racoceanu, D., Deep learning in the biomedical applications: Recent and future status. *Appl. Sci.*, 9, 8, 1526, 2019.
19. Zhang, Y.D., Dong, Z., Gorriz, J.M., Jiang, Y., Yang, M., Wang, S.H., IEEE Access Special Section Editorial: Emerging Deep Learning Theories and Methods for Biomedical Engineering. *IEEE Access*, 9, 74038–74043, 2021.
20. Zhou, S.K., Le, H.N., Luu, K., Nguyen, H.V., Ayache, N., Deep reinforcement learning in medical imaging: A literature review. *Med. Image Anal.*, 73, 102193, 2021.

5

Application of Deep Reinforcement Learning in Adversarial Malware Detection

Manju* **and Chhavi Rana**

Computer Science and Engineering (UIET), Maharshi Dayanand University, Rohtak, India

Abstract

The rapid evolution of computer technologies has heightened the operational vulnerability to malicious software attacks, leading to frequent data breaches for companies. Despite innovative techniques proposed to counter various cyber threats, their efficacy diminishes with the introduction of adversarial attacks, making malware identification increasingly challenging. This study underscores the critical need for a reliable malware detection algorithm to safeguard Internet data. Furthermore, it highlights the enhanced efficacy achieved in Android malware detection through machine learning and deep reinforcement learning (DL) approaches. Notably, the research provides a comprehensive overview of DL techniques, machine learning methodologies, feature selection approaches, datasets, and the DL implementation framework, with a specific focus on their application in malware detection. The study also demonstrates the generation and utilization of adversarial samples using diverse deep reinforcement learning techniques, showcasing their effectiveness in augmenting training data.

Keywords: Deep, reinforcement, adversarial, attacks, malware, machine, learning, algorithms

5.1 Introduction

Feature selection plays a pivotal role in the effectiveness of adversarial malware detection, especially in the context of integrating deep reinforcement learning (DRL). As DRL combines deep neural networks with reinforcement learning for optimal decision-making in malware detection, selecting relevant features becomes crucial. Features are specific aspects or characteristics of data that the model uses for identification and decision-making. In the realm of cybersecurity, the choice of features directly influences the model's ability to recognize malicious patterns and adapt to evolving threats.

Cybersecurity, malicious software, commonly known as malware, constitutes a diverse range of harmful programs like viruses, worms, trojan horses, ransomware, and spyware, designed with the intent to disrupt, damage, or illicitly access computer systems. Specifically, adversarial malware emerges as a category intentionally crafted to outsmart

**Corresponding author*: dhull.manju91@gmail.com; ORCID ID: 0000-0001-6129-7475

Shubham Mahajan, Pethuru Raj and Amit Kant Pandit (eds.) Deep Reinforcement Learning and Its Industrial Use Cases: AI for Real-World Applications, (91–114) © 2024 Scrivener Publishing LLC

traditional detection methods. Characterized by a high degree of sophistication, adversarial malware employs advanced techniques to evade detection, making it challenging for conventional mechanisms like signature-based or heuristic-based methods to identify its dynamic behavior or appearance alterations. Detection strategies involve pattern recognition, where identifying malicious code patterns, and behavioral analysis, examining software actions for malware-like characteristics, play crucial roles. Adversarial malware detection faces challenges due to its polymorphic nature, constantly changing code, and utilization of zero-day exploits, targeting vulnerabilities unknown to security experts. To counter these challenges, advanced detection techniques such as machine learning and artificial intelligence are employed, allowing systems to dynamically learn and adapt to new threats. The continuous evolution of the threat landscape necessitates constant updates in detection strategies, emphasizing the importance of staying informed about the latest threat intelligence to enhance capabilities and address emerging adversarial techniques.

Artificial intelligence, deep reinforcement learning (DRL) intricately merges the realms of deep learning and reinforcement learning to address complex challenges. Deep learning, a subset of machine learning, employs artificial neural networks with multiple layers to unravel intricate patterns in data. Reinforcement learning, on the other hand, is a machine learning paradigm where an agent learns to make sequential decisions through interactions with an environment, receiving feedback in the form of rewards or penalties. DRL seamlessly integrates these approaches by utilizing deep neural networks to automatically discern hierarchical representations from raw input data, while reinforcement learning establishes the framework for making optimal sequential decisions based on feedback. This integration allows the model to navigate complex problems over time, particularly in tasks with high-dimensional input such as image or sequence data. DRL's versatility is evident in its applications across diverse fields, from robotics to gaming, and notably in cybersecurity, where it proves valuable for intricate tasks like adversarial malware detection.

Machine learning, the role of feature selection emerges as a critical determinant in enhancing model performance and efficiency. Features, representing specific aspects or characteristics of data, serve as the bedrock for a model's decision-making, allowing it to learn intricate patterns and relationships. The information content carried by each feature and the synergy among them significantly impact the overall model performance. Feature selection, defined as the process of choosing the most relevant and informative features, aims to optimize model effectiveness by focusing on discriminative features while discarding irrelevant or redundant ones. This process proves essential for addressing challenges such as the curse of dimensionality, where datasets with numerous features can pose computational complexities. Various methods, including filter, wrapper, and embedded methods, offer diverse approaches to feature selection, catering to different modeling scenarios. Challenges such as handling correlation and redundancy, as well as domain-specific considerations, emphasize the need for a nuanced understanding of the task at hand. Feature selection becomes an iterative journey involving continuous refinement, where initially selected features are honed based on model performance, underscoring its dynamic and evolving nature. Ultimately, the art and science of feature selection strike a delicate balance between domain knowledge, algorithmic techniques, and the specific requirements of the given task, ensuring optimal model generalization and interpretability.

Cybersecurity, the utilization of deep reinforcement learning (DRL) for malware detection emerges as a strategic approach aimed at optimizing decision-making processes.

DRL, known for its prowess in sequential decision-making tasks, proves particularly adept at adapting to the ever-changing threat landscapes inherent in malware detection. The effectiveness of DRL hinges on the quality and relevance of the selected features, acting as the foundational elements that enable the model to comprehend data characteristics and make informed decisions. The discriminating power of the model, crucial for distinguishing benign from malicious patterns, relies heavily on the meticulous selection of features. Relevant features play a key role in identifying anomalies associated with malicious activities, facilitating the model's ability to flag potential threats. As the cybersecurity landscape dynamically evolves with the emergence of new malware variants, effective feature selection contributes significantly to the model's adaptability, allowing it to recognize novel patterns indicative of evolving threats. Risks associated with including irrelevant or redundant features, such as hindrance to discrimination, overfitting, and increased computational overhead, underscore the importance of precision in feature selection. The adaptive learning inherent in DRL involves continuous refinement of features, with the model adjusting its behavior over time based on feedback and evolving patterns. In the dynamic cybersecurity landscape, feature selection in DRL for malware detection becomes an iterative process, ensuring continuous improvement and the maintenance of optimal detection performance in the face of emerging threats. The careful consideration of feature selection emerges as a pivotal factor for the success of DRL in the complex domain of malware detection within cybersecurity.

Cybersecurity domain, the impact of features on model performance is paramount in navigating the ever-evolving threat landscape. The selection of pertinent features directly shapes the model's adaptability to dynamic threats, enabling it to stay responsive to emerging patterns and behaviors associated with the regular influx of new malware variants and attack techniques. Features play a pivotal role in pattern recognition, aiding the model in distinguishing normal system behavior from deviations indicative of malicious activity. The relevance of features is instrumental in the model's capacity to discern subtle differences, achieving higher specificity in detection and reducing false positives. The dynamic selection of features is intricately linked to the model's ability to adapt to evolving threats, allowing it to adjust its understanding of normal and suspicious behavior in real-time scenarios. Feature-based learning, incorporating historical data, facilitates the model's learning process and enables it to generalize knowledge to recognize new threats. Continuous monitoring and updating of features, driven by integration with threat intelligence, enhance the model's proficiency in recognizing features associated with the latest threats. The proactive nature of threat detection, facilitated by relevant features, empowers cybersecurity systems to take preventive measures against emerging patterns, exemplifying the critical role of features in early detection and response to dynamic and sophisticated cyber threats.

In the landscape of cybersecurity, the adaptation to evolving threats is imperative for robust security measures, given the dynamic and constantly changing nature of cybersecurity threats. Malicious actors continuously devise new tactics, techniques, and procedures, requiring cybersecurity models to exhibit agility and responsiveness for effective defense. The adaptability of these models is intricately tied to the selection of properly chosen features during the detection process, capturing both known and novel threat characteristics. Features that align with the evolving threat landscape empower the model to recognize and adeptly respond to emerging attack patterns. Learning from relevant features, especially in models incorporating machine learning and deep learning techniques, enhances the model's understanding of the ever-evolving threat landscape. To stay effective against dynamic malware, models rely on a dynamic and well-selected feature set that enables the identification of subtle changes in behavior associated with new malware variants. Continuous monitoring, real-time integration of threat intelligence into feature selection, and a proactive approach to model updating contribute to a cybersecurity model's ability to stay ahead of the rapidly changing threat landscape. Collaborative security measures, including information sharing among different security systems, facilitate the building of a comprehensive understanding of evolving threats. The ability of models to quickly recognize and respond to new threats emerges as a cornerstone for maintaining resilient and effective cybersecurity defenses.

The research conducted by Wang *et al.* (2017) addresses the vulnerability of deep neural networks to adversarial samples, a common weakness in statistical and machine learning models, particularly in the context of malware detection. The study explores techniques to enhance the adversary resistance of deep neural networks, with a specific application to mitigating the impact of adversarial attacks on malware detection systems [1]. Consider a scenario where the feature selection process involves identifying key characteristics of malware, such as code structure, file behavior, and network traffic patterns. A well-chosen set of features enhances the model's accuracy in distinguishing between benign and malicious activities—for instance, if a malware strain exhibits specific code structures or unusual behavior, these features become critical indicators for the detection model.

In pursuit of enhanced model accuracy, meticulous feature selection is employed to focus on pertinent aspects, thereby refining the model's precision in detecting and countering malware threats. This approach instills adaptability to the dynamic and evolving nature of cyber threats, allowing the model to recognize new patterns and adjust to unforeseen

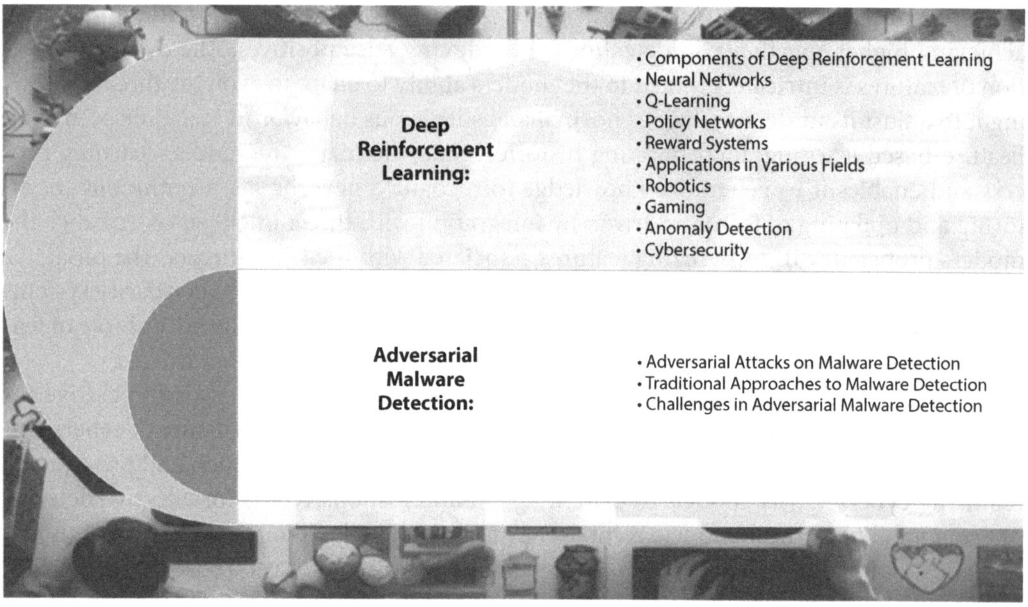

Figure 5.1 Application of deep reinforcement learning in adversarial malware detection.

challenges in the ever-changing cyber landscape. The judicious selection of features optimizes computational resources, preventing the unnecessary processing of irrelevant data. Despite these advantages, challenges persist, particularly in addressing the dynamic nature of adversarial tactics. Adversaries consistently refine their strategies, necessitating ongoing adjustments to the feature selection process for the model to effectively keep pace with evolving threats. Another challenge lies in striking the right balance between relevance and complexity when selecting features, as an excess of features may increase computational demands without a commensurate improvement in model performance.

The importance of feature selection in adversarial malware detection, particularly within the framework of DRL, lies in optimizing model accuracy, adaptability to evolving threats, and the efficient use of computational resources. Thoughtful feature selection ensures that the model remains effective in identifying and mitigating both known and novel malware strains, contributing to the transformative impact of DRL in the ongoing battle against cyber threats [2].

5.1.1 Background

In the rapidly evolving of computer technologies, the escalation of vulnerabilities to malicious software attacks poses a significant threat to the security of companies' confidential data. The intricate interplay between technological progress and cyber threats challenges traditional security measures, particularly with the emergence of adversarial attacks that add complexity, diminishing the efficacy of conventional methods. The evolving sophistication of malware not only exploits vulnerabilities but actively outsmarts detection mechanisms, presenting a formidable challenge for cybersecurity experts and demanding a paradigm shift in defense strategies. Adversarial attacks, designed with a keen understanding of detection system weaknesses, enable bypassing traditional security measures, leaving companies susceptible to data breaches and intellectual property theft. The imperative to develop robust and trustworthy malware detection algorithms becomes paramount for safeguarding sensitive data on the Internet. The ongoing arms race emphasizes the need for advanced, adaptive, and resilient detection mechanisms. The development of such algorithms involves leveraging cutting-edge technologies, particularly artificial intelligence (AI), where machine learning, specifically deep learning, holds promise in enhancing malware detection capabilities. Zhong *et al.* (2023) discuss the integration and perspectives of artificial intelligence (AI) in enhancing head and neck tumor management, scrutinizing cutting-edge algorithms such as machine learning (ML) and neural networks (NNs). The article highlights the significant contributions made by AI algorithms in the detection and characterization of head and neck tumors [3]. By analyzing vast datasets and learning complex patterns indicative of malicious behavior, these algorithms provide proactive defense against emerging threats. Integrating behavioral analysis and anomaly detection strengthens the resilience of malware detection, enabling these advanced systems to adapt to the dynamic nature of cyber threats by recognizing deviations from normal behavior. This proactive approach not only identifies known malware but also detects novel and previously unseen threats, underscoring the necessity for more advanced and adaptive malware detection algorithms to safeguard confidential data in the face of evolving cyber threats.

5.1.2 Significance of Malware Detection

Faced with a surge in malicious software attacks, companies are compelled to establish an effective detection mechanism, not merely as an imperative but as a cornerstone in preserving organizational integrity and reputation. Beyond averting financial losses, the detection and neutralization of malware become fundamental to maintaining the trust of clients, stakeholders, and the wider public. The potential repercussions of a successful malware attack encompass data breaches, the compromise of sensitive information, and severe damage to an organization's reputation. The gravity of this situation underscores the pressing need for innovative approaches in cybersecurity. While conventional methods exhibit effectiveness to a certain extent, they are proving insufficient against increasingly sophisticated malware. Kesan and Hayes (2011) explore the concept of mitigative counterstriking as a strategy for self-defense and deterrence in cyberspace, particularly in response to software vulnerabilities and malicious activities. The authors emphasize the significance of this approach in the context of executive orders addressing cybersecurity, highlighting the gravity of the situation [4]. This realization has instigated a quest for more advanced and adaptive solutions, with a specific emphasis on harnessing technologies such as deep reinforcement learning. Serving as a beacon of promise in cybersecurity, deep reinforcement learning combines the strengths of deep neural networks with reinforcement learning principles, introducing a dynamic and self-improving aspect to malware detection. The capacity of these systems to learn and adapt from experience enables them to evolve alongside emerging threats, offering a level of resilience crucial in today's fast-paced cyber landscape. The integration of deep reinforcement learning not only enhances the accuracy of identifying malicious activities but also enables a more proactive defense. Traditional methods often struggle to keep pace with the rapidly changing tactics of cyber adversaries, necessitating the adoption of innovative solutions capable of learning, adapting, and predicting evolving threats. As organizations recognize the paramount importance of cybersecurity, the adoption of deep reinforcement learning in malware detection becomes a strategic imperative, representing a forward-thinking approach to fortify cybersecurity measures. This acknowledgment reflects an understanding that the battle against malicious software demands a continuous evolution of defense strategies, enabling companies to protect their digital assets, sensitive information, and uphold the trust and confidence of those who depend on the security and reliability of their systems.

5.1.3 Challenges with Adversarial Attacks

The challenges posed by adversarial attacks form a formidable obstacle in the domain of malware detection, particularly when integrating deep reinforcement learning (DRL). Adversarial attacks are meticulously crafted maneuvers designed to deceive traditional detection models, introducing a layer of complexity that hinders accurate malware identification. Aryal *et al.* (2021) conducts a comprehensive survey on adversarial attacks in the context of malware analysis, emphasizing the increasing relevance of this research area as the world moves towards automation. The authors focus on the intricacies of adversarial attacks against malware detection systems, addressing the complex changes in the architecture of files that necessitate precise analysis [5]. The dynamic landscape of cyber threats is

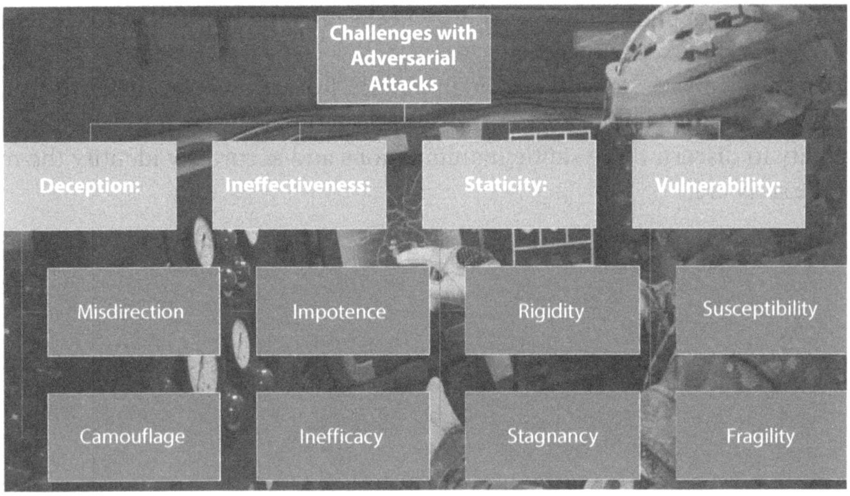

Figure 5.2 Challenges with adversarial attacks.

marked by evolving tactics employed by adversaries, highlighting the inadequacy of existing detection methods in the face of intricate and dynamic threats.

Key challenges in the landscape of malware detection include the utilization of deceptive tactics by adversarial attacks, compelling detection models to discern subtle patterns indicative of malicious activity. The constant evolution of malicious actors renders static detection methods less effective, emphasizing the necessity for adaptive models capable of learning and adjusting in real-time. Recognizing weaknesses in existing detection systems is crucial, highlighting the need for robust models that can withstand and adapt to previously unseen threats. Understanding the intricate dynamics of adversarial attacks is significant in developing effective defense mechanisms, driving the exploration of advanced approaches such as deep reinforcement learning (DRL). Addressing challenges posed by adversarial attacks ensures the development of models capable of preventing evasion and accurately identifying evolving malware threats, contributing to the creation of a proactive defense mechanism that anticipates and counters evolving cyber threats. The strategic response outlined in the introduction focuses on exploring DRL techniques, leveraging their adaptability and learning capabilities to address and mitigate the impact of adversarial attacks on malware detection systems within the context of the ever-changing cybersecurity landscape.

5.2 Foundations of Deep Reinforcement Learning

The integration of deep reinforcement learning (DRL) in the realm of cybersecurity presents its own set of challenges, especially concerning adversarial attacks. Adversarial attacks, accurately identifying malware. As DRL adapts and learns from experiences, it encounters challenges in distinguishing between authentic and adversarial patterns, given the intricate and evolving tactics employed by cyber adversaries. Alshammari *et al.* (2020) explores the status, challenges, and perspectives of cyber deception for cybersecurity, depicting the typical lifecycle of cyber deception for cyber adversaries in their research. The authors provide insights into the use of deception as a strategy in the ongoing battle between defenders and adversaries in the cybersecurity domain [6].

Consider an adversarial attack that manipulates features within malware to resemble benign behavior, aiming to deceive a DRL-based detection system. The challenge lies in the model's ability to discern these subtle manipulations and accurately identify the malicious nature of the software.

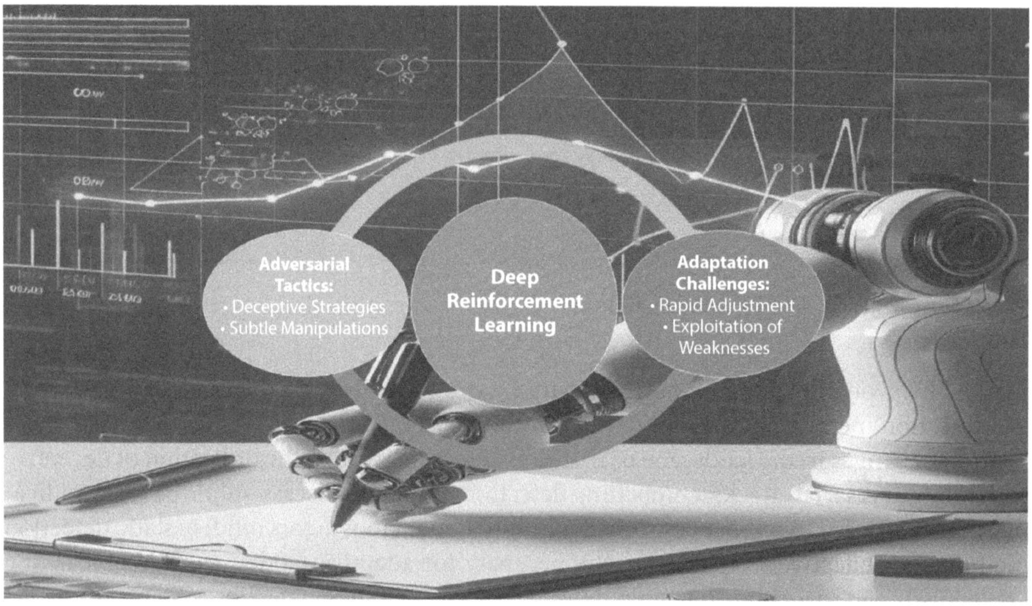

Key challenges in the context of adversarial malware detection using deep reinforcement learning (DRL) involve adversaries employing subtle manipulations of malware features, posing a challenge for DRL models to differentiate between authentic and manipulated patterns. The evolving tactics of cyber adversaries necessitate rapid adaptation from DRL models to discern new patterns indicative of malicious activity. Ensuring robust adaptation to unforeseen adversarial maneuvers is crucial, as adversaries exploit weaknesses to evade detection. Addressing these challenges enhances the resilience of DRL models, enabling them to accurately identify malware even in the face of deceptive tactics. The ability to adapt in real-time allows DRL models to keep pace with dynamic tactics employed by adversaries, ensuring effective and reliable malware detection within the ever-changing cybersecurity landscape. Recognizing these challenges sets the stage for a strategic response, emphasizing the need for ongoing research and development to enhance the robustness of DRL models and offer a more adaptive, resilient, and proactive defense against the elusive tactics of cyber adversaries.

5.2.1 Overview of Deep Reinforcement Learning

Deep reinforcement learning unfolds as a technological marvel, a sophisticated amalgamation of artificial intelligence and machine learning that empowers machines to learn and make decisions in a manner mirroring human cognition. This paradigm represents a pivotal advancement within the broader field of reinforcement learning, distinguishing itself through the integration of deep neural networks. At its core, deep reinforcement learning operates on the principles of reinforcement learning, a framework where an artificial agent interacts with an environment, takes actions, and receives feedback in the form of rewards or penalties. What sets deep reinforcement learning apart is its utilization of deep neural networks, enabling the system to comprehend intricate patterns and representations within complex datasets. This fusion allows the algorithm to discern and learn hierarchies of features, extracting nuanced information that proves invaluable in decision-making processes. The deep neural networks in this paradigm serve as powerful function approximators, capable of understanding and representing intricate relationships within the input data. This characteristic is particularly advantageous in scenarios where the environment is dynamic and the state space is vast, such as in robotics, game playing, or, as in the context of cybersecurity, the identification of malicious software. Kott (2015) presents the science of cybersecurity as a system of models and problems, emphasizing the significance of identifying and characterizing malicious software within the cyber security landscape [7]. One of the key strengths of deep reinforcement learning lies in its ability to learn autonomously through trial and error. The algorithm refines its decision-making processes based on the consequences of its actions, iteratively adapting its strategies to achieve optimal outcomes. This adaptability is especially crucial in dynamic environments, where the system must evolve to navigate novel challenges, making it particularly well-suited for applications like malware detection, which demands constant adaptation to the evolving tactics of cyber adversaries. The versatility of deep reinforcement learning is further emphasized by its applicability in a myriad of domains. From playing complex games like Go and poker to optimizing robotic control and enhancing natural language processing, its reach extends across diverse fields. In the realm of cybersecurity, the paradigm's ability to discern patterns in data and adapt to emerging threats positions it as a potent tool for fortifying defenses against sophisticated cyber-attacks. Deep reinforcement learning stands as a technological frontier, embodying the synergy of artificial intelligence, machine learning, and neural networks. Its adaptability, autonomy, and capacity to discern complex patterns make it a formidable force in addressing challenges in various domains, including the ever-evolving landscape of cybersecurity where it promises to be a key player in advancing the detection and mitigation of malicious software.

5.2.2 Core Concepts and Components

It sheds light on the challenges associated with adversarial attacks, especially in the application of deep reinforcement learning to malware detection. As this educational initiative unfolds the intricate foundations of deep reinforcement learning, it addresses the formidable hurdles presented by adversaries aiming to deceive detection systems. Santos Jr. and Johnson Jr. (2004) contribute to the field of deception detection in intelligent systems by presenting a theory of deception and various types of deception, offering insights into a model for detecting deception [8]. The evolving tactics of cyber adversaries, meticulously crafted to mislead conventional detection models, pose a significant obstacle within the realm of cybersecurity.

Consider an adversarial attack designed to manipulate input features, such as code structures or file behaviors, to deceive a deep reinforcement learning-based malware detection system. The challenge lies in the model's ability to discern between authentic and manipulated patterns, showcasing the need for robust defenses against such deceptive maneuvers.

Key challenges in the domain of adversarial malware detection using deep reinforcement learning (DRL) include adversaries employing subtle manipulations of features, presenting a difficulty for DRL models to accurately distinguish between authentic and manipulated patterns. The dynamic tactics of cyber adversaries necessitate continuous adaptation from DRL models to discern new patterns indicative of malicious activity. Ensuring robust adaptation to unforeseen adversarial maneuvers is crucial, as adversaries exploit weaknesses to evade detection. Addressing these challenges enhances the resilience of DRL models, ensuring accurate identification of malware even in the face of deceptive tactics. The ability to adapt in real-time allows DRL models to keep pace with the dynamic tactics employed by adversaries, ensuring effective and reliable malware detection within the ever-changing cybersecurity landscape. The "core concepts and components" module establishes a foundation for a strategic response, emphasizing the necessity for ongoing research and development to enhance the robustness of DRL models. Recognizing and addressing these challenges enables the evolution of deep reinforcement learning in adversarial malware detection, offering a more adaptive, resilient, and proactive defense against the elusive tactics of cyber adversaries.

5.2.3 Relevance to Malware Detection

The narrative emphasizes the learned decision-making capabilities of artificial agents and their adaptability to dynamic environments as crucial factors, positioning DRL as a compelling paradigm to tackle the intricate challenges posed by malware. The text meticulously delineates how the adaptive decision-making processes in DRL systems contribute to their efficacy in the context of malware detection, surpassing traditional mechanisms by learning from experience and navigating the evolving landscape of cyber threats. Vinod, Laxmi, and Gaur (2012) present "Reform," a non-signature-based method for detecting obfuscated malware using machine learning techniques. The approach involves extracting mnemonic n-grams from malware and benign samples, selecting a subset of features using methods like principal component analysis (PCA) and minimum redundancy and maximum relevance (mRMR) to effectively discriminate between malware and benign samples. The proposed method demonstrates promising results with small features and improved accuracies compared to previous work, making it effective for identifying malicious files [9].

Consider an adversarial attack attempting to deceive a malware detection system utilizing DRL by subtly altering malware features. The challenge lies in the system's ability to autonomously adapt its decision-making strategies based on real-time feedback from the dynamic cybersecurity environment, ensuring accurate identification despite adversarial attempts.

Key challenges in deploying deep reinforcement learning (DRL) systems for malware detection include the need for autonomous adaptation in decision-making strategies, especially given the dynamic nature of cyber threats. DRL models must effectively learn from past experiences to navigate the ever-evolving landscape of malware, ensuring continued adaptability in the face of sophisticated evasion tactics employed by malicious software. The significance lies in the enhanced adaptability of DRL, where its learned decision-making capabilities bolster the resilience of malware detection systems, enabling dynamic responses to emerging adversarial tactics. Emphasizing proactive defense, this section underscores DRL's ability to stay ahead of evolving threats through continuous learning and real-time adaptation, showcasing tangible impacts on fortifying cybersecurity measures. Serving as a strategic response, it acts as a bridge between theoretical knowledge and actionable strategies, illustrating how DRL principles actively contribute to safeguarding digital landscapes against the sophisticated and ever-changing nature of malware.

5.3 Malware Detection Landscape

The historical journey begins with an examination of the earliest techniques employed for malware detection, marking the advent of signature-based approaches. As the narrative unfolds, it traverses through the chronological evolution of detection methods, capturing the shift from static signatures to more dynamic approaches. The exploration delves into heuristics, behavior-based detection, and anomaly detection, showcasing the continuous refinement of strategies to keep pace with the sophistication of malware.

In dissecting the intricacies of adversarial attacks in cybersecurity, the narrative ventures into the clandestine realm where cyber adversaries deploy advanced techniques to evade detection. The arms race between defenders and attackers is laid bare, emphasizing the deceptive and evolving tactics employed by those seeking to exploit vulnerabilities. Adversarial attacks, designed to mislead detection models, introduce a layer of complexity that demands a paradigm shift in defense strategies. Corona, Giacinto, and Roli (2013) conduct a comprehensive survey on adversarial attacks against intrusion detection systems (IDSs). The paper presents a general taxonomy of attack tactics, describes how attacks can exploit IDS weaknesses at different abstraction levels, critically investigates proposed solutions for each attack implementation, and outlines open issues. The survey aims to contribute to the design of adversary-aware, more robust IDS solutions by providing insights into attack strategies and potential countermeasures [10].

The narrative elucidates the imperative for advanced detection strategies in contemporary digital security, emphasizing the inadequacy of traditional methods in the face of sophisticated threats. As technology advances, so do the capabilities of malicious actors, necessitating innovative and adaptive approaches to stay one step ahead. The exploration

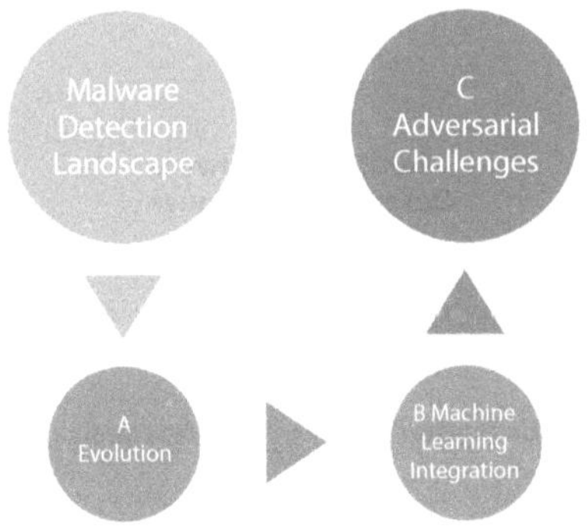

Figure 5.3 Malware detection landscape.

underscores the urgency for proactive defense mechanisms capable of not only identifying known malware but also adapting to the emergence of novel and previously unseen threats.

From the historical evolution of detection techniques to the intricacies of adversarial attacks, the exploration highlights the ongoing battle between defenders and adversaries. The imperative for advanced detection strategies in contemporary digital security becomes evident, signifying the need for a comprehensive and dynamic approach to safeguarding digital landscapes against the relentless and ever-evolving threat of malware.

5.3.1 Evolution of Malware Detection Techniques

The evolution of malware detection techniques, as explored in the narrative, unveils a journey marked by the perpetual struggle between defenders fortifying digital landscapes and malicious actors innovating to exploit vulnerabilities. The historical exploration begins with signature-based methods relying on predefined patterns, progresses through heuristic and behavior-based approaches, and culminates in contemporary strategies integrating machine learning and artificial intelligence. Each phase addresses challenges posed by evolving cyber threats, demonstrating the necessity for dynamic and adaptive detection strategies. Sahay, Sharma, and Rathore (2020) provide an overview of the evolution of malware and its detection techniques in their paper. The work discusses the threat posed by second-generation sophisticated malware to valuable information, ranging from early signature-based detection to contemporary machine/deep learning techniques. The paper aims to enhance understanding of malware evolution and detection methods to address the challenges posed by evolving cyber threats [11].

Consider the challenge faced by signature-based methods when encountering polymorphic and zero-day threats. These threats constantly mutate to evade detection, highlighting the limitation of relying solely on predefined signatures.

Key challenges in contemporary malware detection encompass the need for adaptability to polymorphic threats for signature-based methods, which struggle against mutating and zero-day threats. Heuristic-based techniques face hurdles in keeping pace with rapidly changing cyber adversary tactics, prompting a call for a more flexible detection mechanism. Anomaly identification introduces challenges in accurately discerning deviations from normal system behavior, emphasizing the necessity for a nuanced understanding of baseline norms. The evolution underscores the demand for dynamic and adaptive detection strategies, transitioning from signature-based to heuristic, behavior-based, and machine learning

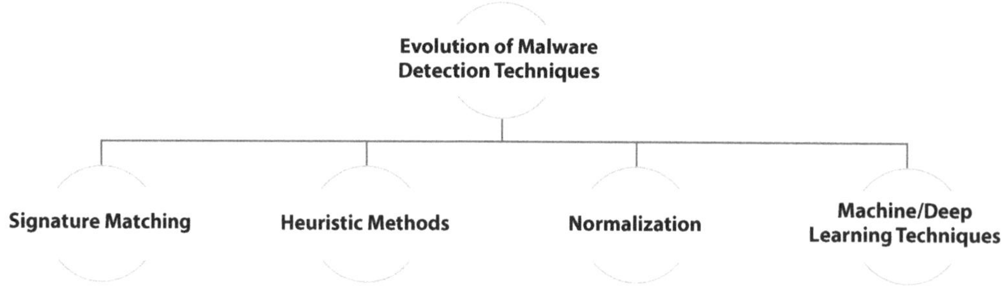

Figure 5.4 Evolution of malware detection techniques.

approaches, enhancing flexibility against evolving cyber threats. An adaptive approach is exemplified by behavior-based techniques, focusing on real-time actions and interactions to combat polymorphic malware seeking to obfuscate signatures. The integration of machine learning and AI technologies, particularly deep learning neural networks, is highlighted for autonomously processing vast datasets and recognizing intricate patterns associated with both known and novel malware. The current landscape advocates for a defense-in-depth approach, combining various methods to create a robust defense mechanism against the adversarial nature of cyber threats. This evolution reflects an ongoing commitment to staying ahead in the perpetual arms race between defenders and cyber adversaries, emphasizing the need for a multifaceted and adaptive approach as the landscape of malware detection remains dynamic.

5.3.2 Adversarial Attacks in Cybersecurity

Adversarial attacks in cybersecurity represent a sophisticated category of maneuvers designed to exploit vulnerabilities in systems, evade detection mechanisms, and compromise the integrity and confidentiality of digital assets. Rosenberg *et al.* (2021) presents a comprehensive survey of adversarial machine learning attacks and defense methods in the cyber security domain. The article characterizes adversarial attack methods, categorizes their applications in cyber security, and highlights the risks associated with machine learning vulnerabilities, particularly in adversarial environments. The work discusses the challenges of implementing end-to-end adversarial attacks in cyber security and provides a unified taxonomy to guide future research directions [12]. These attacks are often orchestrated with a deep understanding of the weaknesses in existing security measures, making them particularly challenging to thwart. This narrative explores the concept of adversarial attacks, providing insights into their nature, tactics, and potential consequences.

Nature of Adversarial Attacks

Adversarial attacks leverage cunning and strategic manipulation to deceive or exploit security systems. These attacks are not limited to a specific vector but can manifest across various layers of a system, including software, networks, and machine learning models. The primary objective is to outsmart and bypass existing defenses, often with the intention of gaining unauthorized access, stealing sensitive information, or disrupting normal system functionality.

In the ever-evolving landscape of cybersecurity, adversaries employ sophisticated tactics, such as obfuscation and polymorphism, to alter the appearance of malicious code,

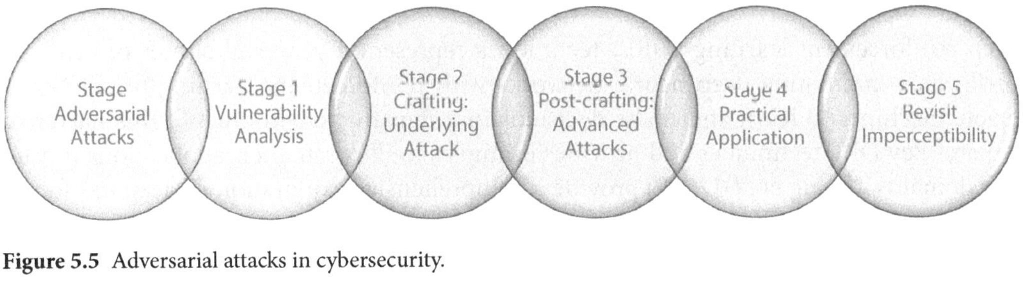

Figure 5.5 Adversarial attacks in cybersecurity.

challenging signature-based detection systems. Adversaries adeptly evade machine learning models by crafting malware specifically designed to mislead the detection process, manipulating features to deceive the system. Zero-day exploits capitalize on unknown vulnerabilities, enabling attacks before security patches are available. Social engineering tactics, exemplified by phishing attacks, exploit human psychology to bypass technical defenses. The potential consequences of adversarial attacks are severe, ranging from data breaches leading to unauthorized access to confidential information, disruption of services through techniques like distributed denial of service (DDoS) attacks, and financial losses incurred in recovery efforts and compensation. Furthermore, successful adversarial attacks can inflict reputation damage on individuals, organizations, or entire industries, eroding trust among stakeholders and the general public. The cybersecurity landscape necessitates a multifaceted and adaptive strategy to effectively counter these nuanced and evolving threats posed by adversaries in the ongoing battle between defenders and malicious actors.

5.3.3 Need for Advanced Detection Strategies

The need for advanced detection strategies in the realm of cybersecurity is paramount as traditional methods prove increasingly inadequate against the evolving sophistication of cyber threats. This narrative delves into the reasons behind the imperative for advanced detection strategies, exploring the limitations of conventional approaches and providing examples to illustrate the pressing demand for innovative solutions.

In the arena of cybersecurity, conventional detection approaches exhibit limitations that hinder their effectiveness. Signature-based detection, reliant on predefined malware signatures, falls short against polymorphic malware that alters its code to evade recognition. Heuristic analysis, identifying potential threats based on predefined rules, may generate false positives or negatives, compromising accuracy. Traditional behavioral detection struggles with rapidly evolving malware tactics, potentially overlooking subtle deviations in behavior. Highlighting the need for innovation are pressing examples like advanced persistent threats (APTs), fileless malware operating in system memory, and zero-day exploits, all challenging traditional detection methods. Moreover, adversarial attacks on machine learning models underscore vulnerabilities, emphasizing the necessity for ongoing model training and resilient algorithms. Recognizing these challenges, there is a clear imperative for advanced detection strategies. Such strategies must be adaptive to evolving tactics, proactively defend against potential threats, enhance precision, and integrate multiple techniques to fortify cybersecurity in the face of a dynamic and sophisticated threat landscape.

5.4 Deep Reinforcement Learning Techniques

Deep reinforcement learning (DRL) techniques represent a powerful subset of artificial intelligence, combining deep neural networks with reinforcement learning principles to enable machines to learn and make decisions in a dynamic environment. This narrative explores key DRL techniques and provides examples to illustrate their applications in various domains. "Dong *et al.* (2020) provide a comprehensive exploration of deep reinforcement learning in their book. Tailored for enthusiasts of reinforcement learning, especially deep reinforcement learning, the text offers insights and practical experiences in applying reinforcement learning techniques to real-world applications" [13].

Deep reinforcement learning encompasses various algorithms, each serving specific purposes across diverse domains. Q-Learning, a foundational algorithm, guides agents in decision-making to maximize cumulative rewards, as exemplified in training game-playing programs. Deep Q-Network (DQN) extends Q-Learning by using deep neural networks, achieving human-level performance in Atari 2600 games. Policy gradient methods directly learn optimal policies, showcased in training robots for tasks like walking using deep deterministic policy gradient (DDPG). Proximal policy optimization (PPO) improves stability in natural language processing, such as training conversational agents. Actor–critic methods, combining policy-based and value-based approaches, optimize trading strategies in finance. Trust region policy optimization (TRPO) ensures stable policy updates, applicable in healthcare for optimizing treatment plans. These examples underscore the versatility of deep reinforcement learning in gaming, robotics, finance, healthcare, and other domains, promising advancements in solving complex problems and enhancing decision-making across real-world applications.

5.4.1 Application of Deep Learning in Malware Detection

Deep learning (DL) has emerged as a potent tool in the field of cybersecurity, particularly in the detection and mitigation of malware. By leveraging sophisticated neural networks, deep learning techniques enhance the ability to recognize complex patterns and anomalies indicative of malicious software. This narrative explores the application of deep learning in malware detection, providing a description of the techniques involved and an illustrative example.

Deep learning techniques in malware detection often involve the use of deep neural networks, such as convolutional neural networks (CNNs) and recurrent neural networks (RNNs). HaddadPajouh *et al.* (2018) propose a deep recurrent neural network (RNN) based approach for Internet of Things (IoT) malware threat hunting. Utilizing recurrent neural networks to analyze ARM-based IoT applications' execution operation codes, the study achieves a high accuracy of 98.18% in detecting new malware samples, demonstrating the effectiveness of the LSTM approach in IoT malware detection [14]. These networks excel at automatically learning hierarchical representations of data, allowing them to discern intricate features within files or network traffic associated with malware.

Deep learning techniques, such as convolutional neural networks (CNNs), are adept at analyzing spatial hierarchies, making them effective for discerning patterns in file structures crucial for malware detection. Recurrent neural networks (RNNs), specialized in processing sequential data, prove valuable in identifying temporal patterns in system behaviors or network traffic, offering insights into potential malware activities. Autoencoders, as unsupervised models, excel in anomaly detection by learning to reconstruct input data and flagging deviations from expected reconstructions. The adaptability of deep learning models to diverse malware forms and their capacity for automatic feature extraction enhance resilience to evolving threats. These models, optimized for deployment, enable real-time or near-real-time malware detection, crucial for promptly addressing emerging threats. However, challenges like data imbalance and the explainability of decisions in deep learning models remain pertinent considerations for ensuring robust and trustworthy malware detection systems.

The application of deep learning in malware detection showcases the potential for sophisticated neural networks to enhance cybersecurity measures. By autonomously learning patterns and anomalies within data, these techniques contribute to a more robust defense against the dynamic and evolving landscape of malicious software.

5.4.2 Reinforcement Learning Algorithms

Reinforcement learning (RL) algorithms represent a category of machine learning techniques where an agent learns to make decisions by interacting with an environment. Through a trial-and-error process guided by rewards or penalties, the agent refines its strategies over time. This narrative explores various reinforcement learning algorithms, providing descriptions and illustrative examples for each. Oh *et al.* (2020) presents a novel meta-learning approach for discovering reinforcement learning (RL) algorithms. The method, named learned policy gradient (LPG), autonomously learns both what to predict (e.g., value functions) and how to learn from it (e.g., bootstrapping) by interacting with a set of environments. Empirical results demonstrate LPG's capacity to discover alternatives to fundamental RL concepts and its surprising ability to generalize from toy environments to complex Atari games, showcasing the potential for data-driven discovery of general RL algorithms [15].

Q-Learning serves as a foundational algorithm, enabling agents to learn an action-value function to make optimal decisions in given states—for instance, in grid-based navigation scenarios, Q-Learning can be applied to instruct a robot in maze navigation by determining optimal paths through learned Q-values. Deep Q-Network (DQN) extends Q-Learning by incorporating deep neural networks, enabling efficient handling of complex state-action spaces, as demonstrated in video game playing where the algorithm achieves human-level

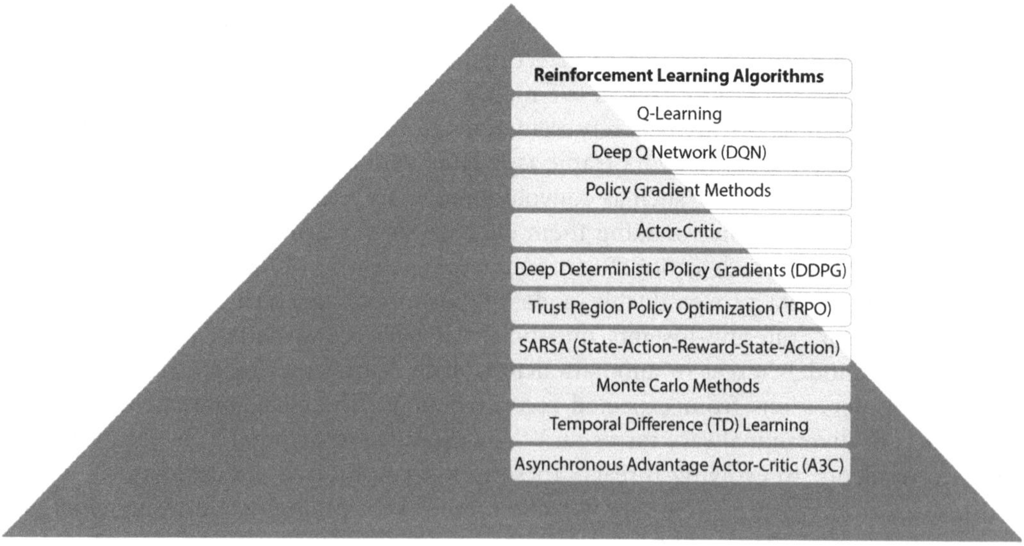

Figure 5.6 Reinforcement learning algorithms.

performance. Policy Gradient methods directly optimize an agent's policy, exemplified in robotic control for teaching complex tasks like walking through trial and error. Actor–critic methods blend value-based and policy-based approaches, making them suitable for continuous control tasks such as robotic arm manipulation. Deep deterministic policy gradient (DDPG) specializes in continuous action spaces, applied to train robotic systems for tasks involving continuous movements. Proximal policy optimization (PPO) ensures stable learning during policy optimization, making it applicable in training conversational agents for natural language processing. Monte Carlo methods estimate state or state-action pair values through episode sampling, beneficial in strategic board games like chess. SARSA (state–action–reward–state–action) represents an on-policy TD learning algorithm, where the agent updates Q-values based on the current policy, demonstrated in training a robot for specific paths in robotic navigation.

These reinforcement learning algorithms showcase the diversity of approaches used to teach agents to make decisions in various environments. Whether navigating mazes, playing games, or controlling robotic systems, these algorithms contribute to the advancement of autonomous decision-making across different domains.

5.5 Feature Selection Strategies

In the context of adversarial malware detection, the integration of deep reinforcement learning (DRL) serves as a strategic enhancement to traditional detection methods. DRL's adaptability and autonomous decision-making capabilities prove crucial in mitigating the evolving tactics of cyber adversaries who aim to deceive and mislead detection models. Steingartner *et al.* (2021) propose a cybersecurity strategy in their paper, focusing on a cyber deception-based approach and education for resilience within a hybrid threats model. Emphasizing the limitations of prevention-only strategies, the authors advocate for a more balanced approach that includes detection and response. The paper explores the use of technologies such as user and entity behavioral analytics, big data, artificial intelligence, and deception to enhance the accuracy of recognizing malicious activity, providing a comprehensive overview of the evolving landscape of threat defense [16].

Consider an antivirus system equipped with DRL for adversarial malware detection. The DRL agent learns from historical data and adapts its strategies to identify novel malware variants designed to evade traditional signature-based detection. Through continuous learning and dynamic decision-making, the system enhances its resilience against adversarial attacks, staying ahead of emerging threats in real-time.

In cybersecurity, the integration of deep reinforcement learning (DRL) brings forth several benefits. DRL exhibits a high level of adaptability, dynamically adjusting to evolving malware tactics, thereby ensuring the system's efficacy against sophisticated adversarial attacks. This adaptability is crucial for maintaining a robust defense in the ever-changing landscape of cybersecurity threats. Additionally, the integration of DRL facilitates real-time decision-making, enabling swift responses to emerging threats. This capability is essential for proactive cybersecurity measures that aim to identify and mitigate potential risks promptly. DRL also contributes to a reduction in false positives, enhancing the accuracy of threat identification by effectively distinguishing between normal and malicious activities. Furthermore, the adversarial training and continuous learning inherent in DRL improve the system's resilience against deceptive tactics employed by malware, further fortifying its defense mechanisms. However, challenges exist, particularly in terms of interpretability, as DRL models may lack transparency in their decision rationale, requiring efforts to ensure a clear understanding of the detection process. Additionally, the effectiveness of DRL relies heavily on the representativeness of training data, necessitating careful curation to avoid biased learning and enhance the overall reliability of the cybersecurity system.

5.5.1 Importance of Feature Selection in Malware Detection

Feature selection is paramount in the realm of adversarial malware detection, particularly when employing deep reinforcement learning (DRL). It involves choosing the most relevant and discriminative features from the dataset to enhance the model's accuracy, interpretability, and resilience against adversarial attacks. Feizollah *et al.* (2015) conduct a comprehensive review on feature selection in mobile malware detection, highlighting the escalating threat of mobile malware and the necessity for effective detection mechanisms. The paper categorizes features into static, dynamic, hybrid, and applications metadata groups, providing insights into the diverse methodologies employed in mobile malware research. The authors emphasize the significance of feature selection in enhancing the efficiency of malware detection systems, shedding light on the evolving landscape of mobile security [17].

In the application of DRL for adversarial malware detection, feature selection might involve identifying critical characteristics in the file or network traffic data. Relevant features could include behavioral patterns, code structures, or frequency of certain operations. By selecting these features judiciously, the DRL model gains a more focused understanding of malicious activities, improving its detection capabilities.

In the context of adversarial malware detection using deep reinforcement learning (DRL), the careful selection of features offers several advantages. Notably, it enhances model efficiency by reducing data dimensionality, thereby expediting the training process and enabling faster inference in DRL models. Moreover, the inclusion of meaningful features contributes to improved model interpretability, allowing security analysts to comprehend the underlying basis of malware detection decisions. This transparency in decision-making is crucial for maintaining a clear understanding of the model's functionality. Additionally, well-thought-out feature selection enhances the model's resilience against adversarial attacks, as the chosen features create a more robust defense system less susceptible to manipulation by deceptive tactics. However, challenges persist, primarily due to the dynamic nature of malware, necessitating continuous adaptation of feature selection strategies to address emerging threats. Furthermore, the interplay with adversarial tactics underscores the importance of ongoing refinement of selection criteria to ensure the effectiveness of feature selection in the face of evolving cybersecurity challenges. In summary, feature selection plays a pivotal role in optimizing the performance of DRL models for effective adversarial malware detection, balancing efficiency, interpretability, and resilience against evolving threats.

5.5.2 Techniques for Feature Selection

Feature selection is vital in the context of adversarial malware detection with Deep reinforcement learning (DRL). Various techniques exist to choose the most informative features, enhancing the efficiency and effectiveness of DRL models in identifying malicious activities while resisting adversarial attacks.

In the application of DRL for adversarial malware detection, techniques like recursive feature elimination (RFE) or information gain analysis may be employed. RFE recursively removes less relevant features, while information gain assesses the significance of each feature based on its contribution to distinguishing between benign and malicious files. These techniques refine the set of features the DRL model considers, improving its ability to discern malicious patterns.

In the realm of adversarial malware detection using deep reinforcement learning (DRL), various feature selection techniques play a pivotal role. Commonly employed methods include filter, wrapper, and embedded approaches, each with distinct characteristics. Filter methods, such as information gain analysis, evaluate feature relevance independently of the learning algorithm by employing statistical measures like correlation. On the other hand, wrapper methods, exemplified by recursive feature elimination (RFE), assess feature subsets based on the performance of a specific learning algorithm through iterative testing. Embedded methods seamlessly integrate feature selection into the model-building process; for instance, dropout layers in a DRL neural network eliminate certain features during training, emphasizing the most informative ones. Principal component analysis (PCA) transforms original features into uncorrelated variables, reducing dimensionality and enhancing information retention. These techniques offer several benefits, including improved model efficiency, enhanced interpretability, and increased robustness against adversarial attacks. Feature selection, however, faces challenges in adapting to the dynamic threat landscape and countering potential exploitation by adversaries. In navigating these challenges, the careful application of feature selection techniques proves crucial for optimizing the efficiency, interpretability, and resilience of DRL models in the context of adversarial malware detection, thereby fortifying cybersecurity defenses.

5.5.3 Optimization for Deep Reinforcement Learning Models

Optimizing deep reinforcement learning (DRL) models is critical in the context of adversarial malware detection. Optimization involves refining the model's parameters, training process, and decision-making strategies to enhance its efficiency, accuracy, and resilience against evolving adversarial tactics. Li *et al.* (2020) introduces a novel framework, DRL-MOA (Deep Reinforcement Learning-based Multi-Objective Optimization Algorithm), for solving multiobjective optimization problems (MOPs) using deep reinforcement learning. The approach decomposes MOPs into scalar optimization subproblems, modeled as neural networks. Through collaborative optimization and a neighborhood-based parameter-transfer strategy, the proposed method achieves Pareto-optimal solutions, as demonstrated in solving the Multi objective traveling salesman problem (MOTSP). The results highlight DRL-MOA's strong generalization ability and efficient solving speed, making it a competitive approach for Multi objective optimization [18].

In adversarial malware detection, optimization might involve tuning the hyperparameters of a DRL algorithm, such as the learning rate or exploration-exploitation balance. This ensures the model learns effectively from data, adapts to emerging threats, and maintains robust decision-making capabilities in the face of adversarial attempts to deceive the detection system.

In the optimizing, deep reinforcement learning (DRL) models for adversarial scenarios, several common techniques prove instrumental. Hyperparameter tuning involves adjusting parameters like learning rates and exploration probabilities to find the optimal configuration for training DRL models, ensuring a delicate balance between rapid learning and stability in the face of adversarial variations. Batch normalization normalizes the inputs of each neural network layer, reducing internal covariate shift and enhancing the model's generalization across diverse scenarios. Experience replay involves storing and randomly sampling past experiences to break temporal correlations in training data, contributing to more stable and efficient learning. Reward shaping modifies the reward structure to guide the model toward desired behaviors or address sparse reward problems, such as encouraging the DRL model to prioritize accurately identifying adversarial patterns in malware. These optimization techniques collectively enhance model performance, increase robustness against adversarial manipulation, and ensure stability in training processes. However, challenges include navigating the complexity of the hyperparameter space and striking a balance between avoiding overfitting to training data while generalizing effectively to adversarial scenarios, highlighting the nuanced nature of optimizing DRL models for cybersecurity defense.

Optimizing deep reinforcement learning models for adversarial malware detection is a crucial aspect of building robust and effective cybersecurity systems. Through careful tuning and application of optimization techniques, DRL models can adapt to the dynamic threat landscape and maintain high-performance standards in identifying and mitigating adversarial attacks.

5.6 Datasets and Evaluation

Selecting appropriate datasets and defining evaluation metrics are pivotal steps in the application of deep reinforcement learning (DRL) for adversarial malware detection. Datasets provide the foundation for model training and testing, while evaluation metrics quantify the effectiveness of the DRL model in identifying and mitigating adversarial threats. Mesaros, Heittola, and Ellis (2018) provide an extensive exploration of datasets and evaluation methodologies in the domain of environmental sound scene and event recognition. The chapter discusses the crucial aspects of designing evaluation datasets, highlights commonly used metrics for system evaluation, and includes a survey of available datasets. The comprehensive insights contribute to the development and assessment of computational systems in the field [19].

In the context of adversarial malware detection, a dataset might consist of a diverse collection of malware samples, both benign and adversarial. Evaluation metrics could include accuracy, precision, recall, and F1 score, providing a comprehensive assessment of the DRL model's performance in distinguishing between malicious and non-malicious entities.

Optimizing deep reinforcement learning (DRL) models for adversarial scenarios, several common techniques prove instrumental. Hyperparameter tuning involves adjusting parameters like learning rates and exploration probabilities to find the optimal configuration for training DRL models, ensuring a delicate balance between rapid learning and stability in the face of adversarial variations. Batch normalization normalizes the inputs of each neural network layer, reducing internal covariate shift and enhancing the model's generalization across diverse scenarios. Experience replay involves storing and randomly sampling past experiences to break temporal correlations in training data, contributing to more stable and efficient learning. Reward shaping modifies the reward structure to guide the model toward desired behaviors or address sparse reward problems, such as encouraging the DRL model to prioritize accurately identifying adversarial patterns in malware. These optimization techniques collectively enhance model performance, increase robustness against adversarial manipulation, and ensure stability in training processes. However, challenges include navigating the complexity of the hyperparameter space and striking a balance between avoiding overfitting to training data while generalizing effectively to adversarial scenarios, highlighting the nuanced nature of optimizing DRL models for cybersecurity defense.

5.7 Generating Adversarial Samples

Adversarial samples are malicious inputs crafted to mislead machine learning models, and in the context of deep reinforcement learning (DRL), understanding and addressing them is crucial. Techniques in DRL for sample generation involve creating adversarial samples to train models robustly. Augmenting training data with adversarial samples enhances model resilience against deceptive tactics. Grosse, Papernot, Manoharan, Backes, and McDaniel (2017) present an innovative approach to crafting adversarial examples specifically tailored for malware detection models. In their work, they address challenges related to discrete and binary input domains, achieving a highly effective attack on a neural network trained for malware detection. The study demonstrates the potential vulnerability of malware detection systems to adversarial manipulation [20].

In adversarial malware detection, an adversary might manipulate a malware sample's code to evade detection. Understanding adversarial samples involves recognizing these manipulations, and DRL techniques can generate new samples with similar deceptive patterns. Augmenting the training data with such adversarial samples ensures the model learns to identify and adapt to evolving threats.

Deep reinforcement learning (DRL) for adversarial malware detection, key concepts encompass adversarial samples—inputs crafted to exploit vulnerabilities in machine learning models. An illustrative instance involves subtly altering malware code to evade traditional signature-based antivirus detection. Techniques in sample generation, employing DRL methods, focus on creating adversarial samples during training to expose models to diverse deceptive tactics. Adversarial training augments the training dataset with manipulated instances, allowing the model to learn and recognize subtly altered malicious patterns. This approach bolsters model robustness, making it more resilient against manipulative tactics employed by adversaries. Furthermore, augmenting training data with adversarial samples ensures adaptability to emerging threats, vital in the face of evolving malware tactics. However, challenges persist, particularly in addressing the dynamic nature of adversarial tactics and balancing false positives to prevent misclassification of benign samples as malicious. In summary, comprehending adversarial samples, leveraging DRL techniques for their generation, and judiciously augmenting training data with such samples are pivotal strategies for fortifying deep reinforcement learning models in adversarial malware detection scenarios.

Conclusion and Future Directions

The exploration of deep reinforcement learning (DRL) for adversarial malware detection reveals promising outcomes. Evaluation metrics such as accuracy, precision, and recall showcase the model's effectiveness in identifying and mitigating threats. Comparative analyses demonstrate the advantages of DRL over traditional methods, emphasizing its adaptability to dynamic cyber threats. The study's implications suggest that DRL models, with their ability to handle adversarial samples and evolving tactics, hold substantial promise in fortifying cybersecurity measures. Insights gained from the results inform decision-makers about the strengths and weaknesses of the DRL approach in the specific context of malware detection.

Future Directions

The path forward involves addressing challenges related to model interpretability and generalization. Future research should focus on enhancing the transparency of DRL models and ensuring their applicability to diverse cybersecurity scenarios. Continuous refinement of DRL techniques for adversarial sample generation and model optimization remains crucial for staying ahead of evolving threats.

The integration of deep reinforcement learning in adversarial malware detection marks a significant step forward in cybersecurity. As the field advances, ongoing research and development efforts will contribute to the continued effectiveness and adaptability of DRL models, ultimately bolstering the resilience of cybersecurity frameworks against emerging adversarial tactics.

References

1. Wang, Q., Guo, W., Zhang, K., Ororbia, A.G., Xing, X., Liu, X., Giles, C.L., Adversary resistant deep neural networks with an application to malware detection, in: *Proceedings of the 23rd ACM SIGKDD International Conference on Knowledge Discovery and Data Mining*, pp. 1145–1153, August 2017.
2. Romanovsky, A., Troubitsyna, E., Gashi, I., Schoitsch, E., Bitsch, F., *Computer Safety, Reliability, and Security*, Springer Nature, Finland, 2019.
3. Zhong NN, Wang HQ, Huang XY, Li ZZ, Cao LM, Huo FY, Liu B, Bu LL. Enhancing head and neck tumor management with artificial intelligence: Integration and perspectives. *Semin Cancer Biol.*, 2023 Oct;95:52-74. doi: 10.1016/j.semcancer.2023.07.002. Epub 2023 Jul 18. PMID: 37473825.
4. Kesan, J.P. and Hayes, C.M., Mitigative counterstriking: Self-defense and deterrence in cyberspace. *Harv. J. Law Technol.*, 25, 429, 2011.
5. Aryal, K., Gupta, M., Abdelsalam, M., A survey on adversarial attacks for malware analysis, vol. 2, pg. 08223, 5 Jan. 2022, arXiv preprint arXiv:2111.08223.
6. Alshammari, A., Rawat, D.B., Garuba, M., C. A. Kamhoua, Njilla, L.L., Deception for cyber adversaries: status, challenges, and perspectives, in: *Modeling and Design of Secure Internet of Things*, pp. 141–160, 2020.

7. Kott, A., Science of cyber security as a system of models and problems. *Network Science and Cybersecurity*, Springer, New York, 2015, arXiv preprint arXiv:1512.00407.
8. Santos Jr., E., and Johnson Jr., G., Toward detecting deception in intelligent systems, in: *Enabling Technologies for Simulation Science VIII*, Univ. of Connecticut, USA, vol. 5423, pp. 130–141, SPIE, August 2004.
9. Vinod, P., Laxmi, V., Gaur, M.S., Reform: Relevant features for malware analysis, in: *2012 26th International Conference on Advanced Information Networking and Applications Workshops*, IEEE, pp. 738–744, March 2012.
10. Corona, I., Giacinto, G., Roli, F., Adversarial attacks against intrusion detection systems: Taxonomy, solutions and open issues. *Inf. Sci.*, 239, 201–225, 2013.
11. Sahay, S.K., Sharma, A., Rathore, H., Evolution of malware and its detection techniques, in: *Information and Communication Technology for Sustainable Development: Proceedings of ICT4SD 2018*, Springer Singapore, pp. 139–150, 2020.
12. Rosenberg, I., Shabtai, A., Elovici, Y., Rokach, L., Adversarial machine learning attacks and defense methods in the cyber security domain. *ACM Comput. Surv. (CSUR)*, 54, 5, 1–36, 2021.
13. Dong, H., Dong, H., Ding, Z., Zhang, S., Chang, *Deep Reinforcement Learning*, Springer Singapore, Singapore, 2020.
14. HaddadPajouh, H., Dehghantanha, A., Khayami, R., Choo, K.K.R., A deep recurrent neural network based approach for internet of things malware threat hunting. *Future Gener. Comput. Syst.*, 85, 88–96, 2018.
15. Oh, J., Hessel, M., Czarnecki, W.M., Xu, Z., van Hasselt, H.P., Singh, S., Silver, D., Discovering reinforcement learning algorithms, in: *Advances in Neural Information Processing Systems*, vol. 33, pp. 1060–1070, 2020.
16. Steingartner, W., Galinec, D., Kozina, A., Threat defense: Cyber deception approach and education for resilience in hybrid threats model. *Symmetry*, 13, 4, 597, 2021.
17. Feizollah, A., Anuar, N.B., Salleh, R., Wahab, A.W.A., A review on feature selection in mobile malware detection. *Digit. Invest.*, 13, 22–37, 2015.
18. Li, K., Zhang, T., Wang, R., Deep reinforcement learning for multiobjective optimization. *IEEE Trans. Cybern.*, 51, 6, 3103–3114, 2020.
19. Mesaros, A., Heittola, T., Ellis, D., Datasets and evaluation, in: *Computational Analysis of Sound Scenes and Events*, pp. 147–179, 2018.
20. Grosse, K., Papernot, N., Manoharan, P., Backes, M., McDaniel, P., Adversarial examples for malware detection, in: *Computer Security-ESORICS 2017: 22nd European Symposium on Research in Computer Security, Oslo, Norway, September 11-15, 2017, Proceedings, Part II*, Springer International Publishing, pp. 62–79, 2017.

6

Artificial Intelligence in Blockchain and Smart Contracts for Disruptive Innovation

Eashwar Sivakumar[1,2*], Kiran Jot Singh[1] and Paras Chawla[3]

[1]Department of ECE, Chandigarh University, Punjab, India
[2]Department of ECE, Jain (Deemed-to-be) University, Bengaluru, India
[3]School of Engineering, Ajeenkya DY Patil University, Pune, India (iNurture Education Solutions Private Ltd., Bengaluru)

Abstract
Blockchain represents a burgeoning technological advancement that is transforming a number of different organizations. The purpose of this study is to examine bibliographic aspects related to blockchain in digital identity, financial security, supply chain management, insurance, healthcare, agriculture, construction, and education published between 2016 to 2022. This study summarizes and evaluates the present level of research on smart contracts by conducting an analysis of 104 publications written on the subject of smart contracts. Based on the findings of the study, the top 10 countries amount to 69.35% of total contribution, out of which United States, China, and India had a greater number impact compared to those in other nations contributing to 43.12%. The investigation further revealed the use of artificial intelligence in blockchain amounting to almost 9%. The combination of these technologies has significant advantages and are presented here.

Keywords: Blockchain, smart contracts, artificial intelligence, applications of blockchain, bibliographic analysis

6.1 Introduction

A blockchain [1] is a replicated and shared data structure that operates within a network. It was built utilizing bitcoin in order to overcome the problem of double spending. Blockchain refers to a series of linked records that are very difficult to alter and are kept secure *via* the use of cryptographic protocols. Once the blockchain is established, it will be impossible for hackers to manipulate the data for the reason that every user shall have a replica of the ledger and the block data is encrypted. A conventional database management method often necessitates the engagement of a third party and comes with a number of drawbacks, including expensive transaction costs, the possibility of double spending, inadequate data recovery, and online frauds.

**Corresponding author*: eashwarsivakumar@gmail.com

Shubham Mahajan, Pethuru Raj and Amit Kant Pandit (eds.) Deep Reinforcement Learning and Its Industrial Use Cases: AI for Real-World Applications, (115–136) © 2024 Scrivener Publishing LLC

Conventional databases may be utilized if shared write access is not desired. Blockchain technology offers a solution in situations in which the authors are unknown or untrustworthy, our interests do not coincide, and we do not want to go *via* a third party that can be trusted.

The database is unparalleled when compared to other options in terms of its use, speed, and precision. Blockchain technology is a clear winner when it comes to the categories of innovation, verification, and automation. Databases should be selected for applications with broad scopes. Blockchain is the technology to use if you are looking for trust, transparency, and verification. Blockchain has four fundamental characteristics:

1. Consensus: Every single transaction is validated and approved by each and every member in the chain. A standard illustration of this would be the use of blockchain technology to reach an agreement about payment, storage, and logistics in a supply chain.
2. Provenance: Everyone who took part in the activity is aware of where the asset came from—for instance, in a blockchain for a supply chain, the assets in question may include food products, currencies, equipment, intellectual property rights, etc.
3. Immutability: Each participant's contribution to the distributed ledger is immutable and cannot be altered in any way. It is impossible to forge any supply chain blockchain transaction, inventory data, delivery time, etc.
4. Finality: Finality is a statement made after the transaction. The blockchain transaction that has been made cannot be undone. This condition is known as finality. Various blockchain networks have various finality times—for instance, it takes 60 min for Bitcoin and 2 s for EOS.

6.1.1 Smart Contract

Nick Szabo, an American computer scientist, came up with the idea of "smart contracts" in 1997 [2], which seemed like a far-fetched idea at the time. Smart contracts are digital transaction mechanisms created to ensure the compliance of contractual requirements. At that time, the economy and information systems were not strong enough to support these methods. Today, the necessary technology is in place, and smart contracts are being created, tried, and used in various industries around the globe. The workflow of smart contracts is given in Figure 6.1.

Smart contracts are helpful because they are fast, accurate, efficient, trustworthy, clear, safe, and cost effective. When the condition is met, the contract is put into effect. Smart contracts have become digital so that people do not have to deal with the paperwork and waste time that can come with filling out paper forms. Smart contracts are carried out without any help from a third party. Each transaction is subjected to encryption and subsequently distributed to all parties, assuring trust and transparency. The data and records on a blockchain global system are safe and hard to change because they are protected and linked together in a chain. Hackers would have to change every record in the chain to change just one, which is impossible. Smart contracts are executed independently, without any requirement for intermediaries.

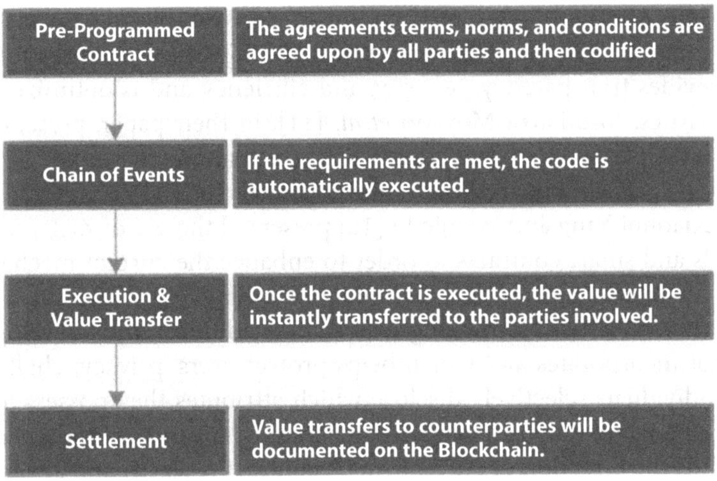

Figure 6.1 Smart contracts workflow [3].

6.2 Literature Review

The most important results from a number of studies are collected and presented here. The pervasive adoption of recent advancements in blockchain technology and smart contracts across multiple sectors such as digital identity, financial security, supply chain management, insurance, healthcare, agriculture, construction, and education are incorporated in the literature study.

6.2.1 Blockchain and Smart Contracts in Digital Identity

Affan Yasin and Lin Liu [4] presented a framework to aggregate digital identity by implementation in the blockchain. The TURS "Tsinghua University User Reputation System" combines the Windhover principle of digital identity and trust framework. Nikolaos Kapsoulis *et al.* [5] developed two variants of smart contracts for protecting the confidentiality of user information on enterprises blockchains. It is developed with the help of IPFS "Interplanetary File System" and Quorum blockchain. Ashok Kumar Yadav and Ramendra Kumar Bajpai [6] proposed an improved KYC system utilizing blockchain smart contract technology. Each entity in the consortium can only validate, rather than tamper, the specifics. Jose Parra-Moyano and Omri Ross [7] presented a system to reduce expenses associated with basic KYC verification procedure in financial institutions and enhances the client experience. Clients only need to go through the basic KYC verification procedure one time, regardless of how many financial institutions they want to use. The key aspect is that the customer owns the smart contract information and not the institutions. Alex Biryukov *et al.* [8] proposed an Ethereum-based private KYC method. The system will allow financial service companies to utilize blockchain technology to improve client onboarding efficiency while adhering to regulations and safeguarding customers' privacy. Wazen M. Shbair *et al.* [9] proposed a framework which operates on Grid'5000 platform. The technology gives developers working on private and consortium blockchains insights that allow them to discover challenges and assess the performance of their apps under a variety of scenarios.

N. Sundareswaran *et al.* [10] presented an enhanced KYC blockchain system based on Ethereum utilizing symmetrical AES encryption and LZ-based compression mechanism. The system provides transparency, security, and efficiency and is optimized using blockchain characteristics. José Parra-Moyano *et al.* [11], in their paper, presented a dynamic blockchain-based KYC system that decreases KYC costs, allows these expenses to be equally distributed among participating Fis, and removes the need of TTP to manage authorizations in the system. Xiaohui Yang and Wenjie Li [12] presented the use of ZKP (zero-knowledge proof) methods and smart contracts in order to enhance the current mechanism of identifying claims in blockchain and achieve identity unlinkability, thereby preventing the ownership of characteristics from being exposed. In addition, they developed BZDIMS, a prototype that incorporates and which helps protect users' privacy; challenge-response protocol lets individuals selectively disclose which attributes they possess to service providers. Sara Rouhani and Ralph Deters [13], in his study, covers the fundamental principles and recommends the direction of current smart contract research and advancements. The research findings were categorized into three distinct groups: security methodologies and tools, performance optimization approaches, and smart contract-based decentralized apps. Ahmad Sghaier Omar and Otman Basir [14] developed a blockchain-based digital identity and a semi-decentralized IoT identity management system that assures a worldwide and unique device identification. Yogita Borse *et al.* [15] presented a method for protecting the confidentiality of sensitive user information by employing a zero-knowledge protocol for interval membership based on the Pedersen commitment scheme. This study describes in depth how some user characteristics may be concealed in a blockchain environment.

6.2.2 Blockchain and Smart Contracts in Financial Security

Hemang Subramanian [16] defines the security token framework as a smart contract application. Additionally, the author uses the proposed security token framework and smart contract capabilities to show the design and implementations of the Simple Agreement for Future Equity (SAFE), a widely used financial instrument. Engin Demirel and Seda Karagoz Zeren [17] investigates the idea of "smart contracts" and the most important things to think about when building a financial payment system. This research also examines the potential advantages of using smart contracts as a legal, technical, and innovative part of the new business environment. Omar Ali *et al.* [18] provides a taxonomy that divides blockchain-enabled financial advantages, difficulties, and capabilities into three distinct categories. Implications for future studies and applications are identified within the blockchain paradigm. Mallikarjun Reddy Dorsala *et al.* [19] proposed a cryptography paradigm based on blockchain, which is used to construct protocols and validate their safety using universally composable theory. Hao Wang *et al.* [20] proposed a digital account paradigm for transferring assets between centralized and decentralized ledgers and proposed locking and unlocking algorithms for smart contracts. Evaluations of unlocking codes and chain code reveal that the proposed technology is appropriate in a real-world financial lending environment. Yue Li [21] proposed that SafePay analysis methodology was created to detect unfair payments in Ethereum smart contracts. Compared to previous analysts, SafePay is able to locate blockchain transactions that have vulnerabilities that might be exploited, hence preventing the misleading reports. Haneffa Muchlis Gazali *et al.* [22] presented a proof-of-concept for handling student loan payments using smart contracts and blockchain. Borrowers have unrestricted access to their financial records, and with the assistance of smart contracts, corporate filings and management systems are kept up to date in real time [23]. In order to ensure consistent and equitable payment between retailers, customers, and logistics firms, three distinct forms of smart contracts have been developed. Blockchain's immutability and auditability make it a useful tool for verifying assets and data sharing throughout the transportation industry. Saurabh Ahluwalia [24] demonstrates how blockchain technology might enhance the startup financing process and develop a model based on transaction cost economics. Sabyasachi Chakraborty *et al.* [25] suggested two credit scoring systems, one of which scores an individual's credit for loans or credit invoicing. In the second framework, we suggest using blockchain to handle client credit orders for pushing outblocked transactions. Sabyasachi Chakraborty [26] built a transaction logic-based financial product management platform using Hyperledger Fabric. This paper's BNA platform helps financial institution players effectively manage life cycles.

6.2.3 Blockchain and Smart Contracts in Supply Chain Management

Angwei Law [27] proposed a prototype using Ethereum platform to develop decentralized applications (DApps) to identify provenance, monitor items, and maintain an open database; thus, three separate smart contracts were created. The contracts provide a standardized structure for supply chain parties to use in recording, sharing, and accessing information, hence reducing uncertainty [28]. The blockchain platform eliminates the risk involved in supply chain, but it requires investment for implementation as well as recurring cost. Pietro De Giovanni investigates the cases in which smart contracts are economically feasible. Paolo Bottoni [29] offer a new form of smart contract that aims to address two major issues that impede supply chain efficiency and effectiveness: trust and coordination. Pietro De Giovanni [30], in his work, compares the solutions of centralized and decentralized versions of the game in static and dynamic circumstances. It also looks at the decentralized supply chain by assessing two smart contracts: shared revenue and wholesale pricing contracts. Ilhaam A. Omar *et al.* [31] proposed automating GPO contracting using blockchain smart contracts and presented a generalized HCSC contracting framework with algorithms illustrating stakeholder interactions. Ilhaam A. Omar *et al.* [32] presented a general architecture utilizing decentralized storage systems and Ethereum smart contracts to automate operations and information sharing and comprehensive algorithms that capture supply chain interactions between stakeholders. Gunnar Prause [33], in their study, examines how smart contracts and blockchain technology might enable collaborative business models for sustainable entrepreneurship activity in smart supply chains. Thomas Bocek *et al.* [34] presented modum.io, a startup that employs IoT sensor to reduce operating expenses by incorporating blockchain technology into the pharmaceutical supply chain to establish data immutability and open access to temperature records. The information is then sent to the blockchain, where a smart contract evaluates it based on the characteristics of the product. Hajar Moudoud *et al.* [35] presented a blockchain framework, LC4IoT, tailored specifically for usage in supply chains, consisting of a variety of distributed IoT entities. Sofia Terzi *et al.* [36] illustrates how two real-world supply chain scenarios utilize BC technology. The first one covers logging and tracing products, while the second one covers authentication procedures for BC-identity-holding individuals. Haya Hasan *et al.* [37] presented a supply chain management system that would increase efficiency through the utilization of blockchain technology and smart containers. The proposed approach governs and manages interactions between the sender and recipient by using smart contract functionalities in the Ethereum network. Lu Wang [38] presented consortium and smart contract-based system for monitoring and tracing the process of agrifood supply chain and implementing traceability and shareability. Margin Indicator (MI), a revolutionary design feature [39], was developed to generate trustworthy predictive analytics outcomes using conventional machine learning algorithms. The suggested solution promotes cost and energy management while retaining high transparency in managing decentralized AV supply chain operations, monetary consequences, and environmental sustainability.

6.2.4 Blockchain and Smart Contracts in Insurance

Veneta Aleksieva *et al.* [40] examines the use of public and private blockchains for the provision of insurance services by means of the experimental deployment of smart contracts using Hyperledger Fabric and Ethereum. Valentina Gatteschi *et al.* [41] seeks to help decision-makers by giving a general introduction of blockchain technology, emphasizing its advantages and disadvantages, and demonstrating a variety of real-world applications in the insurance sector that can be easily transferred to other industries. Fabrizio Lamberti *et al.* [42] presented a working model that incorporates a mobile app and a carry-on electronic gadget. The motorist may dynamically modify the coverage status of various insurance policies using the mobile app. Alpen Sheth and Hemang Subramanian [43] presented a conceptual framework of smart contracts and explore how they were used to build Etherisc, a decentralized insurance exchange built on the Ethereum blockchain. Veneta Aleksieva [44] describes an experimental smart contract system for insurance services that use the Ethereum network. Based on the ERC20 standard for smart contracts, a decentralized crypto-token is built. Mayank Raikwar *et al.* [45] focused on building a solid infrastructure using BC technology to handle insurance transactions. The careful consideration and selection of parameters during the establishment of a blockchain network is crucial, given the substantial impact that they have on network latency, as evidenced by scalability experiments. Veneta Aleksieva *et al.* [46] developed a blockchain application that is intended to serve the insurance industry. It can be accomplished to develop an insurance policy using smart contracts as well as to calculate the insurance risk and to carry out insurance claims. The Hyperledger Fabric Blockchain serves as the foundation for the proposed smart contract. Utilizing the capabilities of both machine learning and BC technology, Anubhav Elhence *et al.* [47] created an efficient framework for the health insurance sector. The application of smart contract eliminates the involvement of third party and enhances the overall process. Efthymios Chondrogiannis *et al.* [48] proposed a decentralized application built on blockchain technology that facilitates consensus between people and healthcare insurance providers as they carry out the healthcare insurance policies outlined in individual contracts. Houyu Zheng *et al.* [49] presented an innovative plan for filing medical insurance claims using blockchain technology, smart contracts, and zero-knowledge proof. The authors primarily concentrated on two different processes: acquiring medical insurance and making a claim on medical insurance.

6.2.5 Blockchain and Smart Contracts in Healthcare

Asma Khatoon [50] proposed the use of blockchain technology numerous processes that are engaged in the healthcare ecosystem in order to enhance data management. Ilhaam A. Omar [51] illustrates how BC technology can be used to solve problems in clinical study data management and provide a blockchain-based framework built on top of ESC. The following steps of a CT procedure are represented in the framework: phases of new medication submission, clinical trial launch, and enrollment of patients. Ilhaam A. Omar [52] proposed a solution using blockchain technology in order to tackle challenges connected with the maintenance of CTs data. In particular, the study presents a proof-of-concept solution that tackles the issues associated with handling data, protocol conformity, data integrity, and transparency in CTs. Timothy Nugent [53] expands the notion by utilizing smart contracts that exist at a particular blockchain address and are cryptographically approved by the network to show how clinical trials may ensure trust and avoid data tampering. Mengyi Li [54] presented preliminary research on incorporating the FHIR standard's semantics into smart contracts. Using these standardized ontological concepts, they also address the appropriate data to extract from transaction records in decentralized ledgers to detect irregularities and information abuse. Mehdi Sookhak [55] outlines a comprehensive survey of blockchain-based authorization techniques in the healthcare domain as a premise for classifying extant and prospective advancements in the access control area. Raifa Akkaoui [56] proposed a role-based authorization system referred to as RBAC-HDE, leveraging the robust qualities of a blockchain, including its immutability and decentralization, to be developed, which will allow safe sharing of data for healthcare purposes. AaYusH [57] presented an ESC-based system for securely and effectively completing the TS process and demonstrated how ESC with IPFS provides privacy, security, and affordable data storage by removing third-party organizations. Nghia Duong-Trung *et al.* [58] proposed a patient-focused healthcare system which is built on top of smart contracts. The patients are in charge of their own medical records, which means that they have the ability to decide what information, if any, pertaining to their medical history should be shared with medical facilities, regardless of whether it is highly confidential. Shekha Chenthara *et al.* [59] primarily focused on the implementation of an effective referral system that makes use of sophisticated smart contracts for the purpose of facilitating the efficient exchange of healthcare information across various stakeholders in the healthcare sector. This referral system is based on a model that prioritizes the needs of patients, and it is restricted to authorized providers that are part of the health data network. Huanrong Tang *et al.* [60] presents a solution for the safe exchange of medical images that is based on the use of credit ratings and smart contracts generated by blockchain. Aleksandr Kormiltsyn *et al.* [61] identified problematic procedures in the healthcare industry and then provided a mapping that is built on smart contracts with the intention of addressing these processes. Yusen Wu *et al.* [62] proposed Soteria, a privacy smart contract architecture for managing, sharing, and analyzing clinical trials using fabric private chain code (FPC).

6.2.6 Blockchain and Smart Contracts in Agriculture

Based on blockchain and IoT technology, the authors [63] provide a reliable, independent, open, and eco-friendly food monitoring system that includes all parties involved in the ecosystem of smart agriculture, some of whom may not trust each other. Tahmid Hasan Pranto *et al.* [64] proposed a framework wherein smart contracts oversee transactions among participants, Internet of Things (IoT) devices gather field data, and blockchain technology serves as the central ledger. Mark Kim *et al.* [65] presented a Harvest Network, a Ethereum blockchain-based food traceability application with Internet of Things devices exchanging GS1 message standards. Mohsin Ur Rahman *et al.* [66], in their paper, presented an access-controlled, scalable data exchange system for smart agriculture. Yorghos Voutos *et al.* [67] examines the tantalizing possibility of integrating Internet of Things and smart contracts into smart agriculture to not only produce agrifood goods of greater quality but also to improve the distribution network and agricultural logistics linked with such products. Hiren Patel and Bela Shrimali [68] proposed a prototype AgriOnBlock, with the goal of preventing or minimizing monetary loss, crop contamination, and spoiling, which would ultimately lead to increased profitability in a shorter amount of time. Anusha Vangala *et al.* [69] presented a novel authentication model for smart agricultural infrastructure using smart contracts that employ hybrid blockchain and edge computing. Banupriya Sadayapillai and Kottilingam Kottursamy [70] presented a revolutionary way to automate and regulate the supply chain process using blockchain and smart contracts while keeping the trust of all parties involved. The authors specifically build smart contracts with two criteria, Percentage Profit Share and Minimum Purchase Rate, to control product costs and enhance farmer revenue.

6.2.7 Blockchain and Smart Contracts in Real Estate

Ioannis Karamitsos *et al.* [71] shows how blockchain technology and smart contract can be used in real estate. An in-depth smart contract architecture is presented, and its potential application to the leasing of commercial and residential real estate is discussed. Sandeep Kumar Panda *et al.* [72] examined how blockchain technology can resolve challenges within the current land registration system. The authors focused on a blockchain-based land registration system using Ethereum platform for decentralized application (DApps) development. Priyanka Kumar *et al.* [73] built a buyer–seller protocol on Ethereum utilizing smart contracts and blockchain technologies. They describe and execute the suggested approach and address the digitalization of land registration and the need to utilize blockchain instead of conventional storage systems. The authors' [74] overarching goal is to reduce the risk of fraudulent actions occurring inside land authentication systems by using blockchain technology. A working prototype based on smart contracts has been developed and validated with the participation of real users. SoonHyeong Jeong and Byeongtae Ahn [75] used the ZKP algorithm to construct a virtual real estate contract system based on Ethereum. Online contract administration and the detection of forgeries are made possible by the real estate contract system, with the help of BC technology. Miroslav Stefanović *et al.* [76] presented two case studies providing modifications that would result in a speedier transaction execution process, thereby eliminating the potential of "double spending." These two case studies illustrate some of the potential uses for blockchain and smart contracts in real estate transaction registrations. The authors [77] presented a service model that would connect transactions on the blockchain and exchange data on real estate sales, forming a shared monitoring system between properties to avoid fraudulent sales registration by real estate agents. Iftikhar Ahmad *et al.* [78] proposed a system that would use a private blockchain to safely record transactions involving the purchase and sale of properties. The ubiquitous and risk-free transfer of ownership may be accomplished *via* the use of a hybrid system that consists of public and private keys in addition to smart contracts. Miroslav Stefanović [79] proposed a solution using smart contract that can accommodate a wide range of scenarios often seen in land administration systems, including but not limited to joint ownership, partial transfer of ownership, split or merging of properties, and restrictions on real estate transactions.

6.2.8 Blockchain and Smart Contracts in Education and Research

Shivani Pathak *et al.* [80] proposed a decentralized application called AcadCerti. Using the AcadCerti smart contract, certificates can be stored securely on the blockchain, where they can later be accessed by an organization's recruiting team or enrollment team throughout the course of the admissions or hiring process. Junyao Wang *et al.* [81] focus on the analysis and practical applications of blockchain technology in the domains of intellectual property and scholarly research. Vijay Mohan [82] emphasizes the usefulness of emerging technologies such as blockchain in combating academic publication malpractice. Emilija Stojmenova Duh *et al.* [83] present a blockchain system designed to facilitate decentralized scholarly communication. The system performs the execution of a customizable research paper inside a blockchain smart contract. In their article, Jonathan P. Tennant *et al.* [84] explore the fundamental components of a hybrid model for publishing and peer review as well as the potential ways in which these components might be merged. The most important benefit that is discussed in this article is the transformation of the review into an activity that is inherently communal and social, free from the constraints of any journal-based system. In their study, Gipp *et al.* [85] introduced a method that aims to ensure the secure verification of research proposals, information, or conclusions before the submission of a paper for review. The suggested methodology produces a cryptographic hash, serving as a unique identifier, for the document and its corresponding outcomes, if relevant. Michał R. Hoffman *et al.* [86] presented a series of use case scenarios that illustrate the suitability of blockchains for the field of scientific publication, drawing upon the inherent qualities of this technology. Tim K. Mackey *et al.* [87] proposed a paradigm aimed at revolutionizing scientific publication using blockchain technology. All of the aforementioned concepts, including our own proposal, present challenges in terms of their acceptance and implementation. Consequently, they would need more testing, review, and collaborative engagement with the scientific community in order to be effectively implemented. In the context of the area of scientific publishing, Janowicz Krzysztof *et al.* investigated the potential applications of distributed ledger technology, highlighting the opportunities they provide for research endeavors. The aim of the study was to illustrate the construction of a management process for the Semantic Web journal by using a crypto-coin ecosystem. Novotny *et al.* [89] presented a comprehensive overview of the advancements in permissioned blockchains, focusing specifically on Hyperledger Fabric. The authors also investigate the potential applications of these advances in the field of scientific publishing. Sinclair Davidson *et al.* [90] presents a new way of looking at the economic significance of blockchains: as an emerging institutional technology. Aletheia [91] was created by Kade Morton *et al.*, a decentralized, open access, and open-source journal that allows authors to submit research papers. It requires participants to exchange documents, establishing a P2P network. Sina

Rafati Niya *et al.* [92] introduced a blockchain-based system known as Eureka, designed specifically for the purpose of journal review management. The publishing platform has furthermore provided incentives for writers and reviewers to engage in its activities. Joris van Rossum [93] highlighted the obstacles encountered in the realm of scientific publishing and highlighted the potential of blockchain technology to revolutionize the publication sector, particularly in the context of scholarly communication. Blockchainpeerreview.org [94] is an industry initiative in which multiple publishers are working together to improve the peer review system for academic publications. Antonio Tenorio-Fornes *et al.* [95] propose a framework for a decentralized publishing system based on blockchain and IPFS. The system is equipped with the following features: open access, a distributed reviewer system, and transparency. Orvium [96] was established by Manuel Martin, Antonio Romero, and Roberto Rabasco with the intention of using blockchain technology to bring together all of the parties involved in the process of scientific publication. In addition, they used Orvium coins as an incentive for peer review. US 2018/0323980A1 [97] proposes the use of BC technology in the field of scientific research. The proposal involves the generation of a block to record and store information related to tests, data collection, performance analysis, and outcomes. The blockchain serves as a research endeavor in which all stages are documented in a transparent manner.

6.2.9 Blockchain and Smart Contracts in Other Sectors

Rui Xie *et al.* [98], in their paper, proposed a method for constructing a blockchain and smart contract-based decentralized certificate system, with the end goal of offering certificate services on the blockchain to participants in college-level innovation and entrepreneurship competitions. Rifat Sonmez *et al.* [99] designed and developed a revolutionary BIM-based smart contract payment management system to upgrade the conventional progress payment method for construction assignments. The suggested method is used on a real building project to demonstrate the technique and collect views from construction experts through a survey. Abdul Wahab *et al.* [100] proposed BCT-SmContract using an automated program that uses blockchain to establish contractual agreements among several participants in a building project. Using blockchain technology and smart contracts, this paper outlines a system architecture [101] for the management of information at building sites. Using this approach, numerous autonomous smart contracts are constructed for the various sorts of data. It guarantees that the system is able to dynamically install new smart contracts without disrupting service, which makes it possible to expand scenario services. The proposed system [102] uses both public blockchain and consortia blockchain technology to facilitate RI negotiations, credit issuance, and credit trading using SC. Experimental findings show that the suggested plastic credit system may achieve a lightweight operation in terms of resource needs and system maintenance, as shown by the overall system performance analysis. Rowaid Ibrahim *et al.* [103] presents a prototype ecosystem built on top of a programmable smart contract hosted on a new cryptocurrency blockchain. The prototype's goal is to ensure that the building sector operates in a decentralized, economically independent manner. Debashis Das *et al.* [104] proposed a vehicle-to-vehicle communication system (BVHS) to improve the safety of vehicles and confidentiality of data exchanged between vehicles. The suggested technique has the capability to authenticate user identities, identify unlawful access, and create encrypted communication between vehicles.

6.3 Critical Analysis of the Review

Literature search was done using Scopus and Google Scholar Database because majority of the quality journals are covered by these two databases. The search keywords were smart contract, blockchain and digital identity, blockchain and financial security, blockchain and supply chain management, blockchain and insurance, blockchain and clinical trials, blockchain and agriculture, blockchain and real estate, blockchain on education and research, blockchain and construction, blockchain and cryptocurrency, and 3,259 papers were retrieved. A total of 2,947 papers were excluded after the title and abstract screening. After the exclusion criteria were applied—poor paper quality, improvements in blockchain, improvements in security, cryptocurrency mining, blockchain scalability issues, and review

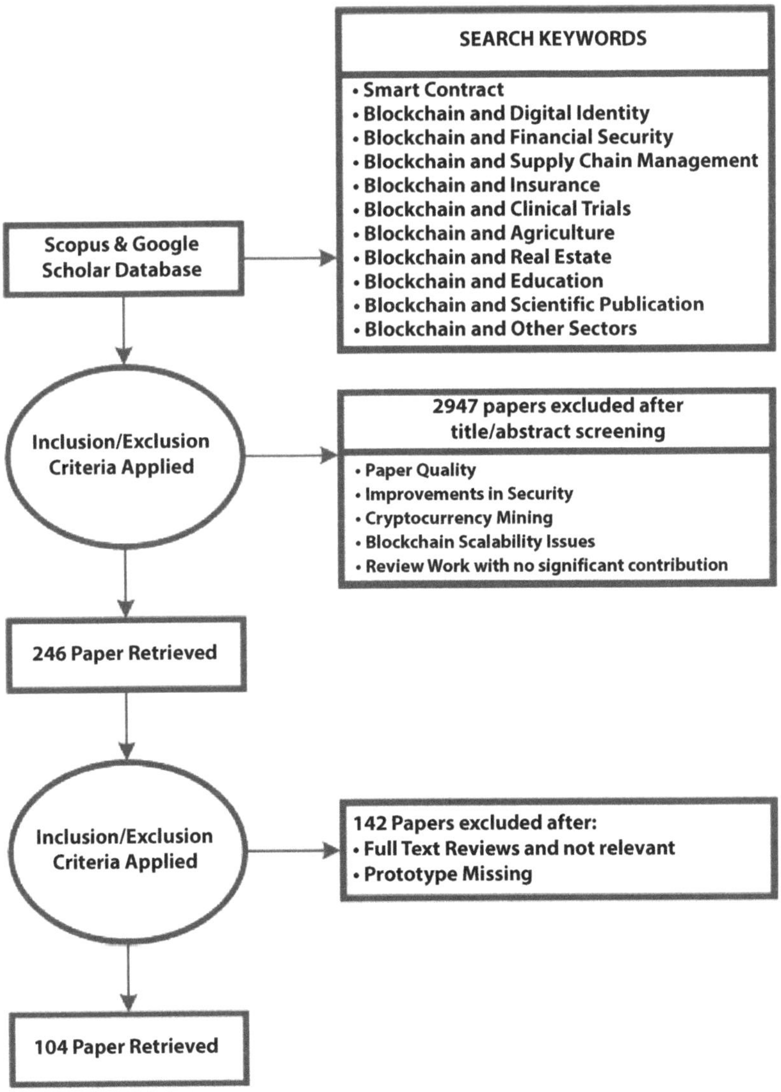

Figure 6.2 Flowchart depicting evidence acquisition.

work with no significant contribution—246 papers were retrieved. The exclusion criteria were applied again based on full text review that are not relevant to the study and prototype was missing. Finally, 104 papers were retrieved and included as references and depicted in Figure 6.2.

The bifurcation of blockchain and smart contracts implemented in various sectors is depicted in Figure 6.3. Out of 104 papers, 12 papers were from digital identity, 11 papers from financial security, 13 papers from supply chain management, 10 papers from insurance, 13 papers from healthcare and clinical trials, eight papers from agriculture, nine papers from real estate, 18 papers from education and research, and 10 papers from construction and cryptocurrency.

The literature review's classification of articles is shown in Figure 6.4. There was a total of 104 papers that were taken into consideration for this literature study. Of those papers, 53% were published in journals (55 papers), 37% were published in conference proceedings (39

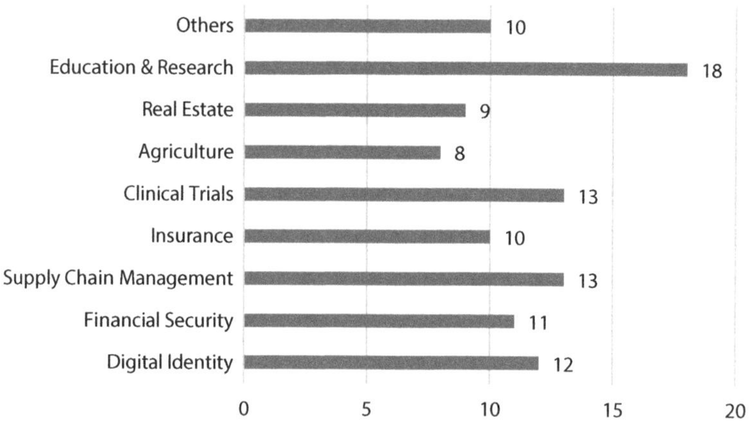

Figure 6.3 Sector-wise bifurcation of papers ($N = 104$).

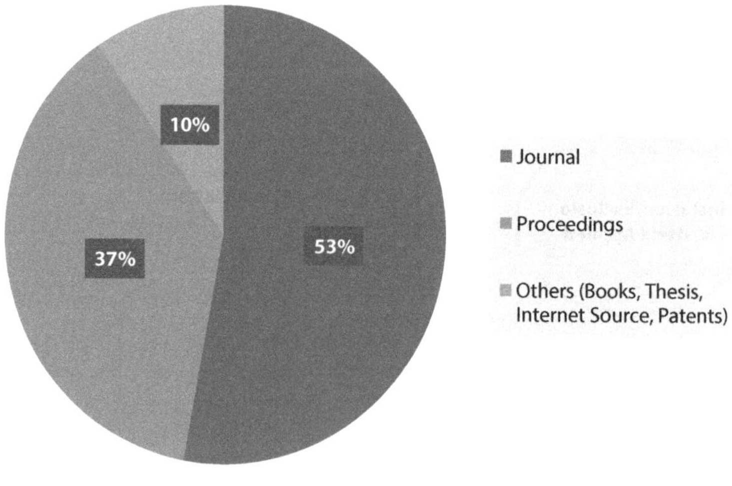

Figure 6.4 Categorization of publication ($N = 104$).

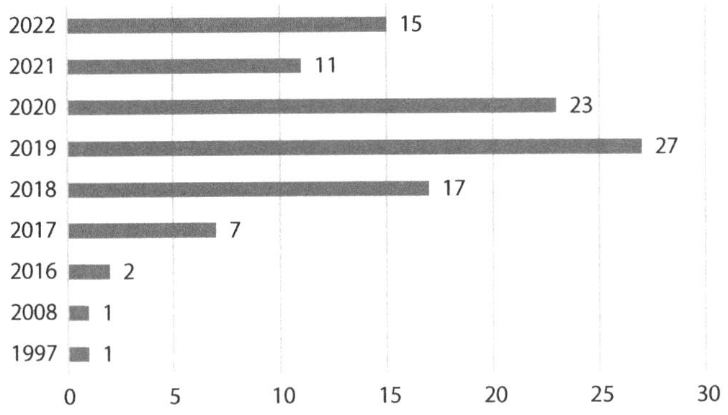

Figure 6.5 Year-wise paper dissemination.

papers), and 10% were published in books, thesis, online sources, and patents (10 papers/articles).

Figure 6.5 illustrates the progression of the number of publications produced each year. It is feasible to verify that research did not start to emerge until 2016, and it remained uncommon up to 2017. However, there was a significant increase in 2018 (17 articles), and the number of publications reached their highest point in 2019 (27 articles). It is worth mentioning that the surge after 2017 corresponds with the formal acknowledgment of cryptocurrencies by governments and financial markets as well as the 2017 cryptocurrency boom. Due to the fact that this graph only takes into account articles that were issued up to the time that the data was received, the number of publications decreased in 2022.

According to the literature review, IEEE and Elsevier have published more papers compared to other publication houses. IEEE and Elsevier publication platforms published 51.92% of the papers (38 and 16 papers, respectively) which are considered in this study. The papers published by various publication houses are presented in Figure 6.6.

Table 6.1 shows a list of the top 10 nations in terms of their author contributions, which account for 69.35% of the total, while research on blockchain conducted in the remaining countries only contributed 30.65% of the total. United States, Republic of China, and India

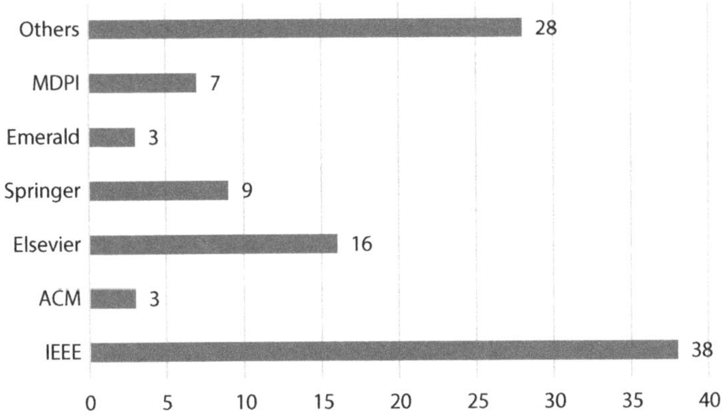

Figure 6.6 Papers published by various publication houses.

Table 6.1 Top 10 countries—author contribution.

Country name	No. of authors	Percentage
United States	60	15.58%
Republic of China	55	14.29%
India	51	13.25%
United Kingdom	19	4.93%
Greece	19	4.93%
United Arab Emirates	17	4.42%
Italy	15	3.89%
Australia	13	3.38%
South Korea	9	2.34%
Canada	9	2.34%

have the largest number of published articles as of the year 2022, followed by the United Kingdom, Greece, and United Arab Emirates.

6.4 Blockchain and Artificial Intelligence

Based on the above-mentioned investigation of 104 distinct publications, we found the concept of artificial intelligence in nine papers, which is almost 9% and presented in Figure 6.7. The recent advancements in combining artificial intelligence and blockchain is gaining popularity and wide acceptance. It was also found that the combination of these technologies is applied to supply chain management, healthcare, real estate, education and research, etc.

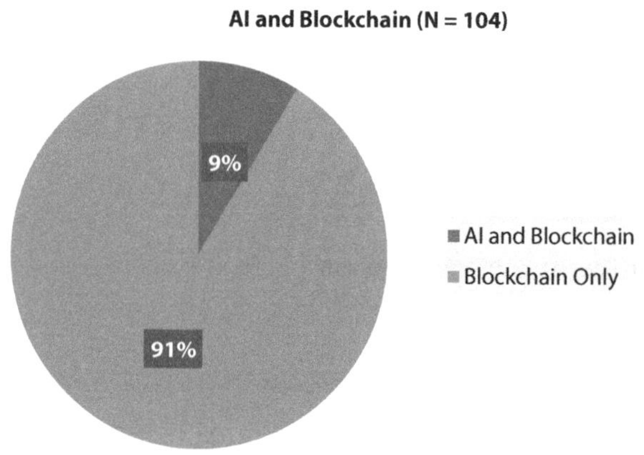

Figure 6.7 Artificial intelligence and blockchain.

Machine learning has the capability to process big data, whereas blockchain technology ensures reliability. We infer that by combining these technologies, data privacy, security, and integration could be preserved.

6.5 Discussion on the Reasoning for Implementation of Blockchain

Digital identification, when carried out using blockchain, ensures protecting the privacy, confidentiality, efficiency, and security. Due to this feature, banking and other financial institutions employ blockchain technologies for the KYC process. Several studies indicated that smart contracts have the ability to widen the scope of financial instruments, financial payment systems, and the overall financial product management. This is because of the trust in blockchain's immutability and auditability features.

Implementing blockchain in supply chain management helps stakeholders to track, record, and share information of goods. Due to the verifiable, transparent, and decentralized features of blockchain, the supply chain management process is reliable, trustworthy, and secure. Integrating blockchain and IoT technologies, smart agriculture system such as Harvest Network, AgriOnBlock, etc., minimizes monetary loss, crop contamination, and spoiling. It enhances the traceability of agri goods and improve the distribution network and trust among the stakeholders.

Smart contracts in the insurance sector pave the way to eliminate third party organizations. Hence, implementation of blockchain technologies in the insurance sector helps in improving the efficiency and cost. In clinical trials, blockchain solves issues related to data integrity, protocol conformity, data management, and transparency. It also avoids irregularity and information abuse. Furthermore, it aids the health data network by providing a solution to exchange medical images safely. In real estate, blockchain is used to cut third parties, reduce fraud, and bring transparency with smart contracts. It has eliminated the need for brokers, lawyers, and title companies. Tokenization helps to own a part of real estate, similar to crowdfunding.

Blockchain, being a decentralized application, is used to maintain the certificates securely by the higher education institutions. Many publication houses are trying to implement blockchain in scholarly communication. The scholarly communication and scientific publication needs to be transparent, decentralized, and unbiased. Based on the survey, it was found that a role-based smart contract between publisher, reviewer, and author helps in automation, avoiding the intermediary. Attestation of publication is completely absent, leading to which journals are hijacked and mushrooming of cloned journals.

The advantages on combining artificial intelligence and blockchain are manifold. Some of them are listed as follows:

- AI can influence blockchain to become intelligent.
- There is reduction of errors due to human interference.
- Smart contracts in blockchain, through continuous learning, allow them to be more efficient and intelligent.
- Security is enhanced and vulnerability gets reduced.

- The efficiency gets significantly improved due to the optimization and speed of data query.
- Transparency and reliability are ensured in blockchain-packed AI applications.
- Due to decentralization in blockchain, AI models' computing capacity and computational complexity problem could be addressed.

6.6 Conclusion

Blockchain and smart contracts are being widely accepted, and the research community is continuously working to bring innovation in this field. Through this study, we have analyzed the research papers using blockchain in various sectors. We found that the education and research sector is the forerunner in adopting this technology, contributing to 17.3% of the total publications. It was also found that the majority of the researchers resort to go for journal publications, and we have also found that IEEE is the major publishing house that publishes research in the field of blockchain. Country-wise analysis based on authors revealed that United States ranks first, followed by the Republic of China and India. The investigation carried out presented the importance of combining AI with blockchain. This analysis will help future researchers to identify the opportunities available in various sectors. Our future research will focus on research dissemination through blockchain and smart contract.

References

1. Nakamoto, S., Bitcoin: A Peer-to-Peer Electronic Cash System, 2008, [online] Available: https://bitcoin.org/bitcoin.pdf.
2. Szabo, N., Formalizing and securing relationships on public networks. *First Monday*, 2, 9, 1–21, 1997.
3. Sivakumar, E. and Chawla, P., Role-Based Smart Contract: An Intelligent System for Scholarly Communication. *SN Comput. Sci.*, 3, 4, 277, 2022, doi: 10.1007/s42979-022-01163-4.
4. Yasin, and Liu, L., An Online Identity and Smart Contract Management System, in: *2016 IEEE 40th Annual Computer Software and Applications Conference (COMPSAC)*, vol. 2, pp. 192–198, 2016, doi: 10.1109/COMPSAC.2016.2.
5. Kapsoulis, N., Psychas, A., Palaiokrassas, G., Marinakis, A., Litke, A., Varvarigou, T., Know Your Customer (KYC) Implementation with Smart Contracts on a Privacy-Oriented Decentralized Architecture. *Future Internet*, 12, 2, 1–13, 2020, doi: 10.3390/fi12020041.
6. Yadav, A. and Bajpa, R., KYC Optimization using Blockchain Smart Contract Technology. *Int. J. Innov. Res. Appl. Sci. Eng.*, 4, 669–674, 2020, doi: 10.29027/IJIRASE.v4.i3.2020.669-674.
7. Moyano, J.P. and Ross, O., KYC Optimization Using Distributed Ledger Technology. *Bus. Inf. Syst. Eng.*, 59, 6, 411–423, 2017, doi: 10.1007/s12599-017-0504-2.
8. Biryukov, A., Khovratovich, D., Tikhomirov, S., Privacy-preserving KYC on Ethereum. *Proceedings of 1st ERCIM Blockchain Workshop*, Amsterdam, Netherlands, 2018, doi: 10.18420/blockchain2018_09.
9. Shbair, W.M., Steichen, M., François, J., State, R., Blockchain orchestration and experimentation framework: A case study of KYC. *NOMS 2018 - 2018 IEEE/IFIP Network Operations and Management Symposium*, Taipei, Taiwan, pp. 1–6, 2018, doi: 10.1109/NOMS.2018.8406327.

10. Sundareswaran, N., Sasirekha, S., Louis Paul, I.J., Balakrishnan, S., Swaminathan, G., Optimised KYC Blockchain System. *2020 International Conference on Innovative Trends in Information Technology (ICITIIT) Kottayam, India*, pp. 1–6, 2020, doi: 10.1109/ICITIIT49094.2020.9071533.
11. Parra-Moyano, J., Thoroddsen, T., Ross, O., Optimized and Dynamic KYC System Based on Blockchain Technology. *SSRN Electron. J., International Journal of Blockchains and Cryptocurrencies*, 1, 1, 85–106, 2018, doi: 10.2139/ssrn.3248913.
12. Yang, X. and Li, W., A zero-knowledge-proof-based digital identity management scheme in blockchain. *Comput. Secur.*, 99, 102050, 2020, doi: https://doi.org/10.1016/j.cose.2020.102050.
13. Rouhani, S. and Deters, R., Security, Performance, and Applications of Smart Contracts: A Systematic Survey. *IEEE Access*, 7, 50759–50779, 2019, doi: 10.1109/ACCESS.2019.2911031.
14. Omar, A.S. and Basir, O., Identity Management in IoT Networks Using Blockchain and Smart Contracts. *2018 IEEE International Conference on Internet of Things (iThings) and IEEE Green Computing and Communications (GreenCom) and IEEE Cyber, Physical and Social Computing (CPSCom) and IEEE Smart Data (SmartData)*, Halifax, NS, Canada, pp. 994–1000, 2018, doi: 10.1109/Cybermatics_2018.2018.00187.
15. Borse, Y., Chawathe, A., Patole, D., Ahirao, P., Anonymity: A Secure Identity Management Using Smart Contracts. *SSRN Electron. J.*, Jan. 2019, doi: 10.2139/ssrn.3352370.
16. Subramanian, H., Security tokens: architecture, smart contract applications and illustrations using SAFE. *Manage. Finance*, 46, 6, 735–748, Jan. 2020, doi: 10.1108/MF-09-2018-0467.
17. Demirel, E. and Karagoz Zeren, S., Developing Smart Contracts for Financial Payments as Innovation, in: *Handbook of Research on Managerial Thinking in Global Business Economics* pp. 96–114, 2021.
18. Ali, O., Ally, M., Clutterbuck, Dwivedi, Y., The state of play of blockchain technology in the financial services sector: A systematic literature review. *Int. J. Inf. Manage.*, 54, 102199, 2020, doi: https://doi.org/10.1016/j.ijinfomgt.2020.102199.
19. Dorsala, M.R., Sastry, V.N., Chapram, S., Fair payments for verifiable cloud services using smart contracts. *Comput. Secur.*, 90, 101712, 2020, doi: https://doi.org/10.1016/j.cose.2019.101712.
20. Wang, H., Guo, C., Cheng, S., LoC — A new financial loan management system based on smart contracts. *Future Gener. Comput. Syst.*, 100, 648–655, 2019, doi: https://doi.org/10.1016/j.future.2019.05.040.
21. Li, Y., Liu, H., Yang, Z., Ren, Q., Wang, L., Chen, B., SafePay on Ethereum: A Framework For Detecting Unfair Payments in Smart Contracts. *2020 IEEE 40th International Conference on Distributed Computing Systems (ICDCS)*, Singapore, pp. 1219–1222, 2020, doi: 10.1109/ICDCS47774.2020.00116.
22. Gazali, H.M., Hassan, R., Nor, R.M., Rahman, H.M.M., Re-inventing PTPTN study loan with blockchain and smart contracts. *2017 8th International Conference on Information Technology (ICIT)*, Amman, Jordan, pp. 751–754, 2017, doi: 10.1109/ICITECH.2017.8079940.
23. Wang, S., Tang, X., Zhang, Y., Chen, J., Auditable Protocols for Fair Payment and Physical Asset Delivery Based on Smart Contracts. *IEEE Access*, 7, 109439–109453, 2019, doi: 10.1109/ACCESS.2019.2933860.
24. Ahluwalia, S., Mahto, R.V., Guerrero, M., Blockchain technology and startup financing: A transaction cost economics perspective. *Technol. Forecast. Soc Change*, 151, 119854, 2020, doi: https://doi.org/10.1016/j.techfore.2019.119854.
25. Chakraborty, S., Aich, S., Seong, S.J., Kim, H. C., A Blockchain based Credit Analysis Framework for Efficient Financial Systems. *2019 21st International Conference on Advanced Communication Technology (ICACT)*, PyeongChang, Korea (South), pp. 56–60, 2019, doi: 10.23919/ICACT.2019.8701926.

26. Chen, B., Tan, Z., Fang, W., Blockchain-Based Implementation for Financial Product Management. *2018 28th International Telecommunication Networks and Applications Conference (ITNAC)*, Sydney, NSW, Australia, pp. 1–3, 2018, doi: 10.1109/ATNAC.2018.8615246.
27. Law, A., *Smart Contracts and Their Application in Supply Chain Management*, Massachusetts Inst. Technol., Cambridge, MA, USA, 2017.
28. De Giovanni, P., Blockchain and smart contracts in supply chain management: A game theoretic model. *Int. J. Prod. Econ.*, 228, 107855, 2020, doi: https://doi.org/10.1016/j.ijpe.2020.107855.
29. Bottoni, P., Gessa, N., Massa, G., Pareschi, R., Selim, H., Arcuri, E., Intelligent Smart Contracts for Innovative Supply Chain Management. *Front. Blockchain*, 3, 1–19, 2020, doi: 10.3389/fbloc.2020.535787.
30. De Giovanni, P., Digital Supply Chain through Dynamic Inventory and Smart Contracts. *Mathematics*, 7, 12, 1–25 2019, doi: 10.3390/math7121235.
31. Omar, I.A., Jayaraman, R., Debe, M.S., Salah, K., Yaqoob, I., Omar, M., Automating Procurement Contracts in the Healthcare Supply Chain Using Blockchain Smart Contracts. *IEEE Access*, 9, 37397–37409, 2021, doi: 10.1109/ACCESS.2021.3062471.
32. Omar, I.A., Jayaraman, R., Salah, K., Debe, M., Omar, M., Enhancing Vendor Managed Inventory Supply Chain Operations Using Blockchain Smart Contracts. *IEEE Access*, 8, 182704–182719, 2020, doi: 10.1109/ACCESS.2020.3028031.
33. Prause, G., Smart Contracts for Smart Supply Chains. *IFAC-PapersOnLine*, 52, 13, 2501–2506, 2019, doi: https://doi.org/10.1016/j.ifacol.2019.11.582.
34. Bocek, T., Rodrigues, B.B., Strasser, T., Stiller, B., Blockchains everywhere - a use-case of blockchains in the pharma supply-chain. *2017 IFIP/IEEE Symposium on Integrated Network and Service Management (IM)*, Lisbon, Portugal, pp. 772–777, 2017, doi: 10.23919/INM.2017.7987376.
35. Moudoud, H., Cherkaoui, S., Khoukhi, L., An IoT Blockchain Architecture Using Oracles and Smart Contracts: the Use-Case of a Food Supply Chain. *2019 IEEE 30th Annual International Symposium on Personal, Indoor and Mobile Radio Communications (PIMRC)*, Istanbul, Turkey, pp. 1–6, 2019, doi: 10.1109/PIMRC.2019.8904404.
36. Terzi, S. *et al.*, Transforming the supply-chain management and industry logistics with blockchain smart contracts. *PCI '19: Proceedings of the 23rd Pan-Hellenic Conference on Informatics*, Nicosia, Cyprus, pp. 9–14, 2019, doi: 10.1145/3368640.3368655.
37. Hasan, H., AlHadhrami, E., AlDhaheri, A., Salah, K., Jayaraman, R., Smart contract-based approach for efficient shipment management. *Comput. Ind. Eng.*, 136, 149–159, 2019, doi: https://doi.org/10.1016/j.cie.2019.07.022.
38. Wang, L. *et al.*, Smart Contract-Based Agricultural Food Supply Chain Traceability. *IEEE Access*, 9, 9296–9307, 2021, doi: 10.1109/ACCESS.2021.3050112.
39. Arunmozhi, M., Venkatesh, V.G., Arisian, S., Shi, Y., Raja Sreedharan, V., Application of blockchain and smart contracts in autonomous vehicle supply chains: An experimental design. *Transp. Res. Part E: Logist. Transp. Rev.*, 165, 102864, 2022, doi: 10.1016/j.tre.2022.102864.
40. Aleksieva, V., Valchanov, H., Huliyan, A., Smart Contracts based on Private and Public Blockchains for the Purpose of Insurance Services. *2020 International Conference Automatics and Informatics (ICAI)*, Varna, Bulgaria, pp. 1–4, 2020, doi: 10.1109/ICAI50593.2020.9311371.
41. Gatteschi, V., Lamberti, F., Demartini, C., Pranteda, C., Santamaría, V., Blockchain and Smart Contracts for Insurance: Is the Technology Mature Enough? *Future Internet*, 10, 2, 1–16, 2018, doi: 10.3390/fi10020020.
42. Lamberti, F., Gatteschi, V., Demartini, C., Pelissier, M., Gomez, A., Santamaria, V., Blockchains Can Work for Car Insurance: Using Smart Contracts and Sensors to Provide On-Demand Coverage. *IEEE Consum. Electron. Mag.*, 7, 4, 72–81, July 2018, doi: 10.1109/MCE.2018.2816247.
43. Sheth, A. and Subramanian, H., Blockchain and contract theory: modeling smart contracts using insurance markets. *Manage. Finance*, 46, 6, 803–814, Jan. 2020, doi: 10.1108/MF-10-2018-0510.

44. Aleksieva, V., Valchanov, H., Huliyan, A., Application of Smart Contracts based on Ethereum Blockchain for the Purpose of Insurance Services. *2019 International Conference on Biomedical Innovations and Applications (BIA)*, Varna, Bulgaria, pp. 1–4, 2019, doi: 10.1109/BIA48344.2019.8967468.
45. Raikwar, M., Mazumdar, S., Ruj, S., Sen Gupta, S., Chattopadhyay, A., Lam, K.-Y., A Blockchain Framework for Insurance Processes. *2018 9th IFIP International Conference on New Technologies, Mobility and Security (NTMS)*, Paris, France, pp. 1–4, 2018, doi: 10.1109/NTMS.2018.8328731.
46. Aleksieva, V., Valchanov, H., Huliyan, A., Implementation of Smart Contracts based on Hyperledger Fabric Blockchain for the Purpose of Insurance Services. *2020 International Conference on Biomedical Innovations and Applications (BIA)*, Varna, Bulgaria, pp. 113–116, 2020, doi: 10.1109/BIA50171.2020.9244500.
47. Elhence, A., Goyal, A., Chamola, V., Sikdar, B., A Blockchain and ML-Based Framework for Fast and Cost-Effective Health Insurance Industry Operations, *IEEE Trans. Comput. Social Syst.*, 2022, doi: 10.1109/TCSS.2022.3219256.
48. Chondrogiannis, E., Andronikou, V., Karanastasis, E., Litke, A., Varvarigou, T., Using blockchain and semantic web technologies for the implementation of smart contracts between individuals and health insurance organizations. *Blockchain: Res. Appl.*, 3, 2, 100049, 2022, doi: 10.1016/j.bcra.2021.100049.
49. Zheng, H., You, L., Hu, G., A novel insurance claim blockchain scheme based on zero-knowledge proof technology. *Comput. Commun.*, 195, 207–216, 2022, doi: 10.1016/j.comcom.2022.08.007.
50. Khatoon, A., A blockchain-based smart contract system for healthcare management. *Electron. (Switzerland)*, 9, 1, 1–23, 2020, doi: 10.3390/electronics9010094.
51. Omar, I.A., Jayaraman, R., Salah, K., Simsekler, M.C.E., Exploiting Ethereum Smart Contracts for Clinical Trial Management. *2019 IEEE/ACS 16th International Conference on Computer Systems and Applications (AICCSA)*, Abu Dhabi, United Arab Emirates, pp. 1–6, 2019, doi: 10.1109/AICCSA47632.2019.9035341.
52. Omar, I.A., Jayaraman, R., Salah, K., Simsekler, M.C.E., Yaqoob, I., Ellahham, S., Ensuring protocol compliance and data transparency in clinical trials using Blockchain smart contracts. *BMC Med. Res. Methodol.*, 20, 1, 224, 2020, doi: 10.1186/s12874-020-01109-5.
53. Nugent, T., Upton, D., Cimpoesu, M., Improving data transparency in clinical trials using blockchain smart contracts. *FResearch*, 5, 2541, Oct. 2016.
54. Li, M., Xia, L., Seneviratne, O., Leveraging Standards Based Ontological Concepts in Distributed Ledgers: A Healthcare Smart Contract Example. *2019 IEEE International Conference on Decentralized Applications and Infrastructures (DAPPCON)*, Newark, CA, USA, pp. 152–157, 2019, doi: 10.1109/DAPPCON.2019.00029.
55. Sookhak, M., Jabbarpour, M.R., Safa, N.S., Yu, F.R., Blockchain and smart contract for access control in healthcare: A survey, issues and challenges, and open issues. *J. Network Comput. Appl.*, 178, 102950, 2021, doi: https://doi.org/10.1016/j.jnca.2020.102950.
56. Akkaoui, R., Hei, X., Guo, C., Cheng, W., RBAC-HDE: On the Design of a Role-based Access Control with Smart Contract for Healthcare Data Exchange. *2019 IEEE International Conference on Consumer Electronics - Taiwan (ICCE-TW)*, Yilan, Taiwan, pp. 1–2, 2019, doi: 10.1109/ICCE-TW46550.2019.8991965.
57. Gupta, R., Shukla, A., Tanwar, S., AaYusH: A Smart Contract-Based Telesurgery System for Healthcare 4.0. *2020 IEEE International Conference on Communications Workshops (ICC Workshops)*, Dublin, Ireland, pp. 1–6, 2020, doi: 10.1109/ICCWorkshops49005.2020.9145044.
58. Duong-Trung, N., Son, H.X., Le, H.T., Phan, T.T., On Components of a Patient-Centered Healthcare System Using Smart Contract, in: *Proceedings of the 2020 4th International Conference on Cryptography, Security and Privacy*, pp. 31–35, 2020, doi: 10.1145/3377644.3377668.

59. Chenthara, S., Ahmed, K., Wang, H., Whittaker, F., A Novel Blockchain Based Smart Contract System for eReferral in Healthcare: HealthChain, in: *Health Information Science*, pp. 91–102, 2020, doi: 10.1007/978-3-030-61951-0_9.
60. Tang, H., Tong, N., Ouyang, J., Medical Images Sharing System Based on Blockchain and Smart Contract of Credit Scores. *2018 1st IEEE International Conference on Hot Information-Centric Networking (HotICN)*, Shenzhen, China, pp. 240–241, 2018, doi: 10.1109/HOTICN.2018.8605956.
61. Kormiltsyn, A., Udokwu, C., Karu, K., Thangalimodzi, K., Norta, A., Improving Healthcare Processes with Smart Contracts, Lecture Notes in Business Information Processing, Vol. 353, pp. 500–513, Springer, 2019, doi: 10.1007/978-3-030-20485-3_39.
62. Wu, Y., Liu, C., Sebald, L., Nguyen, P., Yesha, Y., Apply Trust Computing and Privacy Preserving Smart Contracts to Manage, Share, and Analyze Multi-site Clinical Trial Data, in: *The International Conference on Deep Learning, Big Data and Blockchain (DBB 2022)*, pp. 3–14, 2023, doi: 10.1007/978-3-031-16035-6_1.
63. Lin, J., Zhang, A., Shen, Z., Chai, Y., Blockchain and IoT based food traceability for smart agriculture, in: *Proceedings of the 2nd International Conference on Crowd Science and Engineering*, pp. 38–43, 2017, doi: 10.1145/3126973.3126980.
64. Pranto, T.H., Noman, A.A., Mahmud, A., Haque, A.B., Blockchain and smart contract for IoT enabled smart agriculture. *PeerJ Comput. Sci.*, 7:e407, 1–29, 2021, doi: 10.7717/peerj-cs.407.
65. Kim, M., Hilton, B., Burks, Z., Reyes, J., Integrating Blockchain, Smart Contract-Tokens, and IoT to Design a Food Traceability Solution. *2018 IEEE 9th Annual Information Technology, Electronics and Mobile Communication Conference (IEMCON)*, Vancouver, BC, Canada, pp. 335–340, 2018, doi: 10.1109/IEMCON.2018.8615007.
66. Ur Rahman, M., Baiardi, F., Ricci, L., Blockchain Smart Contract for Scalable Data Sharing in IoT: A Case Study of Smart Agriculture. *2020 IEEE Global Conference on Artificial Intelligence and Internet of Things (GCAIoT)*, Dubai, United Arab Emirates, pp. 1–7, 2020, doi: 10.1109/GCAIoT51063.2020.9345874.
67. Voutos, Y., Drakopoulos, G., Mylonas, P., Smart Agriculture: An Open Field for Smart Contracts. *2019 4th South-East Europe Design Automation, Computer Engineering, Computer Networks and Social Media Conference (SEEDA-CECNSM)*, Piraeus, Greece, pp. 1–6, 2019, doi: 10.1109/SEEDA-CECNSM.2019.8908411.
68. Patel, H. and Shrimali, B., AgriOnBlock: Secured data harvesting for agriculture sector using blockchain technology. *ICT Express*, 9, 2, 150–159, 2023, doi: 10.1016/j.icte.2021.07.003.
69. Vangala, A., Sutrala, A.K., Das, A.K., Jo, M., Smart Contract-Based Blockchain-Envisioned Authentication Scheme for Smart Farming. *IEEE Internet Things J.*, 8, 13, 10792–10806, 1 July, 2021, doi: 10.1109/JIOT.2021.3050676.
70. Sadayapillai, B. and Kottursamy, K., An Agriculture Supply Chain Model for Improving Farmer Income Using Blockchain Smart Contract, in: *Inventive Communication and Computational Technologies*, pp. 587–598, 2022, doi: 10.1007/978-981-16-5529-6_46.
71. Karamitsos, I., Papadaki, M., Barghuthi, N., Design of the Blockchain Smart Contract: A Use Case for Real Estate. *J. Inf. Secur.*, 09, 177–190, 2018, doi: 10.4236/jis.2018.93013.
72. Panda, S.K., Mohammad, G.B., Mohanty, S.N., Sahoo, S., Smart contract-based land registry system to reduce frauds and time delay. *Secur. Privacy*, 4, 5, e172, 2021, doi: 10.1002/spy2.172.
73. Kumar, P., Dhanush, G.A., Srivatsa, D., Nithin, A., Sahisnu, S., A Buyer and Seller's Protocol via Utilization of Smart Contracts Using Blockchain Technology, in: *Advanced Informatics for Computing Research*, pp. 464–474, 2019, doi: 10.1007/978-981-15-0108-1_43.
74. Jayabodhi, L.W.D.C., Rajapakse, C., Senanayake, J.M.D., Minimization of fraudulent activities in land authentication through Blockchain-based system. *2020 International Research*

Conference on Smart Computing and Systems Engineering (SCSE), Colombo, Sri Lanka, pp. 68–74, 2020, doi: 10.1109/SCSE49731.2020.9313057.
75. Jeong, S. and Ahn, B., Implementation of real estate contract system using zero knowledge proof algorithm based blockchain. *J. Supercomput.*, 77, 10, 11881–11893, 2021, doi: 10.1007/s11227-021-03728-1.
76. Stefanović, M., Ristić, S., Stefanović, D., Bojkić, M., Pržulj, D., Possible Applications of Smart Contracts in Land Administration. *2018 26th Telecommunications Forum (TELFOR)*, Belgrade, Serbia, pp. 420–425, 2018, doi: 10.1109/TELFOR.2018.8611872.
77. Huh, J.H. and Kim, S.K., Verification Plan Using Neural Algorithm Blockchain Smart Contract for Secure P2P Real Estate Transactions. *Electronics*, 9, 6, 1–25, 2020, doi: 10.3390/electronics9061052.
78. A.A.A., L.A., Ahmad, I., Alqarni, M.A., Real Estate Management via a Decentralized Blockchain Platform. *Comput. Mater. Contin.*, 66, 2, 1813–1822, 2021, doi: 10.32604/cmc.2020.013048.
79. Stefanović, M., Pržulj, Đ., Ristić, S., Stefanović, D., Nikolić, D., Smart Contract Application for Managing Land Administration System Transactions. *IEEE Access*, 10, 39154–39176, 2022, doi: 10.1109/ACCESS.2022.3164444.
80. Pathak, S., Gupta, V., Malsa, N., Ghosh, A., Shaw, R.N., Smart Contract for Academic Certificate Verification Using Ethereum, in: *Advanced Computing and Intelligent Technologies*, pp. 369–384, 2022, doi: 10.1007/978-981-19-2980-9_29.
81. Wang, J., Wang, S., Guo, J., Du, Y., Cheng, S., Li, X., A Summary of Research on Blockchain in the Field of Intellectual Property. *Proc. Comput. Sci.*, 147, 191–197, 2019, doi: 10.1016/j.procs.2019.01.220.
82. Mohan, V., On the use of blockchain-based mechanisms to tackle academic misconduct. *Res. Policy*, 48, 9, 103805, 2019, doi: 10.1016/j.respol.2019.103805.
83. Duh, E.S. *et al.*, Publish-and-Flourish: Using Blockchain Platform to Enable Cooperative Scholarly Communication. *Publications*, 7, 2, 1–15, 2019, doi: 10.3390/publications7020033.
84. Tennant, J.P. *et al.*, A multi-disciplinary perspective on emergent and future innovations in peer review [version 3; peer review: 2 approved]. *F1000Research*, 6, 1151, 1–66, 2017, doi: https://doi.org/10.12688/f1000research.12037.3.
85. Gipp, B., Breitinger, C., Meuschke, N., Beel, J., CryptSubmit: Introducing Securely Timestamped Manuscript Submission and Peer Review Feedback Using the Blockchain. *2017 ACM/IEEE Joint Conference on Digital Libraries (JCDL)*, pp. 1–4, 2017, doi: 10.1109/JCDL.2017.7991588.
86. Hoffman, M.R., Ibáñez, L.-D., Simperl, E., Scholarly publishing on the blockchain – from smart papers to smart informetrics. *Data Sci.*, 2, 291–310, 2019, doi: 10.3233/DS-190018.
87. Mackey, T.K., Shah, N., Miyachi, K., Short, J., Clauson, K., A Framework Proposal for Blockchain-Based Scientific Publishing Using Shared Governance. *Front. Blockchain*, 2, 19, 2019, doi: 10.3389/fbloc.2019.00019.
88. Janowicz, K. *et al.*, On the prospects of blockchain and distributed ledger technologies for open science and academic publishing. *Semant. Web*, 9, 545–555, 2018, doi: 10.3233/SW-180322.
89. Novotny, P. *et al.*, Permissioned blockchain technologies for academic publishing. *Inf. Serv. Use*, 38, 159–171, 2018, doi: 10.3233/ISU-180020.
90. Davidson, S., De Filippi, P., Potts, J., Blockchains and the economic institutions of capitalism. *J. I. Econ.*, 14, 4, 639–658, 2018, doi: 10.1017/S1744137417000200.
91. Morton, K., Aletheia: blockchain for scientific knowledge with a community management framework, GitHub, 2021, [Online], https://github.com/aletheia-foundation/aletheia-whitepaper/blob/master/white-paper.md, [Accessed: 23 May 2021].
92. Niya, S.R. *et al.*, A Blockchain-based Scientific Publishing Platform. *2019 IEEE International Conference on Blockchain and Cryptocurrency (ICBC)*, pp. 329–336, 2019, doi: 10.1109/BLOC.2019.8751379.

93. Van Rossum, J., Blockchain for Research, 2017, [Online], Available: https://www.labxing.com/files/lab_publications/502-1526736820-87OCgwFK.pdf, [Accessed: 02 June 2021].
94. Blockchain Peer Review Project, 2021, [Online], Available: https://www.blockchainpeerreview.org/, [Accessed: 01 Jun 2021].
95. Tenorio-Fornés, A., Jacynycz, V., Llop-Vila, D., Sánchez-Ruiz-Granados, A.A., Hassan, S., Towards a Decentralized Process for Scientific Publication and Peer Review using Blockchain and IPFS, 2019, Available online https://scholarspace.manoa.hawaii.edu/server/api/core/bitstreams/f67443e9-96d2-40e7-8675-aff625526c7b/content.
96. Orvium, Whitepaper, Accelerated Scientific Publishing (v1.7), 2019b, Available: https://docs.orvium.io/Orvium-WP.pdf.
97. Ahn, J., Chang, M.D., Kokku, R., Watson, P., Blockchain for open scientific research, 2018, US20180323980A1. Available online https://patentimages.storage.googleapis.com/54/52/05/ac22bc22219f63/US10659239.pdf.
98. Xie, R. *et al.*, Ethereum-Blockchain-Based Technology of Decentralized Smart Contract Certificate System. *IEEE Internet Things Mag.*, 3, 2, 44–50, June 2020, doi: 10.1109/IOTM.0001.1900094.
99. Sonmez, R., Ahmadisheykhsarmast, S., Güngör, A.A., BIM integrated smart contract for construction project progress payment administration. *Autom. Constr.*, 139, 104294, 2022, doi: 10.1016/j.autcon.2022.104294.
100. Wahab, A., Wang, J., Shojaei, A., Ma, J., A model-based smart contracts system via blockchain technology to reduce delays and conflicts in construction management processes. *Eng. Constr. Archit. Manage.*, 30, 10, 5052–5072, Jan. 2022, doi: 10.1108/ECAM-03-2022-0271.
101. Zhang, Y., Wang, T., Yuen, K.-V., Construction site information decentralized management using blockchain and smart contracts. *Comput.-Aided Civ. Infrastruct. Eng.*, 37, 11, 1450–1467, 2022, doi: 10.1111/mice.12804.
102. Zhang, X., Liu, C., Medda, F., A Smart-Contract-Aided Plastic Credit Scheme. *IEEE Syst. J.*, 17, 1, 1703–1713, March 2023, doi: 10.1109/JSYST.2022.3205266.
103. Ibrahim, R., Harby, A.A., Nashwan, M.S., Elhakeem, A., Financial Contract Administration in Construction via Cryptocurrency Blockchain and Smart Contract: A Proof of Concept. *Buildings*, 12, 8, 1–22, 2022, doi: 10.3390/buildings12081072.
104. Das, D., Banerjee, S., Chatterjee, P., Ghosh, U., Biswas, U., A Secure Blockchain Enabled V2V Communication System Using Smart Contracts. *IEEE Trans. Intell. Transp. Syst.*, 24, 4, 4651–4660, April 2023, doi: 10.1109/TITS.2022.3226626.

7

Clinical Intelligence: Deep Reinforcement Learning for Healthcare and Biomedical Advancements

Keerthika K.[1]*, Kannan M.[2] and T. Saravanan[3]

[1]*Department of Computer Science, School of Computing, Amrita Vishwa Vidyapeetham, Mysuru Campus, India*
[2]*Department of Computer Science, School of Sciences, CHRIST University, Bengaluru, India*
[3]*Department of Computer Science and Engineering, GITAM School of Technology, GITAM (Deemed to be University), Bengaluru, India*

Abstract

Deep reinforcement learning (DRL) is showing a remarkable impact in the healthcare and biomedical domains, leveraging its ability to learn complex decision-making policies from raw data through trial-and-error interactions. DRL can effectively extract the characteristic information in the environment, propose effective behavior strategies, and correct errors that occurred during the training process. Targeted toward healthcare professionals, researchers, and technology enthusiasts, this chapter begins with notable applications of DRL in healthcare, including personalized treatment recommendations, clinical trial optimization, disease diagnosis, robotic surgery and assistance, mental health support systems, chronic disease management and scheduling, and a few more. It also delves on challenges such as data privacy, interpretability, regulatory compliance, validation, and the need for domain expertise to ensure safe and effective deployment. Next, the chapter seamlessly transitions into DRL algorithms contributing to the biomedical field which are gaining traction due to their potential to provide timely and personalized interventions. Over time, the research community has proposed several methods and algorithms within the field of deep reinforcement learning that help agents learn optimal policies from rich data. Healthcare data is often complex, high-dimensional, and unstructured, such as medical images, genomics data, and patient records. The healthcare-suitable DRL algorithms such as Q-learning, SARSA, Bayesian, actor–critic, reinforcement learning (RL), Deep-Q-Networks (DQN), and Monte Carlo Tree Search (MCTS) are highlighted. In addition, the section offers guidelines for the application of DRL to healthcare and biomedical problems, aiming at providing indications to the designer of new applications in order to choose among different RL methods. Furthermore, a case study is included to fully realize the revolutionary benefits of DRL in healthcare environments, aiming to bridge the gap between theory and practice. The case study presents a remarkable impact on categories such as precision medicine, dynamic treatment regime, medical imaging, diagnostic systems, control systems, chat-bots and advanced interfaces, and healthcare management systems.

Keywords: Deep reinforcement learning, healthcare, biomedical, SARSA, deep-Q-networks, Monte Carlo

**Corresponding author*: kikeerthika@gmail.com

Shubham Mahajan, Pethuru Raj and Amit Kant Pandit (eds.) Deep Reinforcement Learning and Its Industrial Use Cases: AI for Real-World Applications, (137–150) © 2024 Scrivener Publishing LLC

7.1 Introduction

In the domain of artificial intelligence (AI), two powerful paradigms—deep learning and reinforcement learning—have risen to prominence, promising to reshape industries and drive innovation in unprecedented ways. Deep learning, a subset of machine learning, has demonstrated extraordinary prowess in tasks like image processing, natural language processing, automated vehicles, and even healthcare diagnosis. Reinforcement learning, on the other hand, has shown remarkable abilities in making intelligent decisions in complex, dynamic environments. Together they form a formidable duo that holds the keys to unlocking AI's true potential. Deep learning trains artificial neural networks to act like the human brain's ability by learning the patterns from vast amounts of data. Deep neural networks, which can capture hierarchical data representations, are at the heart of it. There are several layers of connected nodes in deep neural networks. In several disciplines, this architecture has been at the forefront of innovative advancements. Reinforcement learning is a science that teaches agents to make decisions sequentially in order to maximize a cumulative reward. It draws on behavioral psychology as inspiration. It mimics real-world scenarios where people have to deal with their surroundings and decide how to go around them.

The synergy between deep learning and reinforcement learning is where the real magic happens. Deep reinforcement learning (DRL) adheres the computational power of deep learning with the decision-making capabilities of RL. This amalgamation has led to groundbreaking achievements. From healthcare and gaming to robotics and finance, these paradigms have demonstrated their versatility and power. The future promises even greater advancements as researchers continue to bridge the gap between these complementary technologies. As we stand on the cusp of a new era driven by AI, the fusion of deep learning and reinforcement learning represents a beacon of hope, pointing toward a world enriched by intelligent, data-driven decision-making.

7.2 Deep Reinforcement Learning Methods

Deep reinforcement learning (DRL) methods are a class of machine learning techniques that enable agents to learn and make decisions in complex environments. An agent learns to make a sequence of actions in an environment to maximize a cumulative reward. DRL methods have gained significant attention and success in various fields, including robotics, autonomous vehicles, gaming, and more. Here are some key reinforcements learning methods:

1. Model-free methods
2. Policy gradient methods
3. Model-based methods

7.2.1 Model-Free Methods

Model-free methods are preferred when there is a wealth of data available for learning. They can leverage large datasets of patient information, electronic health records (EHRs), and medical images to make data-driven decisions without relying on explicit models of the environment. Healthcare and biomedical settings often feature non-stationary conditions, where patient health and treatment protocols change over time. Model-free methods can adapt to these changing environments without the need for constant model updates. The most commonly used model-free methods are as follows:

- Q-learning: A widely used model-free RL algorithm that learns the optimal action-value function by iteratively updating the Q-values based on the Bellman equation. Q-values can be used to understand the reasoning behind the algorithm's decisions, enhancing transparency in clinical decision-making.
- SARSA (State–Action–Reward–State–Action): Another model-free algorithm that updates Q-values based on the current state, action, immediate reward, and the next state–action pair.
- Deep Q-networks (DQN): Combines Q-learning with deep neural networks to learn Q-values from high-dimensional state spaces, making it suitable for complex environments.

7.2.2 Policy Gradient Methods

Policy gradient methods excel at handling healthcare tasks with continuous and high-dimensional action spaces, enabling precise adjustments in treatment and intervention. These methods optimize policies to provide continuous actions in a principled way. They can adapt to dynamic healthcare environments and evolving patient needs over extended periods, aligning with the long-term nature of healthcare and biomedical applications. Policy gradient methods are model-free, which means that they do not require a detailed model of the environment, which can be challenging to develop in healthcare due to the complexity and variability of patient responses. The healthcare-supportive policy gradient methods are as follows:

- Reinforce: A simple policy gradient algorithm that uses Monte Carlo estimation to update the policy parameters in the direction that increases expected rewards.
- Proximal policy optimization (PPO): A more advanced policy gradient algorithm that ensures more stable updates by imposing a constraint on policy changes between iterations.
- Actor–critic: Combines policy-based and value-based methods by using two networks—an actor (policy) network and a critic (value) network—to estimate policy gradients and value functions simultaneously.

7.2.3 Model-Based Methods

Model-based methods require a reasonably accurate model of the environment, which may be available in cases where the underlying physiology or disease dynamics are well understood and can be effectively modeled. In scenarios like treatment planning and drug dosage optimization, a highly accurate model is essential to make decisions that directly impact a patient's health. Therefore, in healthcare scenarios where transparent and interpretable decisions are crucial, model-based methods can offer more insight into decision logic as the model's predictions are often easier to understand than deep neural networks. The efficient model-based algorithms are discussed below.

- Monte Carlo tree search (MCTS): An approach often used for planning in games and sequential decision-making tasks. It builds a search tree by simulating sequences of actions and outcomes.
- Dynamic programming: Techniques like policy iteration and value iteration solve RL problems by iteratively refining policies and value functions based on Bellman equations.

7.3 Applications of DRL in Healthcare

DRL offers a powerful toolset for solving intricate problems in healthcare and medicine. The diverse applications of DRL in biomedical research emphasize its potential to significantly impact patient care, drug development, and healthcare management. Table 7.1 presents more specific applications of reinforcement learning algorithms, including DQN, Q-learning, fitted Q iteration, actor–critic, dynamic programming, Bayesian RL, and Montecarlo, along with their use cases in healthcare. Each algorithm has unique strengths and applications within the healthcare domain, contributing to more effective and personalized healthcare solutions.

Why DRL in Healthcare?

DRL excels in learning the optimal sequences of actions, making it ideal for healthcare tasks that involve a series of decisions over time, such as treatment planning and patient monitoring. DRL models can be updated and improved over time, i.e., continuous learning, aligning with the evolving healthcare landscape. DRL models are capable of handling high-dimensional action spaces, uncertainty, and making decisions in the presence of imperfect or noisy data, which is common in healthcare. The evident applications of DRL are as follows:

- Tailored treatment recommendations
- Optimization of clinical trials
- Disease diagnosis support
- Accelerated drug discovery and design
- Enhanced robotic surgery and assistance
- Health management system

7.3.1 Tailored Treatment Recommendations

DRL algorithms have the potential to create personalized treatment regime for patients by analyzing their medical history, hereditary profile, and other relevant data. These algorithms learn how to make treatment decisions that maximize long-term patient outcomes

Table 7.1 Use cases in healthcare.

Deep reinforcement learning algorithm	Use cases in healthcare
DQN [8]	• Drug discovery • Personalized treatment • Medical imaging
Q-learning [9]	• Clinical decision support • Resource allocation • Chronic disease management
Fitted Q iteration [10]	• Treatment optimization • Disease modeling • Rehabilitation
Actor–critic [11]	• Real-time patient monitoring • Surgical robots • Personalized care plans
Dynamic programming [12]	• Treatment sequencing • Resource optimization • Clinical pathways
Bayesian reinforcement learning [13]	• Clinical trials • Patient risk assessment • Adaptive Interventions
Monte Carlo methods [14]	• Patient outcome prediction • Treatment simulation • Resource simulation
SARSA [15]	• Adaptive drug dosage • Rehabilitation therapy • Treatment sequencing

while considering the possible risks and benefits. DRL is being involved [1] to estimate the severity of Parkinson's disease in individual patients. The model utilized patient data to simulate disease trajectories and tailor treatment plans, highlighting the potential for personalized disease management.

7.3.2 Optimization of Clinical Trials

DRL can be utilized to optimize the design and implementation of clinical trials. RL algorithms can be used to identify potential clinical trial candidates from electronic health records (EHRs) and patient databases. These algorithms can learn from historical patient data and prioritize potential participants based on specific trial criteria, ultimately speeding up the recruitment process. In terms of trial parameters such as patient recruitment strategies and dosing schedules, DRL algorithms can assist researchers in identifying optimal trial designs that yield faster and more accurate results. A study [2] employed DRL to personalize insulin dosing for diabetes patients. By continuously adapting insulin delivery based on blood glucose measurements, the model achieved better glycemic control compared to traditional approaches.

7.3.3 Disease Diagnosis Support

DRL models can aid in medical image analysis for disease diagnosis. By learning from extensive datasets of medical images, DRL algorithms can help identify patterns and irregularities that may be overlooked by human experts, thus improving the accuracy of early disease detection [3]. DRL-based CNN achieved state-of-the-art performance in breast cancer detection using mammography images, outperforming radiologists. In radiology, DRL models have been developed to assist in diagnosing various diseases, including lung diseases (e.g., tuberculosis), cardiovascular conditions, and neurological disorders. These models use medical imaging data, such as X-rays, CT scans, and MRIs, to improve disease detection and localization.

7.3.4 Accelerated Drug Discovery and Design

DRL techniques can accelerate the process of drug discovery by generating innovative molecular structures with desired properties such as efficacy and safety. These algorithms explore vast chemical spaces and optimize molecular structures tailored for specific drug targets. DRL has been applied to drug discovery and target identification. Reinforcement learning algorithms can optimize chemical processes, such as reaction conditions and synthesis pathways, to produce drug compounds more efficiently. In the research published in nature communications article [4], DRL models were used to generate autonomous molecules to develop inhibitors, demonstrating the capability of DRL in accelerating drug development for rare diseases. Recent advancements in DRL have expedited drug discovery pipelines, potentially leading to faster and more cost-effective drug development processes.

7.3.5 Enhanced Robotic Surgery and Assistance

DRL holds promise in enhancing surgical procedures by enabling robotic systems to perform intricate tasks with greater precision. These algorithms learn from surgical data as well as simulations to optimize surgical trajectories and techniques. Studies have shown that DRL-based robotic surgery systems can reduce surgical complications and improve patient outcomes. By optimizing surgical actions and minimizing errors, DRL contributes to safer surgeries. A study published in *Science Robotics* [5] employed DRL to develop an automated system to assess the skill levels of robotic surgeons. The DRL model, trained on a large dataset of surgical videos, demonstrated the ability to evaluate surgical skills accurately and objectively, which can aid in training and credentialing. Studies have shown that DRL-based robotic surgery systems can reduce surgical complications and improve patient outcomes. By optimizing surgical actions and minimizing errors, DRL contributes to safer surgeries. DRL has been used to improve teleoperation systems for remote surgery.

These systems allow expert surgeons to perform procedures from a remote location. DRL algorithms help mitigate issues like communication delays, enabling more responsive and accurate control of robotic surgical instruments. In the article of *JAMA Surgery* [6], DRL was used to develop a gesture-based control system for robotic surgical instruments. Surgeons could control the robot's movements using hand gestures, offering a more intuitive and precise interface for performing minimally invasive surgery.

7.3.6 Health Management System

DRL algorithms have the potential to assist in the field of health management system by analyzing patient information and offering personalized interventions. These interventions can include suggesting drug scheduling strategies, optimal selection of clinical tests, schedule appointments, supporting physical- or cognitive-disabled patients with a tutoring system in challenging environments. Researchers have explored the use of DRL to recommend personalized mental health interventions. A study in *Journal of Medical Internet Research* [7] demonstrated how DRL models can adapt interventions for individuals with depression based on their progress and responses during therapy. In order to construct sequential decision support systems and to determine the best medicine scheduling approach for patients who are HIV-positive, a hybrid method combining model predictive control and RL is proposed. In a system developed for health monitoring, Q-learning is applied for scheduling the clinical dosage of erythropoietin to maintain patients' hemoglobin count to range from 11.5 to 12.5 g/dl, thereby assuring optimal treatment cost and pharmacotherapy.

7.4 Challenges

Certainly, deep reinforcement learning (DRL) in healthcare and biomedical applications faces several challenges. The exploration–exploitation dilemma is a key idea in reinforcement learning. It refers to the trade-off between exploring new states to learn more information and leveraging experienced states to maximize rewards. This dilemma is formally articulated in the multi-armed bandit problem and extends to more complex RL scenarios. Many RL algorithms, including deep reinforcement learning methods, grapple with this dilemma when interacting with an environment. Agents must decide how much exploration is needed to discover optimal policies while avoiding excessive exploration that may hinder performance. Empirical studies have shown that the choice of epsilon-greedy policies significantly impacts the learning process. Too much exploration may lead to slow convergence, while too little exploration may result in suboptimal policies. Q-learning, one of the foundational RL algorithms, addresses exploration–exploitation through epsilon-greedy policies.

These challenges underscore the complexity of applying DRL in healthcare and biomedical contexts and the need for interdisciplinary collaboration between machine learning researchers, healthcare professionals, ethicists, and policymakers to address them effectively. Based on the studies conducted, Table 7.2 highlights the several challenges in employing DRL in biomedical applications.

Table 7.2 Challenges of DRL in healthcare domain.

Challenges	Description
Clinical validation [16]	DRL models need extensive validation and testing in clinical settings to ensure safety and efficacy.
Data integration [17]	Collaborating data from disparate sources, such as EHR and medical images, requires overcoming data heterogeneity.
Interpretable models [18]	Interpreting DRL model decisions is crucial in healthcare for gaining trust and clinical adoption.
Data scarcity and quality [19]	Healthcare data can be scarce, noisy, and imbalanced, which poses challenges for training robust DRL models.
Patient privacy and ethical concerns [20]	Protecting patient privacy while using sensitive medical data is a critical concern.
Regulatory compliance [21]	Ensuring that DRL systems comply with healthcare regulations like HIPAA is essential.
Robustness to covariate shift [22]	Healthcare data often exhibit a covariate shift, which can affect model performance.
Clinical acceptance [23]	Convincing healthcare professionals to trust and adopt DRL-based decision support tools can be a hurdle.
Long-term consequences [24]	DRL models may make decisions that have long-term consequences, requiring consideration of cumulative effects.

7.5 Healthcare Data Types

Healthcare data comes in various types and formats, each serving a specific purpose in monitoring patients, medical research, and healthcare operations. Here are some common types of healthcare data that DRL can learn and interpret according to the use cases: electronic healthcare records (EHRs), biomedical imaging information, laboratory data, and sensor data.

7.5.1 Electronic Healthcare Records (EHRs)

EHRs are extensive computerized records that contain a patient's clinical history, analysis, medications, sensitivities, treatment plans, research facility results, and other clinical data [25]. This category of data includes details about the drugs supplied to patients, such as their names, dosages, frequency, and dates of dispensing. It is necessary for tracking and managing medications. They are used by healthcare providers for patient care and management. DRL leverages EHR data to enhance healthcare applications, improving treatment personalization, clinical decision-making, patient monitoring, and resource optimization. Patient outcomes and healthcare system effectiveness [26] could be dramatically impacted by integrating DRL with EHR data.

7.5.2 Laboratory Data

DRL can use laboratory data to predict disease risk, optimize medication dosages, and recommend personalized treatment plans [28]. DRL algorithms can take as input laboratory data from blood tests, urine, and other biomedical measurements to diagnose diseases. DRL models, for instance, can be trained to interpret test data to detect the presence of particular diseases like diabetes, anemia, or liver ailments. DRL models can optimize treatment plans, doses, and medication selections based on a patient's test results, medical history, and current state in order to maximize well-being while minimizing negative effects.

7.5.3 Sensor Data

Deep learning (DL) and machine learning (ML) methods are significantly reliant on biomedical devices and sensors in both healthcare and biomedical research. These gadgets and sensors gather useful information from subjects and testing environments, which can then be processed and analyzed by DL algorithms to yield insightful conclusions. Data from devices like fitness trackers, heart rate monitors, and glucose meters can be gathered to remotely monitor patients' health thanks to the growth of wearable technology and the IoT (Internet of Things) in healthcare. A few typical biomedical gadgets and sensors utilized for DRL applications are presented in Table 7.3.

7.5.4 Biomedical Imaging Information

It is important to note that while DRL holds great promise in clinical imaging, there are challenges related to data privacy, model interpretability, and training of rich labeled dataset. DRL models, especially convolutional neural networks (CNNs), are employed to process various types of medical images, such as X-rays, MRIs, CT scans, and histopathological slides. These models [27] can automatically detect abnormalities and segment structures and identify patterns that may not be apparent to human observers—for example: DRL can identify tumors in radiological images, helping radiologists and oncologists in early cancer diagnosis.

DRL models can extract high-dimensional features from medical images, which are then used as input to downstream reinforcement learning algorithms. These extracted features provide a more compact and informative representation of the image data for decision-making. In radiation therapy and surgery, DRL can optimize treatment plans based on medical imaging data. It can adapt treatment parameters in real time, accounting for changes in the patient's anatomy and optimizing dose delivery to the target while minimizing damage to the surrounding healthy tissue.

DRL models can assist in diagnosing diseases such as Alzheimer's disease, diabetic retinopathy, and cardiovascular conditions. They can identify suitable patterns or anomalies indicative of these diseases. It assists in performing image registration, aligning multiple

Table 7.3 Healthcare sensor data.

Type of sensors	Purpose	DRL use case
Electrocardiography (ECG) sensors	Monitoring heart activity, arrhythmia detection, and cardiac health assessment	Analyzing ECG data to detect abnormalities, predict heart disease risk, and provide real-time cardiac monitoring
Electroencephalography (EEG) sensors	Recording electrical activity in the brain for epilepsy diagnosis, brain-computer interfaces, and neurological research	Processing the EEG data for brainwave pattern recognition, seizure prediction, and decoding brain signals for controlling devices
Electromyography (EMG) sensors	Capturing electrical activity in muscles for prosthetic control, muscle disorder diagnosis, and physical rehabilitation	EMG signals can be decoded to control prosthetic limbs and assistive devices with greater precision and natural movement
Medical imaging sensors	Medical image capturing, including X-rays, CT scans, MRIs, and ultrasounds	DRL models are used for image analysis, enabling automated disease detection, organ segmentation, and medical image enhancement
Biometric sensors	Capturing biometric data for identification, authentication, and security	Biometric data from fingerprint sensors, facial recognition cameras, and voice recognition sensors can be used for authentication and access control
Glucose monitors	Monitoring glucose levels in diabetic patients for glycemic control	Analysing CGM data to predict blood glucose trends, detect hypo- and hyperglycemic events, and personalize insulin dosing recommendations
Optical sensors (e.g., pulse oximeters)	Measuring blood oxygen levels and heart rate	Pulse oximeter data supports the early detection of respiratory and cardiovascular problems, such as sleep apnea and arrhythmias
Wearable health and fitness sensors	Collecting data on physical activity, heart rate, sleep patterns, and other health-related metrics	Processing of wearable sensor data for personalized fitness recommendations, sleep quality assessment, and early detection of health issues

medical images captured at different time intervals or with different imaging techniques. Furthermore, DRL can guide surgeons during minimally invasive procedures by providing real-time feedback based on intraoperative imaging, ensuring the accurate placement of instruments and implants.

7.6 Guidelines for the Application of DRL

Choosing an efficient deep reinforcement learning (DRL) algorithm for healthcare use cases involves a careful consideration of various factors. Here are some guidelines that will assist to make the right selection in specific scenarios where certain algorithms may be more suitable:

- If the problem involves sequential decision-making with a clear state, action, reward, and termination conditions, consider model-free DRL methods like Q-learning or policy gradient optimization (PPO).
- If you have limited data, consider using model-based DRL algorithms like model predictive control (MPC), which can be more data-efficient.
- If interpretability is crucial, consider using rule-based or tabular DRL approaches like SARSA, which offer more transparent decision-making.
- For resource-constrained environments, prioritize DRL algorithms with low computational demands, such as Q-learning.
- For critical care applications where safety is paramount, consider using safe exploration techniques like categorical DQN (a variant of DQN).
- For dynamic healthcare settings, consider DRL algorithms that support continuous learning, such as actor–critic and dynamic programming.
- For maintaining detailed documentation for transparency and reproducibility, documented DRL algorithms, like DQN or A3C, with well-established codebases can simplify reproducibility efforts.
- For conducting pilot testing in controlled environments before clinical deployment, algorithms like DDPG or SAC that are known for stable and reliable training may be preferred for real-world validation.

7.7 A Case Study: DRL in Healthcare and Biomedical Applications

7.7.1 Optimizing Radiation Therapy Dose Distribution in Cancer Treatment

Radiation therapy is a common treatment modality for cancer patients. However, the challenge lies in delivering a therapeutic dose to the tumor while sparing adjacent healthy tissues. Traditional treatment planning often relies on manual optimization, which can be time-consuming and suboptimal. In the study performed [29], action-based rewards are employed to optimize the radiation therapy dose regime. It has the potential to revolutionize the precision and safety of radiation treatment for cancer patients, ultimately leading to improved clinical outcomes.

Methodology:

- Patient data: Real patient data, including CT scans and organ delineation, are used for training and validation.
- State representation: Patient-specific anatomical data, mean tumor diameters (MTD), and position are used to define the state space.
- Action space: The action space comprises adjustments to the radiation beam angles, intensities, and durations.
- Reward function: A reward function is designed to balance tumor control and healthy tissue sparing. It rewards tumor dose escalation and penalizes dose to critical structures.

The trained RL model resulted with effective dosage formulation to reduce the mean tumor diameters (MTD) in patients treated with simulated temozolomide, procarbazine, and vincristine and chemo- and radiotherapy clinical trials.

7.7.2 Dose Strategy Model in Sepsis Patient Treatment

Sepsis is a life-threatening condition where timely and personalized treatment is critical. Optimizing the distribution of medication doses in sepsis management can significantly impact patient outcomes. The study has been conducted [30] to explore the application of deep deterministic policy gradient (DDPG) to optimize dose distribution in sepsis patient treatment. The study highlights how DDPG can adapt and improve treatment policies, reducing mortality rates and enhancing patient care.

Methodology:

- Patient data: Real patient data, including electronic health records and physiological monitoring data, are used for model training and validation.
- State representation: Patient-specific data, including vital signs, laboratory results, and disease progression, are used to define the state space. Time-series data capture the dynamic nature of sepsis.
- Action space: The action space comprises medication dosage adjustments, fluid administration rates, and vasopressor titration.
- Reward function: A reward function is designed to balance treatment goals, such as stabilizing blood pressure, improving oxygenation, and minimizing adverse effects. Rewards are assigned based on treatment effectiveness.
- DRL algorithm: DDPG is utilized to learn the optimal dosing policies for individual sepsis patients. It leverages continuous action spaces and actor–critic architectures.

This study highlights the potential of deep deterministic policy gradient (DDPG) in optimizing dose distribution for sepsis patient treatment. By adapting treatment strategies based on real-time patient responses, DRL offers a promising avenue to improve sepsis management and enhance patient care.

References

1. Ahmed, S., Komeili, M., Park, J., Predictive modelling of Parkinson's disease progression based on RNA-Sequence with densely connected deep recurrent neural networks. *Sci. Rep.*, 12, 21469, 2022.
2. Zhu, T., Li, K., Kuang, L., Herrero, P., Georgiou, P., An Insulin Bolus Advisor for Type 1 Diabetes Using Deep Reinforcement Learning. *Sensors (Basel)*, 20, 18, 5058, 2020.
3. Shen, L., Margolies, L.R., Rothstein, J.H. *et al.*, Deep Learning to Improve Breast Cancer Detection on Screening Mammography. *Sci. Rep.*, 9, 12495, 2019.
4. Jeon, W. and Kim, D., Autonomous molecule generation using reinforcement learning and docking to develop potential novel inhibitors. *Sci. Rep.*, 10, 22104, 2020.
5. Yang, G.-Z., Cambias, J. *et al.*, Medical robotics- Regulatory, ethical, and legal considerations for increasing levels of autonomy. *Sci. Robot.*, 2, 4, 2017.
6. Luongo, F., Hakim, R., Nguyen, J.H., Anandkumar A, Hung AJ. Deep learning-based computer vision to recognize and classify suturing gestures in robot-assisted surgery. *Surgery*, 169, 5, 1240–1244, 2021 May.
7. Piera-Jimenez, J., Etzelmueller, A., Kolovos, S., Folkvord, F., Lupianez-Villanueva F., Guided Internet-Based Cognitive Behavioral Therapy for Depression: Implementation Cost-Effectiveness Study. *J. Med. Internet Res.*, 23, 5, 2021.
8. Silver, D., Hubert, T., Schrittwieser, J. *et al.*, A general reinforcement learning algorithm that masters chess, shogi, and Go through self-play. *Science*, 362, 6419, 1140–1144, 2018.
9. Ponsen, M., van Hoof, H., Neerincx, M., A survey of intelligent decision support systems for decision making in healthcare. *J. Decis. Syst.*, 28, 4, 295–320, 2019.
10. Antos, A., Busoniu, L., Ernst, D., De Schutter, B., Approximate dynamic programming for solving the repetitive manufacturing problem. *IEEE Trans. Autom. Sci. Eng.*, 4, 3, 364–371, 2007.
11. Daskalaki, E., Diem, P., Mougiakakou, S.G., An actor-critic based controller for glucose regulation in type 1 diabetes, 2012. *Comput. Methods Programs Biomed.*, 109, 2, 116–25, 2013. https://doi.org/10.1016/j.cmpb.2012.03.002.
12. Gade, P.R. and Nemat-Nasser, S., Optimal treatment strategies for septic shock incorporating heterogeneous patient populations. *Ann. Oper. Res.*, 137, 1, 269–283, 2005.
13. Chen, S., He, D., Liu, J., A Bayesian learning-based approach to predictive modeling of patient trajectories. *IEEE Trans. Biomed. Eng.*, 66, 7, 1978–1987, 2019.
14. Zhao, P., Li, S., Jiang, J. *et al.*, Monte Carlo simulations in the evaluation of the optimal operational scheme for an urban sewage system. *Environ. Sci. Pollut. Res.*, 23, 18, 18254–18263, 2016.
15. Sutton, R.S. and Barto, A.G., *Reinforcement Learning: An Introduction*, 2nd ed., The MIT Press, Cambridge, Massachusetts, London, England, 2018.
16. Oakden-Rayner, L., Dunnmon, J., Carneiro, G., Ré, C., Hidden Stratification Causes Clinically Meaningful Failures in Machine Learning for Medical Imaging. *NeurIPS*, 2020.
17. Sun, J., Reddy, C.K., Liu, J., A survey of big data architectures and machine learning algorithms in healthcare. *J. King Saud Univ. Comput. Inf. Sci.*, 2017.
18. Lipton, Z.C., Elkan, C., Naryanaswamy, B., Optimal thresholding of classifiers to maximize F1 measure, in: *Machine Learning for Healthcare*, pp. 49–58, 2016.
19. Miotto, R., Wang, F., Wang, S., Jiang, X., Dudley, J.T., Deep learning for healthcare: review, opportunities and challenges. *Briefings Bioinf.*, 19, 6, 1236–1246, 2017.
20. Choudhury, A., Hong, L., Li, L., AI and Privacy. *Annu. Rev. Stat. Appl.*, 6, 19–36, 2019.
21. Obermeyer, Z., Powers, B., Vogeli, C., Mullainathan, S., Dissecting racial bias in an algorithm used to manage the health of populations. *Science*, 366, 6464, 447–453, 2019.

22. Shickel, B., Tighe, P.J., Bihorac, A., Rashidi, P., Deep EHR: A Survey of Recent Advances in Deep Learning Techniques for Electronic Health Record (EHR) Analysis. *J. Biomed. Health Inf.*, 22, 5, 1589–1604, 2018.
23. Rajkomar, A., Dean, J., Kohane, I., Machine Learning in Medicine. *New Engl. J. Med.*, 380, 14, 1347–1358, 2019.
24. Chen, I.Y., Szolovits, P., Ghassemi, M., Can AI help reduce disparities in general medical and mental health care? *AMA J. Ethics*, 19, 9, 881–888, 2017.
25. Rajkomar, A., Oren, E., Chen, K., Dai, A.M., Hajaj, N., Hardt, M., Lungren, M.P., Scalable and accurate deep learning with electronic health records. *NPJ Digit. Med.*, 1, 1, 1–10, 2018.
26. Raghu, A., Maalouf, R., Vosoughi, S., Reinforcement learning for clinical decision support. *JAMA Netw. Open*, 3, 7, e2011642, 2020.
27. Esteva, A., Kuprel, B., Novoa, R.A., Ko, J., Swetter, S.M., Blau, H.M., Thrun, S., Dermatologist-level classification of skin cancer with deep neural networks. *Nature*, 542, 7639, 115–118, 2017.
28. Zhao, Y., Kosorok, M.R., Zeng, D., Reinforcement learning design for cancer clinical trials. *Stat. Med.*, 28, 26, 3294–3315, 2009, doi:10.1002/sim.3720.
29. Yauney, G. and Shah, P., Reinforcement Learning with Action-Derived Rewards for Chemotherapy and Clinical Trial Dosing Regimen Selection. *Proceedings of the 3rd Machine Learning for Healthcare Conference, in Proceedings of Machine Learning Research*, vol. 85, pp. 161–226, 2018, Available from https://proceedings.mlr.press/v85/yauney18a.html.
30. Lin, T., Zhang, X., Gong, J. *et al.*, A dosing strategy model of deep deterministic policy gradient algorithm for sepsis patients. *BMC Med. Inf. Decis. Mak.*, 23, 81, 2023, https://doi.org/10.1186/s12911-023-02175-7.

8

Cultivating Expertise in Deep and Reinforcement Learning Principles

Chilakalapudi Malathi and J. Sheela*

School of Computer Science and Engineering, VIT-AP University, Vijayawada, Andhra Pradesh, India

Abstract
This abstract provides an insightful exploration of these foundational concepts, beginning with reinforcement learning, which addresses dynamic decision-making and emphasizes agent–environment interaction to optimize cumulative rewards. It uncovers the core principles, underlying mechanisms, key algorithms, and practical applications in various industries while shedding light on the societal implications and ethical considerations linked to this technology. Shifting focus to deep learning, this abstract showcases the use of neural networks with multiple layers to extract intricate patterns from complex data, resulting in transformative advancements in image recognition, natural language processing, and autonomous systems. By comprehending these principles, researchers and practitioners can harness the power of reinforcement and deep learning, thereby contributing to the continual evolution of AI and its real-world problem-solving capabilities.

***Keywords*:** Reinforcement learning, Q-learning, value iteration algorithm, policy evaluation, policy improvement

8.1 Introduction

A vast number of industries, including robots, gaming, banking, healthcare, and several others, have been profoundly altered by the two prominent subdomains of artificial intelligence known as reinforcement learning and deep learning. To fully understand both deep learning and reinforcement learning, readers should leave this chapter with a solid understanding of the basic ideas and ideas that underlie each of them. It also examines how these ideas differ from one another and looks at how they can work together in brilliant systems. An agent is taught how to take action in particular lush surroundings to provide a tranquil and energizing atmosphere to intensify cumulative recompense or accomplish a certain objective using a type of machine learning known as reinforcement learning. The agent engages in environmental interaction by acting, obtaining input (rewards or penalties), and revising its approach in response to the observed results.

**Corresponding author*: sheela.j@vitap.ac.in

Shubham Mahajan, Pethuru Raj and Amit Kant Pandit (eds.) Deep Reinforcement Learning and Its Industrial Use Cases: AI for Real-World Applications, (151–178) © 2024 Scrivener Publishing LLC

The main contribution of this paper is to offer a comprehensive exploration of the core principles, mechanisms, key algorithms, and practical applications of reinforcement learning and deep learning, while also addressing the associated societal and ethical considerations. This contribution aims to provide valuable insights for researchers and practitioners seeking to harness the potential of these technologies in the ongoing evolution of AI and real-world problem-solving.

8.1.1 Reinforcement Learning's Constituent Parts

- Agent: The person who interacts with their surroundings to learn or make decisions.
- Environment: The agent's operating environment outside of it, It provides feedback by giving praise or criticism.
- State (s): A picture of how the environment is right now.
- Action (a): The agent's options for influencing the surroundings.
- Policy (π): An approach or map that an individual will use when making judgments.
- Reward (r): Metric through which the effectiveness of an action made by the agent is measured.
- Value function (V(s)): Determines the potential long-term gain for an agent from a specific state.
- Q-function (Q(s, a)): Determines the anticipated overall payoff for carrying out a certain activity under a specific circumstance.

8.1.2 Process of Markov Decisions (MDP)

In cases when results are unclear, decision-making is modeled mathematically using a Markov decision process (MDP). It is made up of conditions, operations, probabilities of transitions, and benefits, and it aids in the analysis and optimization of sequential operations in ambiguous situations. It comprises of the incentive activity, a price reduction variable,

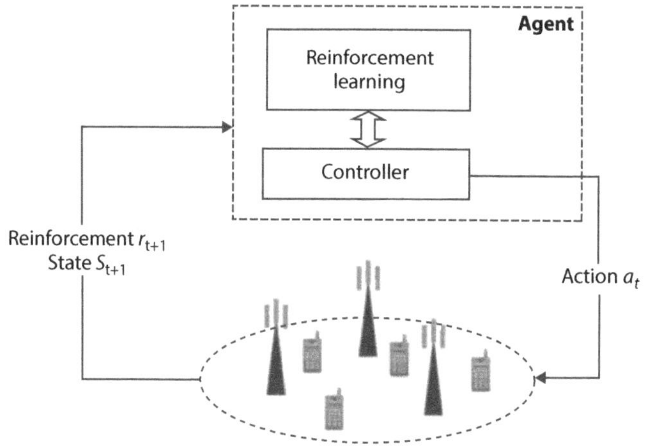

Figure 8.1 Markov decision process.

a change in probability operation, several states, and several procedures. The Figure 8.1 shows the Markov decision process.

The primary elements of an MDP include the following:

- Situations (S): A constrained collection of each potential situation or configuration that an autonomous mechanism could potentially come across.
- Actions (A): A constrained set of decisions and reactions that an individual can undertake in each circumstance.
- Transition probability function (P): It determines the possibility that a state will change when a particular action is carried out. In other words, it describes the dynamics of the environment.
- Reward function (R): It converts each state–activity pair into a numerical reward, showing the agent's immediate gain or loss for doing a specific action in a specific condition.
- Discount factor (γ): A number that weighs the significance of present rewards against those in the future, between 0 and 1. The agent can decide whether to put long-term planning or short-term gains first.

The goal is to determine the appropriate MDP as a course strategy that increases the overall advantages over time [1]. Q-learning, rules revision, and significance loop are just a few of the reinforcement learning techniques used to solve MDPs and decide what the best course of action is for the agent to pursue. MDPs are frequently utilized in many different fields, including robotics, autonomous systems, gaming, finance, and healthcare, to represent decision-making issues involving uncertainty and sequential interactions.

1) Value function: The significance variable (V) figures out the anticipated cumulative advantage a participant can get from a particular condition while putting a particular strategy into practice. Officially, it is referred to as:

$$V(s) = E\left[t * R\frac{(st)}{st} = s\right]$$

2)Bellman optimality equation: The worth of the variable for every condition is altered iteratively in the value-based iteration method. It was developed using the Bellman ideal outcome Formula as its primary discovery. According to the following formula:

$$R(m,b,m') + {}^*V(m') = \max[b]\, T(m,b,m') * [R(m,b,m') + {}^*V(m')]$$

8.1.3 Learning Reinforcement Methods

RL algorithms come in many forms, such as the following:

1) Valuation iteration algorithm: To solve Markov decision processes (MDPs), value iteration, a well-known strengthening learning (RL) technique, is used. It is a technique for

determining, *via* the use of dynamic programming, the appropriate course of action for an agent interacting with its environment to maximize its total rewards. The "Bellman iteration" algorithm is another name for the value iteration method. The "Bellman iteration" algorithm is described below.

The Bellman iteration technique commences after a rough estimation within a given variable (V), updating it periodically until convergence. The revisions make use of the Bellman optimality equation. The algorithm's steps are outlined below:

Step 1: Create random starting values for the state-value function V(m) in each state m.
Step 2: Continually go forward till convergence.

a. The parameters of every state 's' after applying the Bellman optimality equation must be modified as follows:
   V(s) equals the maximum of $[T(m,b,m')*(R(m,b,m')+*V(m'))$ in all cases b.

Step 3: After the permitted amount of repetitions is completed or the changes in the value function are acceptably small, stop iterating.
Step 4: When the value function converges, choose an action that maximizes the expression inside the bellman operator to get the best course of action π *(m).

Given that conditions and dynamics (T and R) are famous, value iteration will always identify the best value function and accompanying best policy for the MDP [2]. However, computational costs might increase significantly in expansive situations and reaction areas. Alternatively, in these circumstances, strategies for reinforcement learning like Q-learning or deep Q-networks (DQNs) are routinely applied.

a. Algorithm for policy iteration: Another well-known method for dealing with Markov decision processes (MDPs) in reinforcement learning (RL) is policy iteration. Additionally, it acts as a dynamic programming technique that helps determine the best course of action for an agent involved in environment-based interactions to maximize cumulative rewards. In contrast to value iteration, policy iteration updates the policy immediately and then refines it until convergence. An outline of the policy iteration algorithm is provided below.
b. Policy evaluation: A given policy's value function (V) is assessed as the first stage of the policy iteration process. The predicted cumulative payoff for each state by adhering to the existing policy is determined throughout the policy evaluation process. The subsequent statement presents a definition of the value function corresponding to a policy:

$$V\pi(s) = E\left[t * R\frac{(st)}{st} = s\right]$$

This stage may be completed by employing iterative techniques, such as iterative policy assessment, or by resolving the linear system of equations governing the value function.

c. Policy improvement: The strategy enhancement phase aims to enhance the strategy after receiving the appropriate operation for the current policy by choosing enhanced actions inside each state, guided by the updated value function. The refined policy is produced by selecting the strategy that increases the anticipated aggregate advantage in each state.

The expression $'(s) = argmax[T(m,b,m')*(R(m,b,m') + *V(m'))]$ in all cases b.

2) Policy iteration algorithm: The strategy assessment and strategy development steps make up the policy iteration approach, which is iterated until convergence [3]. The following basic steps can be used to break down the algorithm:

Step 1: Make a random policy initialized.
Step 2: Continue until convergence occurs.

a. Policy analysis: For the current policy, the value function V(m) must be calculated.
b. Policy development: By selecting the strategy that increases the anticipated aggregate benefit for each state, the policy should be updated.

Step 3: Keep going through the iterations until the policy converges, stabilizes, and no more modifications are made to the policy.
Step 4: The agent can use this policy to maximize its cumulative rewards in the specified MDP until the policy has reached its convergent point, at which point it is the best policy (*).

The identification of the ideal policy and the accompanying ideal value function for the MDP is guaranteed by policy iteration when the dynamics of the environment (T and R) are understood. It may, however, involve computational costs in situations where the huge condition and movement areas, are similar to computational requirements for Value Iteration.

3) Q-learning: It is not necessary for an agent to explicitly understand the dynamics of the circumstances in tolerating and facilitating the acquisition of best practice strategies within that environment. The model-free strengthening learning methodology known as Q-learning was used to attain this success. The core idea behind Q-learning is the construction of a conduct-value operation, abbreviated as Q(m, b), that demonstrates the anticipated cumulative reward that can be attained while the participant performs movement "b" in state "m" and then pursues the most beneficial course of action. These are some general descriptions relating to the qualitative learning technique.

a. Q-value Update: Q-learning involves changing the quantity value connected with a state–action mixture while taking into account both the rewards earned and the Q-values that are the result of successive state–action combinations. This procedure of updating the Q-values makes use of the Bellman's existence formula:

$$Q(m,b) = Q(m,b) + *\max[R(m,b,m') + *[Q(m',b')] - Q(m,b)]$$

The elements of this equation are the present state m, the action selected inside that state, and the outcome state m′. R(m, b, m′) embodies the instantaneous payoff, whereas the learning rate, a modest positive value, regulates the influence of incoming information [4]. Additionally, the discount factor is given a bigger positive value.

b. Exploration vs. Exploitation: Making certain the agent thoroughly examines surroundings in pursuit of the optimal course of action, Q-learning strikes a balance between exploration and exploitation. The detective utilizes an investigation technique like the "-greedy policy", where actions are chosen at random with a chance of (ε), and the probability of choosing the option having the greatest Q-value is (1-ε).
c. Q-learning algorithm: By operating as an off-policy technique, the Q-learning algorithm can learn the best Q-values without being restricted by the policy being applied. The following list summarizes the key steps in the Q-learning algorithm:

Step 1: All state–action pairings in the Q-value function Q(s, a) should be given arbitrary initial values.
Step 2: Continue doing the following until convergence or a predetermined number of episodes:

a. Take note of the existing condition(s).
b. Use an exploratory technique, such as -greedy, to select an action (a) based on the Q-values.
c. Carry out step (a) in the surroundings.
d. Pay attention to the outcome (R) and the subsequent state (s′).
e. The Q-value update equation should be changed to reflect the current state–action pair.
f. Replace the present state (s) with the following state (s′).

Step 3: Use the Q-values to create the best possible policy π(s), which, once the Q-value function converges or after the agent has had enough learning experiences, selects the course to pursue in every location that has the best Q-value.

Dealing with the issue and reaction zones that are significant or ongoing, Q-learning shows efficacy as an RL challenge-solving algorithm. It has demonstrated success in activities requiring discrete and continuous control and exhibits adaptability in a variety of settings.

4) SARSA: A model-free reinforcement learning technique called SARSA is utilized to ascertain the best course of action for an agent involved in environmental interactions. The process by which Q-values are updated based on real-state transitions is described by the acronym "SARSA" which is the abbreviation for "state–action–reward–state–action". As a policy-compliant algorithm, SARSA picks up Q-values in line with the present policy it is putting into practice [5]. An overview of the SARSA formulation is provided below.

a. Q-value update: Incorporating observed rewards together with the ensuing state–action combination's Q-value, SARSA's major objective is to update its Q-values associated with pairings involving states and operations using the current action-selection policy. The following expression demonstrates how SARSA updates Q-values:

$$Q(m,b) = Q(m,b) + \alpha * [R(m,b,m') + \gamma * Q(m',b') - Q(m,b)]$$

Below are the components that make up the following formula: m, present state; b, action taken in current state; m, the state attained after taking action; and b′, the subsequent action selected based on the existing policy. The instant incentive received is represented by R(m, b, m′), the learning rate, and the discount factor.

b. Inquiry vs. abuse: SARSA incorporates the inquiry vs. abuse compromise in a manner comparable to Q-learning. The agency employs a strategy known as the "-greedy policy," where operations are assigned arbitrarily with a fixed frequency (ε), and the greatest Q-values event is then selected with an expected outcome of (1- ε).

c. SARSA algorithm: The SARSA algorithm's key steps are as follows:

Step 1: For each state–action combination, randomly initialize the Q-value functions Q(m, b).
Step 2: Keep performing these actions until convergence or a predetermined number of episodes:

a. Take note of the existing condition (m).
b. Considering the present Q-values, select a course to take (b) using an exploration strategy (such as ε-greedy).
c. Carry out action (b) in the surroundings.
d. c. Pay close attention to the outcome (m′) and the reward (R).
e. Using the current Q-values, decide the next course of action (b′) using the current strategy, such as the "ε-greedy" method.
f. The SARSA Q-value modification formula allows for the modification of the Q-value of that particular state–action pairing.
g. Update the current when shifting beyond a previous situation to the one that follows.

Step 3: When the Q-value function converges or when the participant reaches the appropriate amount of experience through a predefined number of segments, use the Q-values to build ideal regulations, indicated as $\pi^*(s)$, which chooses the procedure having the highest Q-value in every situation.

When the learning agent's policy needs to be changed frequently, SARSA works as an on-policy technique that is very useful [6]. However, under some circumstances, it may converge more slowly than Q-learning and be sensitive to the exploration rate (ε).

5) Deep Q Networks (DQN): Deep Q Networks (DQN) present a revolutionary reinforcement programming technique by fusing large brain networks with the Q-learning strategy. DQNs were introduced in a seminal study published in 2015 by DeepMind researchers. When dealing with expansive and continuous state spaces, traditional Q-learning has substantial problems, which DQNs seek to solve. DQNs can navigate challenging situations and perform very well across a range of tasks by using the function's Q-value can be approximated using neural systems. The DQN algorithm is described in this overview.

a) Q-value function approximation: In conventional Q-learning, a Q-table is utilized to store the Q-values for any prospective state–action combinations. However, for devices having vast or continuous position locations, this strategy is unworkable due to the constant increase in the variety of position–action permutations. To address this challenge, Deep Q Networks (DQNs) implement a rough estimation that Q-value operates referred to to as a Q-network. The brain's neural network receives its state as input, and as a result, Q-values are generated for every potential response.
b) Experience replay: DQNs employ an approach known as "explore rewind" to increase the consistency and effectiveness of learning. The agent gathers experiences as tuples (state, action, reward, and next state) through interactions with surroundings. Such incidents are recorded by DQNs in a recollected memory rather than refreshing Q-network after every experience [7]. Through the arbitrary sampling of small instances in the playback buffering, the connectivity between succeeding events is broken during training. By encouraging more trustworthy results, this strategy improves the stability of learning.
c) Target network: Deep Q Networks (DQNs) use a specialized objective system to increase the general consistency of the method of learning. Although the weights utilized in this objective system are identical to those in the Q-network, they are fixed. Throughout the training phase, it is used to determine the desired Q-values. To match the weights of the Q-network, the intended network's weights are adjusted on a sporadic (though reduced frequency) basis. This strategy not only makes the learning process more stable but also lessens the problem of a "moving target" that Q-learning frequently faces.
d) Loss function and gradient descent: The loss parameter utilized in DQN is the average squared measurement (MSE) loss, which is calculated from the predicted Q-values generated by the Q-network and the intended Q-values received from the target network using the Bellman equation. Then, during

network training, descent slope optimization is performed to minimize this loss as much as possible.

e) Exploration vs. exploitation: During the initial phase of training, DQNs still require a searching methodology to effectively handle the relationship between both discovery and extraction. The widely used tactic is known as the "epsilon-greedy procedure," in which the agent selects an arbitrary response with an expected outcome of (ε) and alternatively selects the course of action with the greatest Q-value and a probability of (1-ε).

DQN is renowned for its astounding proficiency in playing Atari 2600 games and has frequently outperformed humans in these games. It has been enhanced and modified in several ways to handle more challenging jobs, such as continuous control, robots, and multi-agent scenarios [8]. DQN has served as the foundation for various cutting-edge since its inception, sophisticated reinforcement learning techniques have been regarded as an important concept in the field.

6) Progressive ideology actor–critic: A type of neural network for reinforcement learning known as actor–critic techniques combines the advantages of value-driven and policy-based tactics. The critic, who is in charge of estimating the value function (i.e., the anticipated cumulative reward) connected to the acquired policy, and the actor, who is tasked with learning the policy (i.e., the transforming ideas into movements), comprise each of the key components of various strategies. Comparing separate policy-based or value-based algorithms, actor–critic approaches provide a more reliable and effective means of learning in complicated situations. An overview of the actor–critic algorithm is given below.

a) Learning based on policies and values: Policies are directly learned using policy-based approaches, which aim to maximize the predicted cumulative benefit from a given condition. To update the policy parameters, they frequently employ strategies like policy gradients. In centered around value methodologies like Q-education and DQNs, the worth product V(m) or the action–value relationship Q(m, b) are approximations, and rules are subsequently formed from these approximations (for instance, using a greedy technique).
b) The actor: Learning the rules is the actor's responsibility as in actor–critic. The probability distribution of actions is produced using the situation of the world as the starting point. Implementation from the rule representation frequently takes the form of a neural network with programmable parameters. The actor wants to maximize the expected cumulative reward; therefore, it uses gradient ascent to modify its parameters in response to rewards it has already received.
c) Criticism: The value function, as used in the actor–critic framework, is a projection of the predicted cumulative reward connected to a specific state as defined by the present policy. As its input, the critic generates a value that is estimated for the current situation [9]. This is often accomplished by using a different neural network that has its programmable settings and acts as a representation of the value function. The critic's primary objective is to decrease

the average variance between the actual incentives earned during training and the projected values.

d) Advantage function: An essential part of actor-critic approaches, the advantage function provides a gauge of how much activity has improved or declined in comparison to the normative course of action. This function was discovered as a result of the discrepancy between the state-utility equation and the action-utility equation.

$$\text{Benefits}\,(m,b) = Q(m,b) - V(m)$$

The advantage function, which calculates the degree of profit associated with the major source of direction for the actor's update, is an act that occurs inside a defined phase.

e) Actor–critic algorithm: The basic steps that make up the actor–critic algorithm are described below:

Step 1: Assign random weights to the actor and critic neural networks.
Step 2: Engage with the surroundings, gather experiences, and determine rewards.
Step 3: Calculate the advantage function by using both the rewards and the critic's approximations.
Step 4: The critic network should be updated by reducing the average quadratic difference within the awards' true worth and expectations.
Step 5: The advantage function will be used to direct the use of gradient ascent on the projected total reward to change the actor network.
Step 6: Continue steps 2 through 5 until convergence or a predetermined number of episodes.

Particularly, as they skillfully weigh the benefits of prospecting and extracting resources, actor–critic approaches are helpful in scenarios involving continuous action spaces and high-dimensional state spaces.

7) Optimization of proximal policies (PPO): The latest technology method for reinforcing learning (RL) policy optimization is the optimization of peripheral processes (PPO). As a result of PPO's ease of use, stability, and efficiency in training deep neural network rules for challenging situations, researchers at Open AI first introduced it in 2017 [10]. Since then, it has grown significantly in popularity. By using a more straightforward and computationally effective method, PPO seeks to reduce the drawbacks of earlier policy gradient techniques like TRPO. An advanced summary of the proximal policy optimization formula is given in the description that follows.

a) Policy gradient methods: Based on the anticipated cumulative reward, these RL algorithms adjust the policy parameters *via* gradient ascent after directly learning the policy (mapping from states to actions). These techniques have worked well when dealing with high-dimensional state spaces and are especially suited for continuous action spaces.

b) Surrogate objective function: The fundamental concept of PPO is the use of a substitute objective function for policy updates. This substitute goal is a streamlined version of the conventional policy gradient goal. The clipping makes sure that the updated policy is conservative and does not stray too far from the original one to maintain consistency throughout training.

c) Proximal policy optimization objective (PPO): The PPO objective can be described as follows for each stage of data collection:

$$N(\theta) = E[\min(r(\theta) * A, clip(r(\theta), -1, +1) * A)]$$

The equation includes the PPO objective function, denoted as N(θ), where stands the θ for the policy parameters, r(θ) for the likelihood of the new ones *versus* the old ones for observed action within a particular state, and A for the measurement of how much superior the observed activity is the typical action is called the advantage function [11]. Hyperparameter governing the clipping mechanism is also N(θ), which represents the PPO objective function.

d) Clipped surrogate objective: The PPO objective function's clipping operation makes sure that the policy update is limited to a specific range. It prevents the policy from shifting too radically, preserving a steady learning process and enhancing sample effectiveness.

e) PPO algorithm: The proximal policy optimization algorithm's primary steps are as follows:

Step 1: Assign random weights to the policy neural network.
Step 2: Engage with the surroundings, gather experiences, and determine rewards.
Step 3: Use the rewards and the estimated values to compute the advantage function.
Step 4: For each state–action combination, the probable relation (r()) among the most recent and previous regulations should be determined.
Step 5: Using the clipped surrogate objective, calculate the PPO objective function.
Step 6: Use gradient ascent on the PPO goal to update the policy parameters.
Step 7: Continue steps 2 through 6 until convergence or a predetermined number of epochs.

PPO has successfully been used in a variety of RL benchmarks and has demonstrated outstanding performance and stability.

8) Gradient calculation for deep deterministic policy gradients (DDPG): Deep deterministic policy gradients (DDPG), an off-policy sequential decision-making technique, effectively regulate continuous action spaces by elaborating on the ideas of Deep Q Networks (DQN). Initiated by Google DeepMind researchers in their 2016 paper, DDPG has found widespread use in a variety of continuous control domains, including ones like robotic control, autonomous transportation, and drone navigation. By combining actor–critic techniques with the benefits of deep neural networks, DDPG excels at approximating each of the action–value systems and the regulations [12]. The deep deterministic policy gradients algorithm's workings are detailed in the paragraph that follows.

a) Actor–critical architecture: The DDPG employs an actor–critic architecture made up of two neural networks.
b) Actor–network: The actor–network generates continuous action values using the current state as input. The actor's goal is to create a deterministic policy that can convert states into matching actions and is appropriate for continuous action spaces.
c) Critic network: The critic approximates the action–value function Q(s, a) (representing the expected cumulative reward for acting "a" in state "s") using the present condition and the associated action as parameters.
d) Off-policy learning: A contravening algorithm is DDPG, which means that maintains a distinct target policy (the actor's policy that is updated less frequently) and learns from experiences gathered using a different policy (typically an exploration policy). This unconventional strategy boosts sample effectiveness and keeps the learning process steady.
e) Replay buffer: In a replay buffer, DDPG saves experiences similarly to DQN. The agent keeps the incidents in the playback memory. It gathers during interactions with the environment as tuples (state, action, reward, and future state). Mini-batch samples of experiences are taken at random during training to relate similar experiences and lessen bias.
f) Target networks: The actor and critic target networks used by DDPG improve learning stability. Infrequent updates are made to these target networks, just as the core networks. The participant's goal system develops goal actions, while the critic target network projects goal Q-values during training [13]. The weights of the target networks are often updated by interpolation with the main network's objects, which aids in stabilizing the teaching phase.
g) Search techniques: Providing assistance noise that is often added to the actor's output, DDPG uses an exploration approach to efficiently explore the action space. To promote exploration and keep the action space smooth, it is frequently used to introduce Ornstein–Uhlenbeck noise.
h) DDP group algorithm: The DDPG algorithm is composed of a few crucial actions:

Step 1: Initialize the actor, critic, and matching target networks with random weights.
Step 2: Engage with the surroundings, gather experiences, and then store them in the replay buffer.
Step 3: Take a few interactions at a time out of the replay buffer.
Step 4: Utilizing the average squared deviation loss, which is calculated using the target Q-values obtained from the critic target network, update the critic network.
Step 5: Change the actor–network by utilizing the critic network's grade about the actor's settings. Through this procedure, the actor is encouraged to take actions that will maximize the estimated Q-values.
Step 6: Update the target networks by periodically interpolating the target networks' weights into the main networks' weights.
Step 7: Up until convergence or a predetermined number of episodes are attained, repeat procedures 2 through 6 as necessary.

DDPG has successfully been applied to numerous difficult tasks where precise control is required and has been demonstrated to be particularly effective in continuous action spaces [14]. The combination of DDPG's actor-critic design, and off-policy learning, has a significant advantage when it comes to handling continuous control problems because it uses a recording loop and incorporates targeted networks.

9) Optimizing trust region policies (TRPO): A well-liked policy optimization approach in reinforcement learning (RL) is called Trust Region Policy Optimization (TRPO). It was introduced as an advancement of the proximal policy optimization (PPO) method by Open AI researchers in 2015. The drawbacks of conventional policy gradient approaches, such as their sensitivity to the policy update step size and the requirement for complicated hyperparameter tweaking, are addressed by TRPO. By maintaining a trust region, TRPO offers a principled method to policy updates that makes sure updates stay as close as possible to the current policy while maximizing the projected cumulative benefit. An outline of the Trust Region Policy Optimization algorithm is provided below.

a) Methods using policy gradients: Policy gradient approaches maximize the predicted cumulative reward directly by creating a relationship between states and actions. The desired reward result, which is frequently calculated using Monte Carlo sampling or bootstrapping, is used to guide the gradient ascending process used in these systems to understand the rules.
b) b) Trust region policy optimization: The main goal of Trust Region Policy Optimization (TRPO) is to update existing policies while remaining within a trust zone. Because they guarantee that the new policy stays fairly similar to the old one, TRPO's policy update restrictions make the learning process more stable and increase sample efficiency.
c) c) Objective function: The objective function of TRPO, which is constrained by a maximum KL difference between the recent and current policies, is the projected return based on the new policy, or predicted cumulative reward. When comparing the probability distributions of the two policies, the KL divergence is used to sustain the trust region criterion.
d) d) Surrogate objective: TRPO utilizes a surrogate objective function to deal with the restriction on the KL divergence. This function approximates the original objective while making sure that policy updates stay inside the trust region. Utilizing optimization methods like the conjugate gradient method, the surrogate objective can be obtained.
e) e) Line search: Following the computation of the policy update using a substitute purpose, TRPO conducts a path inquiry to identify the ideal step size that maximizes the anticipated cumulative reward under the new policy while remaining within the trust region [15]. The line search aids in making sure that policy adjustments are sensible and do not significantly worsen performance.
f) TRPO algorithm: The initial phases of the Trust Region Policy Optimization method are as follows:

Step 1: Randomly initialize the policy parameters.
Step 2: To interact with the world, gather experiences, and determine rewards.
Step 3: Calculate the expected return (expected cumulative reward) using Monte Carlo sampling or bootstrapping for each state–action pair.
Step 4: Calculate the surrogate objective and update the policies while remaining inside the trust zone.
Step 5: Use a line search to determine the best step size for the policy update.
Step 6: Continue steps 2 through 5 indefinitely or until convergence.

TRPO is especially well-suited for settings with high-dimensional action spaces and has demonstrated encouraging outcomes in a variety of RL tasks. The necessity for line searches and the resolution of limited optimization problems, however, can make it computationally expensive. As a result, TRPO has been improved upon to provide more effective algorithms, such as Proximal Policy Optimization (PPO), which also upholds the principles of trust region optimization while being more useful for large-scale RL tasks.

8.2 Intensive Learning Foundations

8.2.1 A Definition of Deep Learning

To acquire data representations using supervised or unsupervised methods, An aspect of machine learning called deep learning makes use of artificial neural networks, specifically deep neural networks. For neural networks to pick up structured information representations, they need to include a lot of layers, hence the term "deep" in deep learning.

8.2.2 Deep Learning Elements

Artificial neural networks are used in deep learning, a subfield of machine learning, to model and solve complex problems. The following list includes the essential elements of deep learning:

A neural network made artificially deep learning's basic skeleton is composed of artificial neural networks. They take their cues from how biological neurons in the human brain are organized and operate. Artificial neural networks are constructed by layering and connecting artificial neurons [16]. Every artificial neuron in a layer makes a quick calculation on its input and then transmits the result to the layer below it. To derive meaningful representations from the data, During training, the capacities of relationships between axons are modified.

- Layers: Neural networks, which include several layers, handle the processing and transformation of input data. There are three main categories of layers in a typical deep-learning model.
- Data input layer: Receives data first in this layer.
- Hidden layers: The learning of complicated features and representations is done *via* intermediate layers, which are located in the middle of the two layers of input and output.

- Output layer: The output layer is the last one and is in charge of producing predictions from the model.
- Activation functions: To enable neural networks to recognize complex data patterns and correlations, activation functions add non-linearity to the system. Rectified Linear Unit (ReLU), sigmoid, tanh, and softmax (utilized for multi-class classification) are typical activation functions.
- Loss functions: The gap between the goal outputs' actual and expected values is calculated using the loss function, which is occasionally referred to as the price or goal function. As the network updates its parameters throughout training, it acts as a guide. Depending on the particular task at hand, cross-entropy or average square error (MSE) can be used for either classification or regression, respectively.
- Optimization algorithms: Deep learning models are trained to minimize the decreased value by modifying the weights and prejudices of neural networks using optimization methods. A common optimization strategy in deep learning is gradient descent and its variations, such as stochastic gradient descent and Adam.
- Backpropagation: Training deep neural networks requires the use of the backpropagation technique. Through the optimization process, To modify the settings of the model, it generates the slopes of the impairment value concerning those features.
- Regularization: If a model behaves well on initial information and poorly on new information, it is considered to be overfitted. unknown information is avoided using regularization approaches [17]. L1 and L2 regularization, dropout, and batch normalization are typical deep-learning regularization techniques.

8.2.2.1 Different Kinds of Deep Learning Networks

a) Feedforward neural networks (FNN): The fundamental architecture of deep learning networks is embodied by a feedforward neural network (FNN), also referred to as multilayer perceptron (MLP). It is a particular kind of artificial neural network in which the links between the nodes (neurons) form directed acyclic graphs, meaning that data only flows in one direction—from input to output—without involving any loops or feedback connections. With the help of several hidden layers, the network learns to map input data to corresponding desired outputs for supervised learning tasks like classification and regression.

The feedforward framework of neural networks is described in more detail below:

- Input layer: Data entering the system is represented by synapses in the data entry layer. Because each neuron corresponds to a feature in the input data, the dimensionality of the input layer affects how many neurons are present.
- Hidden layers: The number of hidden layers between the input layer and the output layer might range from one to many. Numerous units, or interconnected neurons, are present in each buried layer, and these units form connections with neurons in both the previous and next layers. These connections have related biases and weights that are discovered during training.

- Activation functions: Activation techniques are utilized inside the hidden layers to introduce irregularity into the system, and are achieved by applying these functions to the output of each neuron. Rectified Linear Unit (ReLU), sigmoid, and tanh are a few of the commonly used activation functions [18]. The neural network may learn intricate representations and patterns in the input thanks to nonlinear activation functions.
- Output layer: The neural network's final predictions or outputs are produced by the neurons available in the result phase. The number of categories or output values produced by classification-related tasks or regression determines, accordingly, the number of transistors in the output region.
- Forward pass: The system receives input data during the forward pass and computes level-by-level events are made till they reach the final layer. The output layer's decisive activations show the neural network's anticipated results.
- Loss function: To determine the difference between the goal outputs and the expected outputs, a loss function is employed. Depending on the objective, a particular loss function may be chosen; To determine the difference between the goal outputs and the expected outputs, a loss function is employed.
- Backpropagation and training: Backpropagation is an optimization method used to establish the strengths and weaknesses of the artificial brain. Whenever backpropagation is used, gradients of the lack of function about network parameters are computed and then adjusted using an optimization technique like spherical gradient descent (SGD) or among its adjustments.

Feedforward neural networks are powerful and adaptable models, demonstrating remarkable successes in a variety of applications. They are less effective for jobs involving time series or sequential data because they have trouble capturing temporal dependencies and sequential patterns. Recurrent neural networks (RNNs) and variations of them are frequently employed for such tasks.

b) Convolutional neural networks (CNN): The field of computer vision tasks which includes recognizing images, recognizing objects, and segmentation of pictures has seen a considerable increase in usage of CNNs, a subcategory of deep learning methods. Because of their exceptional work on these projects, CNNs are now an essential part of many cutting-edge computer vision systems. The main advantages of CNNs for computer vision are their capacity to automatically learn hierarchical representations from images and their parameter-sharing method, which minimizes the number of parameters and increases computing efficiency. "ConvNets." are described in detail below.

- Convolutional layers: The central part of a CNN is the neural layer, which is its basic building block. A convolutional layer slides over the input image while using an array of filters (sometimes referred to as kernels), computing the dot product between the weights of the filters and the small, localized portions of the input [19]. Similar to doing a convolutional operation to apply the filters to the picture, this technique creates a feature map intended to highlight significant structures and patterns within the input image.

- Activation functions: By applying an activation function (usually ReLU) after the convolution process, the network gains non-linearity. ReLU (Rectified Linear Unit) is used to enable CNNs to recognize complex, non-linear patterns within the data by canceling out all negative values.
- Pooling layers: Reducing the spatial precision of the map features is accomplished by pooling layers, which significantly lowers the computational workload and lowers the danger of overfitting. Max pooling, which chooses the highest value within a limited zone, and typical pooling, which establishes the mean of data inside a small region, are common pooling techniques.
- Fully connected layers: CNNs frequently use one or more highly linked layers after several layers that are convolutional and pooling. The product's outcome probabilities for tasks like categorization are produced by these thicker layers. They accomplish this by connecting every neuron and flattening the high-dimensional feature maps.
- Parameter sharing: The exchange of parameters is one of CNNs' main benefits. The entire input image is subjected to the same set of filters, which enables the network to acquire translation-invariant features quickly. By sharing parameters, CNNs become computationally efficient and the number of parameters is greatly reduced.
- Training with backpropagation: The same method used to train other deep learning models is used to train CNNs. When the gradients of the loss function are computed relative to the network's variables, the weights are modified during training using methods of Adam optimization, for instance, or online gradient descent (SGD).
- Transfer learning and trained models: A significant quantity of data is frequently required for training CNNs, which can be a time-consuming and expensive computing procedure. Researchers use methods like pertaining and transfer learning to solve this [20]. This method builds on pre-trained models that have been developed on large datasets like ImageNet and then fine-tunes them using smaller datasets that are unique to particular tasks.
- Constructions: Several CNN constructions, including LeNet, AlexNet, VGG, ResNet, Inception, and others, have been proposed. The depth, width, and design decisions of these architectures vary, and they have each shown promise in a variety of computer vision tasks.

With continued efforts to enhance their performance, interpretability, and generalization skills, they remain an active research area.

c) RNN: RNNs are specialized deep learning models created for manipulating sequences of time, text, audio, and different sequential data formats. RNNs are adapted to take into account the special characteristics of sequential information, unlike feedforward neural networks, which feature feedback connections that enable them to process sequences of varying lengths and preserve memories of prior data [21]. RNNs are a good choice for applications that need an understanding of temporal interdependence and sequential patterns because of their memory characteristic. Recurrent neural networks are described in the following manner.

- Recurrent connections: Because RNNs feature recurrent connections, each hidden layer neuron's output is dependent on both the prior hidden state and the present input. The RNN can keep track of long-range dependencies in the sequential input and retain memory thanks to this feedback loop.
- Time unrolling: During training, RNNs are frequently unrolled through time to handle sequences of various lengths. The recurrent connections are revealed over time steps, and the RNN is modeled as a feedforward neural network with each time step representing a different layer. As a result, training might use BPTT.
- LSTM and GRU: Standard RNNs encounter challenges in capturing long-term dependencies as a result of the vanishing gradient problem. The LSTM and GRU are specialized RNN variations created to overcome this problem [22]. They feature gating mechanisms that control the information and gradient flow, making it easier for them to identify and preserve long-range dependencies.
- Bidirectional RNNs: Bidirectional RNNs analyze sequences in both forward and reverse orientations, accumulating data from both previous and upcoming time steps. By modeling the context on both sides of each point in the sequence, bidirectional RNNs can better interpret the data.
- Applications for RNNs:
  1) Speech recognition, sentiment analysis, text generation, and machine translation are just a few of the many jobs in natural language processing where RNNs have demonstrated their effectiveness.
  2) Prediction of the stock market, finding anomalies, and forecasting using time series analysis.
  3) Planning and managing a robot's motion is called robotics.
  4) Recognizing actions in videos, creating captions for them, and creating videos themselves.
- Training RNNs: One of the gradient-based optimization algorithms employed to train RNNs is the algorithm of BPTT. Nevertheless, vanishing/exploding gradient issues can make training RNNs difficult and cause sluggish convergence or instability.
- Stacking RNN Layers: RNNs can be "stacked," or built up from the bottom up, to catch more intricate patterns in sequential data. In a way that is analogous to deep feedforward networks, stacked RNNs enable a hierarchy of temporal representations.
- Challenges: Long-range dependencies can be difficult for RNNs to handle, and training on large amounts of data is necessary. Processing lengthy sequences might also be expensive computationally.

RNNs continue to be an effective tool for modeling sequential data despite these difficulties and have proven to function at the cutting edge in several applications. Sequence modeling in deep learning has evolved thanks to RNN extensions like attention mechanisms and transformer models.

d) LSTM: RNNs are a specific kind of brain network, whereas LSTMs are another. It was created to solve the standard RNN's disappearing and exploding gradient problem. Since their introduction in 1997 by Hochreiter and Schmidhuber, LSTMs have established themselves as a key component of deep learning models that handle sequential data [23]. Tasks involving long-range interdependence and context preservation across protracted timespan are particularly well-suited for LSTMs. Below is a summary of extended instantaneous memory networks:

- Structure: LSTM cells are more sophisticated than the straightforward neurons seen in conventional RNNs, and make up LSTMs. Three primary components are present in each LSTM cell.
- Cell state (c_t): A memory component called cell state enables long-term memory storage (LSTMs) of data. By using gating methods, it can selectively add or remove information.
- Input gate (i_t): The inclusion of data from the input gate (i_t) regulates the cell phase's secret state and present intake.
- Forget gate (f_t): The information from the previous cell state that has to be deleted is determined by the forget gate (f_t), enabling the LSTM to discard unnecessary data.
- Gating mechanisms: The input and forget gates are implemented using sigmoid activation functions, yielding values between 0 and 1. These parameters regulate the amount of data that should be entered or erased for the cell's current condition. Information can move freely when the value is high (near 1), whereas when it is low (near 0), knowledge cannot be modified.
- Output gate (o_t): The resulting gate determines the elements of the cell's condition to include in both the present covert situation and the output of the LSTM. By employing a sigmoid activation function, the knowledge flow from the cell stage to the concealed state is controlled by the resultant gate.
- Training: LSTMs are educated with the BPTT method, just like traditional RNNs. Gradients are calculated about the loss function during the training phase and used to modify the LSTM's parameters.
- Length-term dependencies: By carefully choosing whatever data should be stored in the cell condition during a protracted period, LSTMs are designed to be able to identify deeper connections in a series of data [24]. For jobs like natural language processing, where comprehending the context of a word may need to take into account words that are dispersed across the phrase, this capacity is essential.
- Applications: A variety of applications involving sequential data have had success with LSTMs, including:
  1) Natural language processing: Language modeling, machine translation, sentiment analysis, and text production.
  2) Time series analysis: Forecasting, anomaly identification, and financial projections using time series analysis.
  3) Speech recognition: Acoustic modeling and speech-to-text systems are two examples of speech recognition.
  4) Video analysis: Recognizing actions in videos and adding captions.

Comparing LSTMs to traditional RNNs, it has been shown that the former is significantly better at collecting long-range dependencies and maintaining context in sequential data. They may still need careful hyperparameter adjustment for optimum performance, though, as handling extremely long sequences can be difficult. Gated recurrent units (GRUs), a type of LSTM, offer comparable capabilities with a more straightforward architecture and may be more effective in some circumstances.

e) GRU: The family of recurrent neural network (RNN) architectures includes gated recurrent units (GRUs), which are comparable to LSTM networks. They created a solution to the problem of disappearing gradients and to handle remote dependencies inside sequential data. Cho *et al.* launched GRUs in 2014, and they have grown in popularity because of how well-suited they are for various sequence modeling problems and how easy they are to use. GRUs are an LSTM variant with fewer parameters, which increases their computational efficiency and facilitates training. Here is a description of gated recurrent units:

- Structure: GRUs are constructed similarly to LSTMs in that they have a memory cell with the capacity to store information over diverse time scales. Among a GRU cell's essential elements are:
- Hidden state (h_t): The data that comes from this moment's input and what was previously hidden is contained in the state that is concealed, which stands in for the GRU cell's output.
- Update gate (z_t): The gate for updates controls the amount of fresh input that has to be implemented in the present-day hidden state and the amount of the prior hidden state that should be kept.
- Reset gate (r_t): The reset button establishes the amount of the unconscious state from the previous cycle that should be disregarded while taking into account the incoming intake.
- Gating mechanisms: LSTM-like gating mechanisms are used by GRUs to regulate the information flow. The update and reset gates, which use sigmoid activation functions, support values between 0 and 1 [25]. The GRU can more successfully detect long-range dependencies thanks to these gates' selective updating and resetting of the concealed state.
- Training: Similar to how other RNN types are trained, GRUs are trained *via* BPTT. To update the parameters of GRU throughout training, the gradients have to be calculated by considering the degradation function.
- Simplicity and efficiency: GRUs feature a quicker design than LSTMs and fewer parameters, which increases their computational efficiency and makes them simpler to train. for resources are scarce or for developing sophisticated models with numerous repeating sections, GRUs can be an intelligent decision.
- Applications: GRUs have shown effective in a variety of tasks involving sequence modeling, such as:
  1) Natural language processing: Automated translation, emotional appraisal, text production, and language modeling are all examples of natural language processing.

2) Speech recognition: Voice assistants, speech-to-text software, and acoustic modeling all fall under speech recognition.
3) Time series analysis: Analysis of time series: forecasting, finding anomalies, and financial forecasting. Recognition of actions in videos and captioning of videos.
4) Video analysis: Action recognition and video captioning.

- Comparing GRUs with LSTMs: It can be seen that while both are intended to address long-range dependencies, GRUs have a more straightforward architecture and often call for fewer parameters, which makes them computationally more effective. With minimal computational cost, GRUs can occasionally approach performance levels on par with LSTMs.

GRUs provide a compromise between ease of use, effectiveness, and efficiency in identifying sequential dependencies. For modeling sequential data, they are a crucial tool that is frequently employed in a variety of deep learning applications. Like any neural network architecture, the decision between GRUs and LSTMs may be influenced by the particular objective, the size of the dataset, and the available processing resources.

f) GAN: In 2014, The GAN group of deep learning techniques was introduced by Ian Good Fellow and his associates. By making it possible to create realistic data samples including photos, audio, video, and text, GANs have transformed the field of generative modeling. The basic principle underlying GAN operation is that an engine and the discriminant neural system are trained inside a game-theoretic structure. The discriminator's task in this case is to differentiate between real samples and fake samples, whereas the generator's goal is to provide authentic data samples [26]. This encapsulates the fundamental notion of GANs.

Architecture

- Generator: The grid of reactors produces artificial data samples, such as pictures, from the source of erratic noise, which is taken from a straightforward distribution, such as the Gaussian. It gradually gains the ability to synthesize data samples that closely reflect the real data distribution by converting random noise into useful representations.
- Discriminator: The discriminator network uses samples of real data (from the true data distribution) as well as samples produced by the generator as input. By giving real data with higher probability and producing data with lower probabilities, It has been taught to distinguish between authentic samples and fake ones.
- Training process: The gen and the detector engage in an online minimax game as part of the GAN training process. While the generator seeks to maximize the possibility that the discriminator will mistake fake samples for genuine. To provide more convincing data samples that can fool the discriminator, the generator has been changed.
- By reducing the classification error, the discriminator is updated to more accurately discriminate between genuine and produced data.

- The training procedure is repeated until the generator generates high-quality data samples that are undetectable by the discriminator from actual data.
- Loss functions: The negative log-likelihood of the discriminator accepting generated samples as actual is often used to define the generator's loss. Maximizing this loss motivates the generator to generate samples with higher realism.
- The negative log-likelihoods associated with correctly classifying produced and real samples are multiplied to calculate the discriminator's loss [27]. The discriminator's capacity to distinguish between produced and real samples is improved by minimizing this loss.
- Challenges: It may be difficult and unstable to train GANs. During training, The synthesizer and critic must be balanced carefully and the learning process occasionally experiences mode collapse or divergence.
- Mode collapse: This occurs when the generator produces only a limited set of data samples, failing to capture the diverse range present in the genuine data distribution.

Applications

- GANs are used in many different disciplines, such as:
- Image generation: High-resolution, photorealistic images are created using image generation.
- Image-to-image translation: The process of converting images from one domain to another (for instance, converting sketches into realistic pictures).
- Style transfer: Transfer of creative styles to pictures or movies.
- Super-resolution: Enhancing image resolution.
- Data augmentation: Expansion of training datasets.
- Text-to-image synthesis: Creating visuals from word descriptions
- Video generation: Making realistic videos out of random noise.

A strong method for developing accurate data samples, generative adversarial networks have considerably advanced the science of generative modeling. In producing diversified and top-notch content across numerous domains, they have produced outstanding achievements. However, to further enhance their performance, study into their training and stability is still ongoing.

8.3 Integrating Deep Learning and Reinforcement Learning

8.3.1 Deep Reinforcement Learning

The powerful expressive capacities of deep neural networks are combined with trial-and-error learning methods in deep RL. Both value function (V) and Q-function (Q) can be estimated in deep RL by using function approximators. Examples of these approximators include neural networks. Because of its capability to handle multidimensional and continuous events and event locations, it is appropriate for demanding real-world applications.

The practical use of RL in complicated and high-dimensional tasks has been made possible by combining Reinforcement Learning (RL) and Deep Learning (DL), which has significantly advanced artificial intelligence [28]. The value function or policy in RL algorithms is represented by deep learning, more especially extensive neural systems, It serves as an approximator for functions. In numerous fields, this RL and DL combination is especially efficient. Here are some examples of how to combine RL with DL.

- Deep Q Networks (DQNs): DQNs, that utilize deep learning networks, when using Q-learning, consider the action–value function. The DQN architecture successfully collects and learns the Q-values connected to discrete action areas when put to use in difficult activities like playing challenging games like Atari 2600 games.
- Deep deterministic policy gradients (DDPG): The policies for continuous action spaces need to be estimated, action–value, policy functions and DDPG uses DNNs. DDPG can manage high-dimensional action spaces with effectiveness thanks to actor-critic architecture and deep learning.
- Policy gradients with neural networks: To describe the policy using function approximators, neural networks are used in policy gradient approaches like proximal policy optimization (PPO) and Trust Region Policy Optimization (TRPO). To learn complicated policies in difficult situations, these techniques make use of deep learning's expressive capacity.
- Methods for actor criticism: These techniques integrate policy- and value-based learning. In most cases, a deep neural network serves as the critic's representation. While the character is learning the rules, the critic is evaluating the value system [29]. Through their mutually reinforcing effects, this dual strategy improves stability and learning.
- Deep neural network topologies for RL: Researchers are investigating different deep neural network topologies designed specifically for reinforcement learning tasks within the context of using deep learning (DL) for approximation functions. For instance, in RL, sequential and time-series data have been handled using recurrent neural networks (RNNs).
- Advances in transfer learning and pretraining: Deep learning methods like transferable learning and pretraining have been implemented in RL to draw on knowledge gained from similar activities or contexts. This strategy enhances the generalization of RL agents and sampling efficiency.
- Hierarchical reinforcement learning: To learn higher-level policies or alternatives that can direct the agent's behavior over a longer period, hierarchical reinforcement learning (HRL) uses deep learning. Deep neural networks use these higher-level regulations that they have learned to give the agent a more hierarchical and systematic approach to solving challenging problems.

Learning representations: Deep learning has been applied to representation learning in RL, where neural networks are trained to derive valuable representations from unprocessed sensory input. The performance of the agent can be increased and the dimension of the state space can be decreased by utilizing these gained representations.

Reinforcement learning and deep learning's convergence have opened up new opportunities for solving complex issues that cut across numerous areas. It has sped up development in many practical fields, including robots, autonomous systems, gaming, natural language processing, and many more. Sample efficiency, stability, and safe learning remain important research topics as a result of the difficulties in merging RL with DL.

8.3.2 Deep Reinforcement Learning Complexity Problems

DRL has proven to be incredibly adept at a variety of tasks, such as directing robots, driving autonomously, and navigating challenging video games, to mention a few. Academics and practitioners need to solve a number of the problems it raises if it is to be successful and widely adopted. Numerous significant challenges, such as confronting deep reinforcement learning.

- Sample efficiency: To minimize the number of contacts with the environment required for learning, it is essential to improve sample efficiency. Training Deep Reinforcement Learning (DRL) agents frequently require a large quantity of data, this can cost money and take time, specifically in actual-life situations.
- Exploration–exploitation trade-off: Exploration in balance (experimenting with different methods to find more effective tactics) and exploitation (using the learned policy to exploit known good actions) is challenging. Agents need to explore sufficiently to find optimal policies without getting stuck in local optima.
- Reward function design: The development of a proper reward function is crucial to the effectiveness of deep reinforcement learning (DRL). The reward function must provide meaningful feedback to guide the agent towards achieving the desired task. Sparse or poorly designed reward functions can make learning difficult or lead to suboptimal policies.
- Credit assignment: In tasks with delayed rewards, determining which actions contributed to specific rewards becomes challenging [30]. Assigning credit accurately to the relevant actions is crucial for effective learning.
- Generalization: DRL algorithms may struggle to generalize effectively to unseen states and tasks. Overfitting to specific training scenarios and difficulties in adapting to new environments can hinder the agent's performance within practical applications.
- Exploration in high-dimensional spaces for action: The challenge of effective exploration gets increasingly difficult in scenarios requiring action areas with multiple dimensions. Selecting appropriate actions from a large set of possibilities can be computationally intensive and require specialized exploration techniques.
- Convergence and stability: Deep neural network training in DRL can be unstable. Poor solutions may result from parameter divergence or convergence. To guarantee stable and convergent learning, strategies like adequate initialization, regularization, and learning rate scheduling are crucial.

- Tuning of hyperparameters: deep reinforcement: Many hyperparameters, including learning rates, network designs, exploration rates, and others, are frequently involved in learning. The fine-tuning of these hyperparameters frequently requires significant investments of time and money to achieve optimal performance.
- Health and security: The secure operation of DRL agents is a major problem in real-world applications. A significant difficulty, particularly in contexts where safety is crucial, is ensuring that RL agents behave appropriately and do not engage in unsafe or dangerous behavior.
- Sample complexity: In complicated systems with high-dimensional state spaces, identifying effective policies may necessitate a large number of samples, which lengthens learning time and increases computational cost. To overcome these obstacles, it is necessary to combine algorithmic improvements with effective exploration methods, improved reward engineering, transfer learning, and the incorporation of domain knowledge to make DRL more durable, useful, and scalable for practical applications.

Conclusion

In this book chapter, deep and reinforcement learning principles constitute the bedrock of contemporary artificial intelligence, offering unparalleled capabilities to address complex problems in diverse domains. Deep learning's ability to glean intricate patterns from massive datasets has revolutionized fields like image recognition, natural language processing, and autonomous systems. On the other hand, reinforcement learning has opened doors to intelligent decision-making in dynamic environments, with applications ranging from robotics to finance. Moreover, it has highlighted the critical importance of ethical considerations and societal implications associated with the deployment of deep and reinforcement learning technologies. As we move forward, a deep understanding of these principles will be pivotal in pushing the boundaries of AI research and applications. By leveraging the insights gained from this exploration, researchers and practitioners can harness the transformative potential of deep and reinforcement learning to address pressing challenges and drive innovation in an increasingly AI-driven world.

References

1. Buduma, N., Buduma, N., Papa, J., *Fundamentals of deep learning*, O'Reilly Media, Inc, Sebastopol, CA, 2022.
2. Ramsundar, B. and Zadeh, R.B., *TensorFlow for deep learning: from linear regression to reinforcement learning*, O'Reilly Media, Inc, Beijing, China, 2018.
3. Dong, H., Dong, H., Ding, Z., Zhang, S., Chang, *Deep Reinforcement Learning*, Springer Singapore, Singapore, 2020.
4. Kim, P., *Matlab deep learning. With machine learning, neural networks, and artificial intelligence*, Apress Berkeley, CA, vol. 130, no. 21, pg. 156, 2017.

5. Arulkumaran, K., Deisenroth, M.P., Brundage, M., Bharath, A.A., Deep reinforcement learning: A brief survey. *IEEE Signal Process. Mag.*, 34, 6, 26–38, 2017.
6. Morales, M., *Grokking deep reinforcement learning*, Manning Publications, Canada, 2020.
7. Singh, P. (Ed.), *Fundamentals and Methods of Machine and Deep Learning: Algorithms, Tools, and Applications*, John Wiley & Sons, 2022.
8. Bhardwaj, A., Di, W., Wei, J., Deep Learning Essentials: Your hands-on guide to the fundamentals of deep learning and neural network modeling, Packt Publishing Ltd, U.S.A., 2018.
9. Samudrala, S., *Machine Intelligence: Demystifying machine learning, neural networks, and deep learning*, Notion Press, IndiaGrand Rapids, MI, U.S.A., 2019.
10. Majumder, A., *Deep Reinforcement Learning in Unity: With Unity ML Toolkit*, Apress, Berkeley, CA, 2021.
11. Arulkumaran, K., Dilokthanakul, N., Shanahan, M., Bharath, A.A., Classifying options for deep reinforcement learning, 2016, arXiv preprint arXiv:1604.08153.
12. Van Hasselt, H., Guez, A., Silver, D., Deep reinforcement learning with double q-learning, in: *Proceedings of the AAAI conference on artificial intelligence*, vol. 30, no. 1, 2016, March.
13. Stadie, B.C., Levine, S., Abbeel, P., Incentivizing exploration in reinforcement learning with deep predictive models, 2015, in *Proceeding of ICLR 2016*, arXiv preprint arXiv: 1507.00814.
14. Mnih, V., Kavukcuoglu, K., Silver, D., Rusu, A.A., Veness, J., Bellemare, M.G., Hassabis, D., Human-level control through deep reinforcement learning. *Nature*, 518, 7540, 529–533, 2015.
15. LeCun, Y., Bengio, Y., Hinton, G., Deep learning. *Nature*, 521, 7553, 436–444, 2015.
16. Li, Y., Deep reinforcement learning: An overview, 2017 in *Synthesis Lectures in Artificial Intelligence and Machine Learning*, arXiv preprint arXiv: 1701.07274.
17. Garnelo, M., Arulkumaran, K., Shanahan, M., Towards deep symbolic reinforcement learning, 2016, arXiv preprint arXiv:1609.05518.
18. Gu, S., Lillicrap, T., Sutskever, I., Levine, S., Continuous deep q-learning with model-based acceleration, in: *International conference on machine learning*, PMLR, pp. 2829–2838, 2016, June.
19. Heess, N., Hunt, J.J., Lillicrap, T.P., Silver, D., Memory-based control with recurrent neural networks, 2015, arXiv preprint arXiv:1512.04455.
20. Ho, J. and Ermon, S., Generative adversarial imitation learning, in: *Advances in neural information processing systems*, p. 29, 2016.
21. Kulkarni, T.D., Saeedi, A., Gautam, S., Gershman, S.J., Deep successor reinforcement learning, 2016, arXiv preprint arXiv:1606.02396.
22. Shuford, J. (2024). Deep reinforcement learning unleashing the power of AI in decision-making. *Journal of Artificial Intelligence General Science (JAIGS)* ISSN: 3006-4023, 1(1).
23. Ng, A.Y. and Russell, S., Algorithms for inverse reinforcement learning, in: *ICML*, vol. 1, p. 2, 2000, June.
24. Wahlström, N., Schön, T.B., Deisenroth, M.P., Learning deep dynamical models from image pixels. *IFAC-PapersOnLine*, 48, 28, 1059–1064, 2015.
25. Wulfmeier, M., Ondruska, P., Posner, I., Maximum entropy deep inverse reinforcement learning, 2015, arXiv preprint arXiv:1507.04888.
26. Piao, H. Y., Yang, S., Chen, H., Li, J., Yu, J., Peng, X., ... & Chang, Y., Discovering Expert-Level Air Combat Knowledge via Deep Excitatory-Inhibitory Factorized Reinforcement Learning. *ACM Transactions on Intelligent Systems and Technology*, 15, 4, 1-28, 2024.
27. Salimans, T., Ho, J., Chen, X., Sidor, S., Sutskever, I., Evolution strategies as a scalable alternative to reinforcement learning, 2017, arXiv preprint arXiv:1703.03864.
28. Vezhnevets, A.S., Osindero, S., Schaul, T., Heess, N., Jaderberg, M., Silver, D., Kavukcuoglu, K., Feudal networks for hierarchical reinforcement learning, in: *International Conference on Machine Learning*, pp. 3540–3549, p. PMLR, 2017, July.

29. Nachum, O., Norouzi, M., Xu, K., Schuurmans, D., Bridging the gap between value and policy-based reinforcement learning, in: *Advances in neural information processing systems*, p. 30, 2017.
30. Rummery, G.A. and Niranjan, M., Online Q-learning using connectionist systems, vol. 37, p. 14, University of Cambridge, Department of Engineering, Cambridge, UK, 1994.

9

Deep Reinforcement Learning in Healthcare and Biomedical Research

Shruti Agrawal[1] and Pralay Mitra[2]*

[1]Centre for Computational and Data Sciences, Indian Institute of Technology Kharagpur, Kharagpur, West Bengal, India
[2]Department of Computer Science and Engineering, Indian Institute of Technology Kharagpur, Kharagpur, West Bengal, India

Abstract
Advances in biological research have generated a wealth of data, which prompted the exploration of various machine learning techniques, notably reinforcement learning (RL) and deep reinforcement learning (DRL), for applications in the healthcare and biomedical domain. The utilization of Q-learning, a foundational RL algorithm, and its diverse variants, coupled with the integration of deep neural networks as seen in deep Q-networks (DQN), along with techniques like experience replay, has demonstrated remarkable efficacy in tackling intricate challenges such as protein folding. RL strategies, grounded in proximal policy optimization (PPO), have been extensively employed to meet the protein folding challenges, where the utilization of Markov decision processes aids in formulating intricate protein conformations. The intricate challenge of the protein–ligand docking finds a solution in the innovative application of DRL, employing the advantage asynchronous actor–critic model. Additionally, this complex problem can be effectively tackled using the deep deterministic policy gradient approach, enhanced by the incorporation of a graph neural network to adeptly represent the intricate interactions within the protein–ligand complex. Within the healthcare and biomedical sectors, the integration of deep learning techniques holds the potential for a range of applications. These include the analysis of medical imagery, the prediction of patient outcomes, the identification of intricate patterns within extensive datasets, and support in the field of drug discovery. These techniques harness the capabilities of neural networks, which excel in their ability to discern intricate patterns and relationships from voluminous and complex datasets. DRL has emerged as a transformative force in healthcare and biomedical domains, revolutionizing biomolecular structure prediction, decoding intricate molecular interactions, and elevating therapeutic strategies. Multiple use cases of reinforcement learning techniques are delineated within the realm of healthcare and biomedical applications. The analysis spans diverse applications from protein engineering to healthcare domains, showcasing the transformative impact of these techniques in advancing biomedical research, diagnostics, drug discovery, and the utilization of biological data.

***Keywords*:** Reinforcement learning, deep reinforcement learning, protein engineering and modeling, drug discovery, medical imaging, brain/body machine interfaces

**Corresponding author*: pralay@cse.iitkgp.ac.in

Shubham Mahajan, Pethuru Raj and Amit Kant Pandit (eds.) Deep Reinforcement Learning and Its Industrial Use Cases: AI for Real-World Applications, (179–206) © 2024 Scrivener Publishing LLC

9.1 Introduction

9.1.1 Reinforcement Learning

Deeply rooted in the principles of psychological behavior, a distinct part of the larger machine learning (ML) family is reinforcement learning (RL). In RL, the solution to a problem is achieved by the agent learning from fresh experiences via a trial-and-error methodology. The RL agent undergoes training, enabling it to take actions that aim to maximize the cumulative reward derived from its interactions with the environment. Typically, RL problems are formulated and addressed using the theoretical framework of Markov decision processes (MDPs). This entails the application of Monte Carlo (MC) methods and dynamic programming (DP).

The agent's learning process unfolds continuously through interactions occurring at discrete time steps. In a standard RL cycle, the agent is presented with the current state (s_t) of the environment at time step t and proceeds to select an action (a_t) for interaction. Subsequently, this action makes the environment react, transitioning to a new state (s_{t+1}), at the next time step $t + 1$. It is also decided if the agent will receive the reward (r_{t+1}) regarding the chosen action, which is related to the transition (s_t, a_t, s_{t+1}) [1]. As a result, the agent changes value and action–value function, i.e., V(s), Q(s, a) respectively after each cycle following a predetermined policy. This policy denoted as π is a function that translates states ($s \in S$) to actions ($a \in A$). One effective approach to addressing the RL problem involves modeling a Markov decision process (MDP) for the environment. While following policy π for identifying the agent in state s, the state–value function evaluates the reward. The total of incentives earned at subsequent time steps serves as the basis for this evaluation, each discounted by a factor Υ within the interval [0, 1]. Mathematically, we express this as:

$$V^{\pi}(s) = E_{\pi}\left(\sum_{k=0}^{\infty} \Upsilon^{k} r_{t+k+1} \mid s_t = s\right). \tag{9.1}$$

While adhering to policy π, the desired reward of being in state s and taking action a is evaluated which represents the action–value function. Here we again consider the accumulation of rewards for each state–action pair, similarly discounting future rewards with γ. Formally, this is expressed as:

$$Q^{\pi}(s,a) = E_{\pi}\left(\sum_{k=0}^{\infty} \Upsilon^{k} r_{t+k+1} \mid s_t = s, a_t = a\right). \tag{9.2}$$

This structured approach allows us to systematically analyze and address RL problems, providing a foundation for informed decision-making within dynamic and uncertain environments. To achieve an optimal policy in an MDP, DP starts either by initiating an initial policy or progressively refining it through iterations (policy iteration), or, by commencing with an arbitrary value function and determining the best course of action and its value, an estimate of an improved state–value or action–value function must be refined iteratively known as value iteration [2].

For a given policy, the Bellman expectation equation can be used to estimate the state-value function in the simplest case:

$$V^{\pi}(s) = E_{\pi}(r_{t+1} + \Upsilon V^{\pi}(s_{t+1}) \mid s_t = s). \tag{9.3}$$

This procedure can be thought of as policy evaluation, and by choosing actions greedily to maximize the state–action value, it is possible to achieve an enhanced and eventually ideal policy (π*). However, model-free approaches are required when dealing with situations with uncertain dynamics, bypassing the need for a predefined MDP structure. In these cases, the state–value function is substituted by the action–value function, and the best policy (π*) is determined by optimizing the action–value function:

$$Q^{\pi}(s,a) = E_{\pi}(r_{t+1} + \Upsilon Q^{\pi}(s_{t+1}, a_{t+1}) \mid s_t = s, a_t = a). \tag{9.4}$$

Numerous learning strategies, including Monte Carlo, temporal difference, and state-action–reward–state–action (SARSA), cover various facets of the model-free policy review and improvement process [3]. Since the state–action spaces in real-world RL issues are usually very large, it is impractical to keep a different value function for every possible state. To generalize the value function in such intricate contexts, function approximation techniques are required. By computing a function for a given state–action pair (s, a), designated as $\hat{Q}$ and expressed as $\hat{Q}\,.\ (s,\ a,\ w) \approx Q^{\pi}(s,a) = x(s,a^{T}w)$, one example of a generalization to undiscovered states is the Q-value function approximation. In simpst terms, a feature vector (x) that represents the state–action pair is used to derive an approximation Q-function. (s, a), along with a parameter vector (w) that evolves through Monte Carlo or temporal difference learning [4]. This approximation minimizes the difference between the real and estimated values, which is frequently accomplished using optimization strategies, like gradient descent, to enhance the Q-function. This loss minimization is mathematically represented as:

$$J(w) = E_{\pi}(\mathrm{Q}^{\pi}(s,a) - \widehat{\mathrm{Q}}(\mathrm{s,a,w}))^2. \tag{9.5}$$

Various differentiable function approximators can be employed for this purpose, including neural networks (NN), nearest neighbors, decision trees, linear combinations of features, and Fourier bases [5]. These versatile tools enable RL systems to efficiently navigate and learn from complex environments with vast state–action spaces.

9.1.2 Deep Reinforcement Learning

The inherent ability of RL to autonomously acquire knowledge without requiring intricate feature engineering renders it a potent tool with broad applicability across various domains. However, its effectiveness can be limited when dealing with high-dimensional data and dynamic, nonstationary environments [6]. Furthermore, deep learning's (DL) capacity to discern intricate patterns can occasionally lead to misclassification issues [7]. To address these challenges, recent years have witnessed the successful integration of RL algorithms with deep neural networks (DNN), fostering the development of creative learning methods. This fusion has found use in both training DNN with RL methods and approximating RL functions using DNN architectures [8].

DQN is the leading illustration of such integration, it successfully integrates DNN and Q-learning [9]. The DQN agent exhibits remarkable competence in learning policies through RL, even when confronted with high-dimensional inputs. The action–value function is estimated by deep convolutional neural networks for optimality.

Moreover, approximation of the Q-function may result in instability and divergence problems when tried to solve through the deep CNN where designs of NN are shallow. Then, such problems are solved by methods like experience replay.

The double DQN is a more complex version of the DQN method [10] that uses a double Q-learning algorithm [11] to handle DQN's overestimations. The DQN simultaneously learns two sets of value functions, and experiences are randomly assigned to update one of them. As a result, there are two different sets of weights, one of which chooses the greedy strategy for each update and the other of which chooses the value for it. Additionally, the realm of deep reinforcement learning (DRL) boasts various other notable algorithms, including asynchronous advantage actor–critic, AC with experience replay, dueling network DQN, continuous DQN, asynchronous N-step Q-learning, deep SARSA, and deep deterministic policy gradients [8].

9.2 Learning Methods in Bioinformatics with Applications in Healthcare and Biomedical Research

9.2.1 Protein Folding

Foldit [12] cleverly transforms protein folding into an online puzzle video game to address the computational challenges of protein folding by crowdsourcing solutions where players' intuitions and creativity are exploited. Players' creativity and collective efforts have contributed to enzyme design and structural elucidation. Foldit's simplicity, competitive nature, and data-driven improvements have transformed protein folding into an accessible, crowd-driven scientific endeavor with over 460,000 participants.

Alternatively, DeepFoldit [13] utilizes the latest DRL NN to model protein tertiary structure from their amino-acid sequence (Figure 9.1). The CNN architecture is used to build the agent and uses the Q-learning with experience replay. It was inspired by the Flappy Bird implementation [3]. Initially, the agent starts with a limited understanding of its environment and only a few potential actions. The agent quickly gains an awareness of the game dynamics by exploring the environment, which starts with random move execution and is motivated by the costs it experiences after each action.

The Rosetta algorithm is utilized as a black box problem, an agent learns optimal policies for protein folding by considering energy differences as input. It strives to minimize energy (maximize scores) while penalizing energy growth. Through training episodes, the agent evolves strategies for diverse proteins. Average rewards guide model selection, and parameters are fine-tuned for optimal performance. The agent employs CNN and Q-learning for decision-making, ensuring efficient adaptation and maximizing future rewards in protein folding tasks.

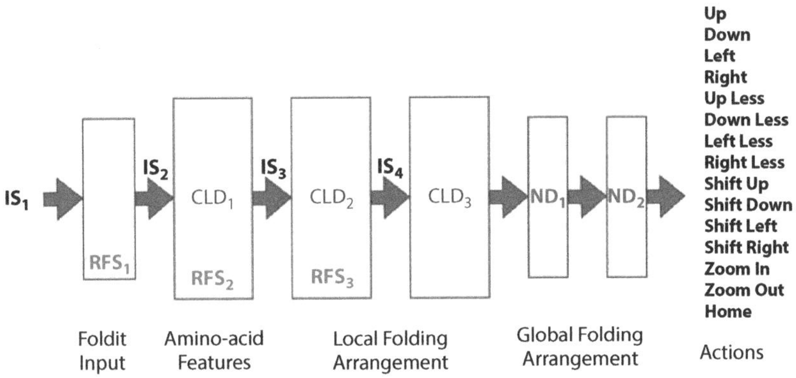

Figure 9.1 Model architecture where IS_i = input size of layer i, RFS_i = receptive field size of layer i, CLD_i = convolutional layer i depth, and ND = dense size and output is any of actions.

Overall, the study's findings indicate the application of DeepFoldit, a DRL framework, for the protein folding game Foldit is shown in this research. It has not quite reached superhuman levels of performance, but its ability to improve protein structures within the limit of 200 moves is encouraging. The study discusses difficulties, such as using "zoom out" sequences and protein replacements to deal with scenarios when the molecule fills the full window. The authors propose that continuous action reinforcement learning may be used to improve the resilience and flexibility of protein folding. They go over parallelization and hyperparameter optimization as ways to increase training effectiveness, and they emphasize how DeepFoldit combines the ease of use of Foldit with the strength of DRL.

9.2.2 Protein Docking

DRLCOMPLEX [14] utilizes 3D atom coordinates and their 2D inter-chain contacts as input to model the protein complex structure through self-play geometric reconstruction. In action space, six actions are designed for the agent, involving translations and rotations along the x, y, and z axes to adjust one protein chain's position against the other in 3D space. A deep CNN (Q-network) approximates the Q-function to predict action values, guiding the agent's decisions. Training involves continuous interaction with the environment, selecting actions either based on the network's highest predicted value or randomly with an ϵ-greedy policy. Experience Replay stores agent experiences (S, A, R, S0) for training. The transition from the current to target states (S and S*) is used to calculate the instant reward, based on the dimer's true or predicted structure. A target network (Q_{target}) aids in training by providing reference Q-values for actions, updated every k steps. These reference Q-values incorporate immediate rewards and predicted future values, using a discount factor γ. It was rigorously evaluated on homodimeric and heterodimeric protein complexes datasets, achieving high accuracy with true contacts and demonstrating its dependence on contact prediction accuracy. When using predicted contacts, DRLCOMPLEX performed competitively, outperforming other optimization methods such as simulated annealing, Markov chain Monte Carlo (MC), gradient descent (GD), crystallography, and NMR systems (CNS), making it a promising advancement in the drug discovery field.

Two well-known protein dimer datasets have been used for model evaluation in the experiments: CASP CAPRI, which contains 28 homodimer targets [15], and Std 32, which has 32 heterodimer targets [16]. For both datasets, the average values of RMSD, TM-score, f_{nat}, I_RMSD, and L_RMSD for DRLCOMPLEX, GD, MC, and CNS are presented. Real inter-chain connections and authentic tertiary structures are used in the evaluations, and the best outcomes are indicated in bold. RMSD, I_RMSD, and L_RMSD are in Å. Based on the Std_32 heterodimer dataset, Table 9.1 shows the algorithms' performance evaluations. DRLCOMPLEX performs better in the ideal situation (Table 9.1) in every evaluation metric except for f_{nat}. The variation in RMSD is quite noticeable. Table 9.2 indicates the average performance of the approaches on the CASP CAPRI homodimer dataset for the ideal situations. In every evaluation criterion, DRLCOMPLEX regularly performs better than or equal to all other predictors, with GD trailing closely behind. Surprisingly, for 10 out of the 28 targets, DRLCOMPLEX obtains a flawless f_{nat} score of 1.0, demonstrating its high accuracy quaternary structure reconstruction capabilities. When given ideal inputs, DRLCOMPLEX can reliably reconstruct quaternary structures for dimeric proteins, as evidenced by its average f_{nat} score of 99.05%. DRLCOMPLEX has demonstrated success in producing high-quality structural models and employs reinforcement self-learning to modify protein locations, particularly when given precise inter-chain interactions. Additionally, it can generate plausible models with anticipated interactions. Even in challenging environments, DRLCOMPLEX recreates protein dimer structures with state-of-the-art accuracy. One approach to enhance DRLCOMPLEX's prediction accuracy and promote the application of cutting-edge DL algorithms in drug development is pre-training it on various protein dimers.

Table 9.1 Performance metrics for DRLCOMPLEX, GD, MC, and CNS on the Std 32 dataset.

Methods	**TMscore (↑)**	**RMSD (↓)**	f_{nat} **(%, ↑)**	**I_RMSD (↓)**	**L_RMSD (↓)**
DRLCOMPLEX	**0.98**	**0. 88**	90.03	**0.92**	**2.15**
GD	0.95	2.90	**92.43**	1.99	7.16
MC	0.94	3.10	92.24	2.20	7.18
CNS	0.82	10.04	69.13	3.71	14.99

Table 9.2 Performance metrics for DRLCOMPLEX, GD, MC, and CNS on the CASP CAPRI dataset.

Methods	**TMscore (↑)**	**RMSD (↓)**	f_{nat} **(%, ↑)**	**I_RMSD (↓)**	**L_RMSD (↓)**
DRLCOMPLEX	**0.99**	**0.37**	**99.05**	**0.22**	**0.82**
GD	**0.99**	**0.37**	99.03	0.35	**0.82**
MC	0.96	1.21	78.45	1.61	2.89
CNS	0.92	2.00	73.45	3.92	4.68

RL-MLZerD [17]—this innovative method employs RL to construct multiple protein complexes, addressing the challenge of elusive quaternary structures. RL-MLZerD treats the assembly of multi-chain complexes as episode, leveraging RL to select and integrate pairwise docking models. It excels in predicting assembly order by analyzing RL paths. RL-MLZerD demonstrates exceptional performance, outperforming existing methods for protein complexes with three to five chains, except in unbound docking where AlphaFold-Multimer excels. It efficiently identifies correct docking poses, achieving an average RMSD of 2.50 Å and 6.30 Å for bound and unbound docking respectively, surpassing CombDock [18] and Multi-LZerD [19]. In bound cases, it outperforms ColabFold [20] and competes with AlphaFold-Multimer. RL-MLZerD employs a unique two-step process, generating pairwise decoys using LZerD, clustering them based on C_α RMSD, and ranking them using GOAP [21], DFIRE [22], and ITScorePro [23]. The RL-based multiple docking is performed on the top-ranked decoys from each subunit pair that forms the pool.

9.2.3 Protein–Ligand Binding

Greater complexity exists when constructing an efficient RL method where sampling and scoring both require such as protein–ligand docking instead of only using deep learning to address these two aspects separately. Few RL-based DL models exist for protein–ligand binding pose prediction [24], with much of the DRL, such as focusing on ion positioning prediction. The rigid docking approach used by the DRL-based developed method places the ligand far away from the protein at random. There are two models in the DRL framework: Two models are used to evaluate the adequacy of predicted binding sites: i) an actor model, structured as a ProDCoNN (protein design using a convolutional neural network) [25], and ii) a critic model that has a similar architecture to the actor model [26]. With an emphasis on creating ligand-specific models for single atoms and small multi-atom ligands, this work investigates the viability of using RL to solve protein–ligand docking problems. These models can be helpful in the structure-based medication development process. To further enhance model robustness and performance, it would be advisable to train the model using a more extensive dataset derived from additional sources like the Binding MOAD (mother of all databases) [27] and CCDC/Astex databases (a newly created test set for verifying protein–ligand interaction predictions) [28]. This approach would contribute to the versatility and effectiveness of the model.

For predicting copper binding ligands, 20,000 training samples were generated and originally positioned distantly from their binding sites. During training, the actor model discovered how to direct ligands towards their actual binding locations. Model performance was calculated by the RMSD, which demonstrated a decline in training and an enhancement in movement strategies. The average RMSD dropped dramatically from approximately 9 to 2 Å. The training and test datasets RMSD statistics and improvement rates are compiled in Table 9.3. The long-term reward was estimated using the critic model in this work.

Mean and median values obtained during the training and testing stages for the sulfate-ligand dataset are compiled in Table 9.4, with particular attention paid to the final position permutation-distance, center-distance, and their respective improvement rates. The overall performance of the Cu ligand dataset is superior to that of the sulfate-ligand dataset, even if center-distance findings beat permutation-distance. This difference can be attributed to the intricate structure of sulfate ligands and the intricate nature of protein–ligand complexes

Table 9.3 Training and testing evaluation metrics for Cu ligand dataset.

	Training dataset		Test dataset	
	Mean	Median	Mean	Median
RMSD (in Å)	2.06	1.52	3.18	2.52
Improvement rate	76.46%	83.87%	62.13%	73.15%

Table 9.4 Training and testing metrics for SO^{2-} ligand dataset.

	Training dataset		Test dataset	
	Mean	Median	Mean	Median
The last position permutation distance	4.2566	3.8366	4.4347	3.9361
The last position center distance	4.1057	3.7129	4.2810	4.2810
Improvement rate for permutation distance	49.00%	59.22%	48.08%	58.80%
Improvement rate for center distance	50.08%	60.85%	49.12%	60.01%

including SO_4^{2-} ligands. Furthermore, the bigger space surrounding the genuine SO_4^{2-} ligand location is not well aligned with the actor model's predilection for sites around atoms. The SO_4^{2-} ligand dataset's improvement rate suggests that the suggested framework is still successful despite these obstacles. Since it is difficult to see the effect of rotation, critic distance is still a useful stopping criterion.

It would also be advantageous to investigate the incorporation of more complex and sophisticated NN designs, such as DenseNet, ResNet, and different attention techniques. Leveraging the latest deep learning architectures could potentially lead to further enhancements in performance for protein–ligand docking models. It would be worthwhile to consider implementing a replay buffer to collect and utilize multiple historical steps. De-correlated samples are drawn at the time of the training phase, to achieve mini-batch gradient descent. A method like this could significantly speed up learning and increase the overall training efficiency of protein–ligand docking models.

Understanding the connections between particular amino acid residues in proteins and ligand molecules depends on correctly identifying the binding posture in protein–ligand interactions. This knowledge is essential for determining binding energy and affinities. Traditional methods like molecular docking often fail to predict the biologically relevant binding pose due to computational limitations. This method does not require that active sites in proteins be defined in advance [29]. GraphCNNs are proposed as an alternative to 3D CNNs for molecular representation, offering computational advantages. The approach combines node and spatial features. DDPG is chosen as the RL algorithm but may face convergence issues. Discretizing the continuous state space or using other RL models like

DDQN or TD3 are potential alternatives. The implementation involves training the model on various viral protein–ligand complexes, validating the RL algorithm's ability to identify active sites, and assessing the importance of atomic features. Encoding ligand-specific and residue-specific features into the algorithm is also considered. The method eliminates the requirement for manual chores like locating protein pockets and grid layout by proposing an RL-based way to automate the determination of the proper ligand posture and binding site. Antiviral protein–ligand complexes are the initial emphasis, with the possibility of extension to a more dynamic framework trained on numerous datasets.

9.2.4 Binding Peptide Generation

Because MHC Class I proteins have distinct peptide repertoires, they are essential for immunotherapy. However, in vitro investigations cannot describe the binding motifs for hundreds of MHC Class I proteins, so a scalable and effective alternative approach is required. PepPPO (peptide proximal policy optimization) is designed for generating binding peptides that interact with MHC Class I proteins [30]. Binding motifs are efficiently determined for approximately thousands of human MHC Class I proteins, showcase a strong resemblance to experimentally derived motifs. This method addresses the scarcity of experimental data by using RL to create peptides. PepPPO optimizes random initial peptides through stepwise amino acid mutations until they are predicted to bind positively to a specific MHC protein according to MHCflurry2.0 [31]. The binding peptides produced by PepPPO have shown a strong correlation using motifs obtained through experimentation. They have properties like strength and robustness, consistently emerging from different initial peptides, demonstrating PepPPO's reliability. Moreover, these motifs are effective for quickly screening human MHC Class I protein i.e., neoantigens, whether experimental data is available or not. PepPPO outperforms other methods in generating high-quality binding peptides, making it a promising tool for peptide-based vaccine development and immunology research.

To assess PepPPO's efficacy, five baseline models were created and contrasted. The effectiveness of these techniques in producing qualifying peptides for 30 MHC proteins is shown in Table 9.5, which is separated into common MHCs with experimental data and rare MHCs without. PepPPO performs better than other approaches in terms of qualified peptide percentage, average presentation scores, and maximum presentation scores whether it is used with or without a diversity-promoting buffer. PepPPO performs noticeably better for common MHCs than the second-best model, sPWM, proving its capacity to identify more patterns in qualifying peptides. PepPPO works significantly better than sPWM in the case of uncommon MHCs, indicating that it is more resilient in situations where there are not enough qualifying peptides available. The data show the mean and standard deviation values for each of the 30 MHC proteins. The following are shown in the columns: The term "Percentage (%)" denotes the proportion of qualified peptides among all created peptides, "Average score" denotes average presentation scores, and "Maximum score" denotes maximum presentation scores. Bold highlights indicate the highest percentage (%) values. Motifs that are retrieved from peptides produced by PepPPO show robustness and good alignment with motifs that are derived experimentally. These patterns are useful for both common and unusual human alleles, as evidenced by their successful application for quick

Table 9.5 Comparison of PepPPO model performance with the other five models.

MHC	Method	Percentage (%)	Average score	Maximum score
Common	BO-VAE	3.37 ± 2.53	0.13 ± 0.23	0.97 ± 0.03
	BP-VAE	1.85 ± 1.26	0.06 ± 0.16	0.95 ± 0.07
	MCTS	13.90 ± 9.34	0.16 ± 0.29	0.96 ± 0.02
	sPWM	30.60 ± 12.78	0.40 ± 0.39	0.99 ± 0.01
	Random	0.43 ± 0.36	0.02 ± 0.09	0.89 ± 0.10
	PepPPO (without buffer)	**91.48** ± **10.68**	0.83 ± 0.17	0.99 ± 0.01
	PepPPO (with buffer)	91.42 ± 10.89	0.83 ± 0.17	0.99 ± 0.00
Rare	BO-VAE	2.59 ± 3.41	0.11 ± 0.21	0.93 ± 0.06
	BP-VAE	1.34 ± 1.24	0.05 ± 0.14	0.92 ± 0.09
	MCTS	8.11 ± 7.11	0.11 ± 0.23	0.94 ± 0.03
	sPWM	18.80 ± 8.99	0.28 ± 0.35	0.98 ± 0.01
	Random	0.26 ± 0.34	0.02 ± 0.07	0.83 ± 0.15
	PepPPO (without buffer)	85.33 ± 14.09	0.78 ± 0.21	0.98 ± 0.01
	PepPPO (with buffer)	**87.15** ± **9.71**	0.79 ± 0.20	0.98 ± 0

neoantigen screening in rectal cancer. This has the potential for the advancement of quick neoantigen screening techniques in precision medicine.

9.2.5 Protein Design and Engineering

Inspired by the existing evolutionary protein design and engineering methods [32-37], EvoPlay [38] uses a policy-value neural network and a Monte Carlo tree search for guidance and implements a self-play RL framework-based protein design technique where single-site residue mutation is an action that mimics moves on a chessboard. EvoPlay is excellent at jobs requiring in-silico-directed evolution for enhancing protein sequences, even supporting structural design using AlphaFold2 for high-affinity peptide binder development, validated through binding assays. Prospective luciferase engineering with EvoPlay led to 7.8-fold improved bioluminescence beyond the wild type, demonstrating its potential for diverse academic, industrial, and clinical protein design applications.

The technique was performed five times for each of the five sequences that were chosen as initial seeds from the low-fitness pool. The highest score is indicated in bold, and the max value is the average cumulative maximum value derived from these five repetitions. The mean value is also determined by averaging the top 10 produced sequences from the five repeats that the TAPE oracle examined. Table 9.6 displays these findings. Using the PAB1

Table 9.6 Comparing EvoPlay with other methods using the PAB1 dataset. For each method, the first column indicates max values, and the second column indicates mean value.

Series no.	Start fitness	EvoPlay		AdaLead		SAC		Cbas		Evolutionary BO		DyNAPPO	
		Max.	Mean	Max.	Mean	Max.	Mean	Max.	Mean	Max.	Mean	Max.	Mean
S1	0.02	**0.73**	**0.59**	0.63	0.49	0.51	0.33	0.53	0.33	0.32	0.28	0.50	0.33
S2	0.04	**0.55**	**0.48**	0.40	0.28	0.29	0.19	0.34	0.28	0.36	0.31	0.29	0.17
S3	0.01	**0.54**	**0.45**	0.50	0.41	0.39	0.26	0.41	0.30	0.39	0.34	0.39	0.24
S4	0.04	**0.59**	**0.55**	0.50	0.45	0.30	0.15	0.34	0.24	0.40	0.36	0.32	0.18
S5	0.29	**1.09**	0.82	1.05	**0.86**	0.45	0.36	1.00	0.58	0.37	0.34	0.91	0.60

dataset, we benchmarked sequences produced by EvoPlay and five additional techniques AdaLead, evolving BO, Cbas, SAC, and DyNA-PPO.

With the application of advanced technologies, DyNA PPO [39], a DRL model that generates sequences sequentially while treating the procedure as a Markov decision process, uses proximal policy optimization [40] for sequence design (Figure 9.2). The agent receives zero reward before sequence completion, with the final reward determined by sequence fitness measurements from machine learning models approximating surrogate fitness functions at the end of each round. DyNA PPO addresses reward estimation trade-offs by leveraging various models to understand different aspects of sequence fitness, updating its policy with the most suitable and accurate one. Although it has demonstrated superiority in large-scale benchmarking compared to several methods, DyNA PPO's practicality awaits

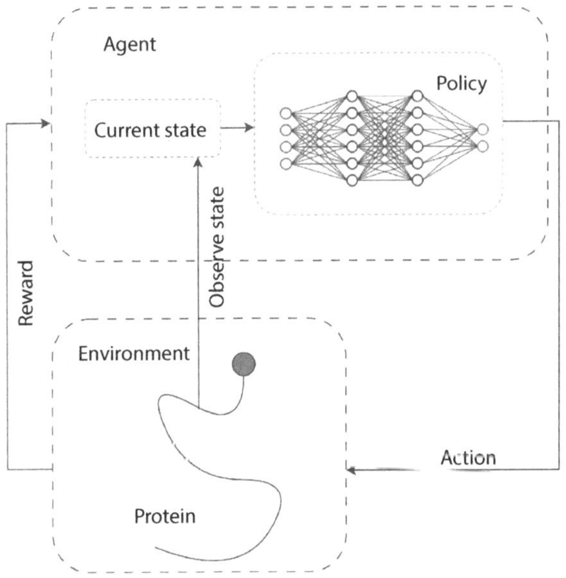

Figure 9.2 DRL-based protein design, where the agent selects the next action by determining the amino acid for mutation in the following phase based on its policy and the current state.

validation through wet lab experiments in future research. Alternatively, pre-trained generative models for designing proteins can be improved using RL, as seen in the tuning of an RNN through policy-based RL to generate desirable compounds. This research offers valuable insights into mitigating the common problem of catastrophic forgetting in protein generative models [41].

Using three *in silico* optimization scenarios intended to emulate expensive wet-lab studies, we compare DyNA PPO with alternative approaches. Additionally, ablation studies are carried out to improve our comprehension of DyNA PPO's functionality. Model-free policy optimization (PPO) and model-based optimization (DyNA PPO) are compared, as well as DbAs, RegEvolution (RegEvol), FBGAN, Bayesopt GP (BO-GP), and Random (refer to Table 9.7). Based on the cumulative maximum reward f(x) for sequences proposed within a certain round, optimization performance is assessed. As a numerical statistic, the area under the cumulative maximum reward curve summarizes the optimization performance. The mean pairwise Hamming distance between sequences proposed at each round is used to measure sequence diversity, and for issues with known optima, we report the fraction of global optima found. A total of 50 different random seeds are used to duplicate these trials. The technique's efficacy is assessed by computing average ranks for each of the 41 transcription factor targets that are unknown. Before being averaged over all objectives, rankings are determined for each goal independently using the average of metrics from optimization rounds. In this ranking system, higher ranks (seven being the highest) denote better performance. DyNA PPO is exceptional in maximizing f(x) and finding a large number of well-separated local optima. A variation of PPO called DyNA PPO uses an exclusive reward function that penalizes identical sequences to promote exploration while improving sampling efficiency through autonomous model selection. The use of surrogate models approximates wet-lab investigations and allows for multiple iterations of optimization in simulation. Although the advantages of DyNA PPO in biological sequence design are showcased, the large-batch, low-round optimization strategy covered here may have wider uses in several scientific and commercial fields.

9.2.6 Drug Discovery and Development

A well-justified hypothesis is formulated for producing novel lead compounds by *de novo* design or choosing compounds from the available data showing that the structure–activity relationship such as synthetically feasible chemical libraries, is a critical step in new drug development efforts [42]. Combining DL and RL techniques, ReLeaSE (Reinforcement

Table 9.7 Average ranking of methods for transcription factor binding targets.

	DyNA PPO	**PPO**	**BO-GP**	**DbAs**	**RegEvol**	**FBGAN**	**Random**
Cumulative max	**6.4**	5.8	5.0	3.7	3.7	2.2	1.3
Fraction optima	**6.8**	5.6	5.4	3.3	3.3	2.5	1.0
Mean Hamming distance	**5.6**	5.4	4.0	2.5	1.0	2.5	7.0

Learning for Structural Evolution) allows the *de novo* synthesis of molecules with certain properties [43]. It incorporates two DNNs, namely, the generative (G) and predictive (P) models (Figure 9.3(a)). Two stages make up the training process: first, with the aid of supervised learning methods, both models are trained independently. The models are then jointly trained using an RL strategy intended to maximize particular target features. An agent is responsible for generating novel SMILES notation which is chemically feasible molecules part of the generative model. It employs a Stack-RNN capable of learning hidden rules for forming legitimate SMILES strings, including correct handling of different bracket sequences, counting ring openings and closures, and valence for atoms. Stack-RNN introduces memory stack operations (PUSH and POP) as differentiable, continuous processes, resembling a pushdown automaton. Simultaneously, the predictive model, responsible for predicting the properties (physical, chemical, or biological) of these new compounds, is done by the critic. It has an embedding layer, two dense layers, and an LSTM layer (Figure 9.3(b)). This DNN calculates user-specified properties (activity) of a molecule directly from its SMILES string, without the need for traditional numerical descriptors. This approach differs from conventional quantitative structure–activity relationship (QSAR) models by directly learning relationships between SMILES strings and target properties. Each created molecule is given a numerical reward or penalty, with the reward depending on the numerical values of the predictive model, and this determines how well the agent performed. Next, training is done to optimize the expected reward for the generative model. ReLeaSE

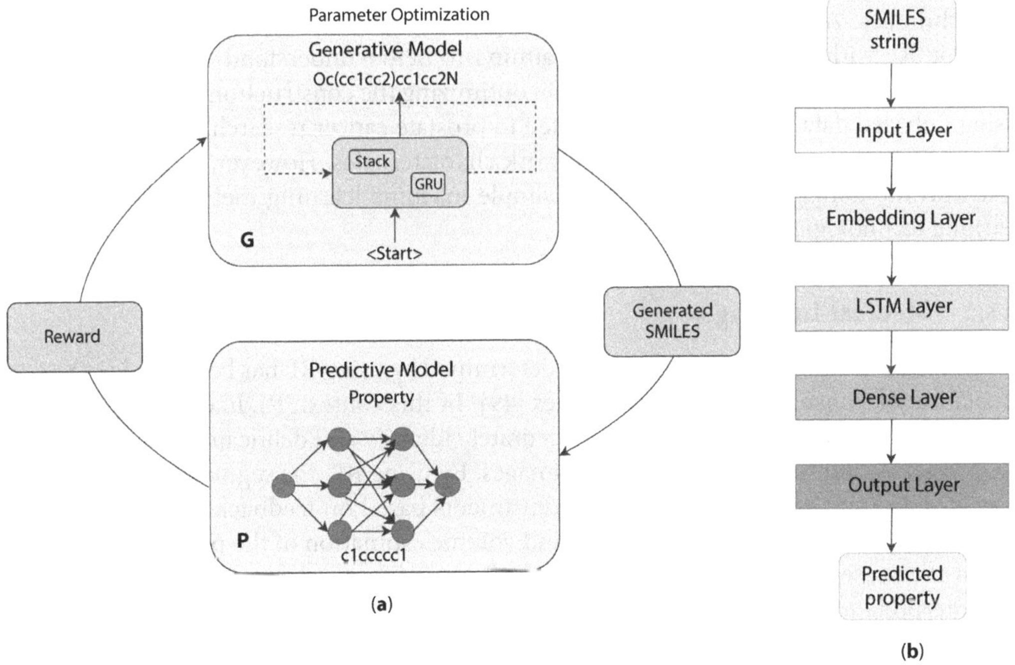

Figure 9.3 DRL-based production of novel SMILES strings of compounds with specified properties. (a) An RL-based framework for generating novel compounds. (b) A SMILES string is supplied to the prediction model as input and outputs an estimated property value as a real number.

has been demonstrated to design chemical libraries with various property biases, making it a versatile tool for generating targeted compounds with specific desired characteristics.

We found that the estimated variation for Tm was 42°C, and the RMSE for log *P* was 0.9 when comparing the experimental and projected data. Interestingly, this accuracy shows that the author's approach is successful in predicting characteristics for an external dataset but slightly less than that of a similar quantitative structure–property relationship (QSPR) model developed using cross-validation. This adds to the proof that the consistent molecules produced by their method have the expected and desired features.

9.3 Applications in Biological Data

9.3.1 Omics Data

Researchers also tried to combine RL and binary particle swarm optimization (BPSO) as an alternative for nontrivial and time-consuming experimental methods for predicting operons in genomes of the bacterial family [44]. It utilizes intergenic distance, metabolic pathway participation, and gene length ratios in *Escherichia coli* to create an RL-based fitness function, demonstrating its prediction performance on three genomes. RL is used in the BioAgent system to improve the precision of biological sequence annotation [45]. RL may have been used to iteratively improve the annotation process based on feedback and reward signals. The challenging fragment assembly problem (in DNA as well as in protein) is crucial in bioinformatics. It introduces a novel RL-based model using Q-learning agents [46], demonstrating promising performance and potential for improving DNA fragment assembly techniques. A hybrid method for building protein-protein interaction networks that combine RL with text mining methods, aiming to better understand complex biological functions [47]. RL may have played a role in optimizing the construction of these networks. Using PubMed data, the approach is applied to prostate cancer research, demonstrating its effectiveness, and revealing scale free network characteristics. However, it is demonstrated that with the domain knowledge even a simple machine learning method can beat deep learning technologies [48].

9.3.2 Medical Imaging

The prostate's location and volume can be determined by using RL has been used in the segmentation of transrectal ultrasound images [49]. In this context, RL likely played a role in optimizing the segmentation process to accurately identify and delineate the boundaries of the prostate gland within the ultrasound images. By using RL, the segmentation algorithm may have learned to make decisions or adjustments based on feedback signals, ultimately improving the accuracy of both location and volume estimation of the prostate. This application showcases how RL can be used in medical imaging tasks to enhance the analysis and interpretation of complex medical data. A DRL method has been created to recognize breast lesions using magnetic resonance imaging, leading to significantly improved accuracy and greatly reduced detection time.

The method introduces a novel approach using DRL to optimize medical image acquisition by leveraging various parameters, including those from medical scanner databases.

A DRL-based technique using trust region policy optimization is used to categorize and segment joint surgical gestures. Deep RL and an actor–critic-based method are used to address the issue of detecting anatomical landmarks. The earlier demonstrates improved accuracy and speed for MRI, ultrasound, and CT images. For 3D landmark detection in real-time in CT images, another way suggests a multi-scale DQN that is tested on a substantial dataset. Additionally, there is a focus on anatomical structure detection in incomplete volumetric data, assessed on a large CT volume dataset. One method employs DQN for automatic view planning in 3D image acquisitions, achieving high accuracy on different MRI planes. Another approach discusses a DRL method for actively locating objects in volumetric scenes, while a different one utilizes DQN to detect multiple landmarks in various medical image datasets. Furthermore, recent research proposes unified end-to-end DRL techniques for 3D medical volume specific tasks such as vascular centerline tracing, multimodal picture registration, and image registration. These innovations significantly enhance medical imaging and analysis across diverse applications [50].

9.3.3 Brain/Body–Machine Interfaces

Research on brain–machine interfaces (BMI) has found uses for RL, especially in controlling robotic or prosthetic devices. Mapping neural activity to intended behavior researchers have used co-adaptive BMI with TD(λ) to map neural activity to intended behaviors, allowing individuals to control devices through their brain signals [51]. Actor–critic techniques have been employed in symbiotic BMI systems, where the brain and the machine work together to achieve desired actions or movements [52]. To produce more accurate BMI control models, a testbed focusing on center-out reaching tasks in monkeys has been established [53]. By translating brain states into actions for prosthetic devices, Hebbian RL approaches have been used to create adaptive control [54]. Using Q(λ) techniques, in multistep goal-directed tracking tasks, BMI systems have been created for the unsupervised decoding of cortical spikes [55]. To provide steady performance over long periods, adaptive BMIs that can adapt to significant changes in brain activity with little training have been created. They frequently use AC architecture [56]. Quantized attention-gated kernel RL is used for mapping the state–action space as an approximation, to accomplish effective nonlinear mapping of brain states to actions [57]. Additionally, to control a robotic device remotely, a BMI can send EEG signals elicited by fictitious movements over the internet [58]. Furthermore, Bayesian models and RL are integrated to choose dynamic thresholds, improving restorative BMI system's performance [59].

9.4 Adaptive Treatment Approach in Healthcare

An Adaptive Treatment Approach (ATA) comprises a series of decision rules used to personalize treatments based on evolving patient data, incorporating data such as diagnostic test results, genetic information, and demographic details, fitting well within the paradigm of personalized medicine. They are especially valuable for managing chronic disorders in time-varying treatment settings. The Sequential Multiple Assignment Randomized Trial (SMART) approach allows for causal inferences at each level of decision and can be used to create ATAs [60]. SMART involves a series of randomized observations and treatments, with patients re-randomized at each stage based on their response to previous treatments. Common goals include comparing predefined ATAs and finding the optimal one. Many statistical techniques have been developed for this purpose, with some leveraging data collected during clinical trials. SMART data naturally align with RL, and Q-Learning is a common RL-based approach to finding optimal ATAs.

Multiple studies have utilized RL methods to optimize ATAs for various medical conditions, including epilepsy and schizophrenia. These ATAs aim to personalize treatment decisions based on evolving patient data. To reduce seizure length and frequency in epilepsy, deep brain stimulation techniques were refined utilizing a fitted Q-iteration strategy in conjunction with extremely randomized trees [61]. An RL-based approach [62] was applied to discover the optimal ATAs for reducing schizophrenia symptoms, monitored over time with the PANSS score (Positive and Negative Syndrome Scale). The SMART study's data were used, and for learning the Q-function by regression, the strategy used a fitted Q-iteration method. Ertefaie *et al.* introduced changes in Q-learning for residual analysis using information from a SMART trial including individuals suffering from schizophrenia [63]. Furthermore, a fitted Q-iteration method can be used to create an efficient DTR for the treatment of schizophrenia when the method is combined with several reward functions.

Sepsis, a potentially fatal infection, is a significant global health concern, and personalized treatment strategies are crucial for its management. Recent studies have explored the application of RL in developing personalized sepsis treatments. These studies investigated clinical treatment strategies using RL agents based on TD-learning and in-place policy iteration, using patient data from the available dataset. Researchers aimed to learn an optimal policy to improve patient survival and evaluate its performance against recorded clinical decisions [64]. In other studies, DRL and continuous state-space models were employed to assess optimal sepsis treatments, emphasizing the potential of RL in tailoring interventions for individual patients. Additionally, neural networks and proximal policy optimization were utilized in the treatment of sepsis patients, highlighting the versatility of RL techniques in healthcare applications [65]. These investigations underscore the growing interest in leveraging RL to enhance sepsis treatment outcomes and reduce its global impact.

Numerous research studies have focused on personalized cancer treatment, a simulation of advanced cancer trials was conducted, employing Q-learning to develop ATAs that continuously adjust drug dosages based on individual patient characteristics and subgroups with varying treatment responses. In a different study, a reinforcement learning framework was used to find personalized lung cancer treatment options [66]. The optimal policies were derived from longitudinal patient trajectories using Q-learning while considering the trade-off between treatment efficacy and toxicity. A DRL approach was employed to determine the optimal treatment regimens for leukemia patients, leveraging medical data collected over a specific timeframe to minimize the risk of graft *versus* host disease [67]. Furthermore, Jalalimanesh *et al.* addressed the challenge of personalizing radiation therapy dosage calculations and utilized Q-learning to tailor radiation doses based on tumor growth phases, highlighting the importance of personalized radiation therapy strategies in cancer treatment [68].

RL methods have found applications in treating anemia, particularly in hemodialysis patients. From medical records, the ideal course of treatment for renal anemia was chosen using the Q-learning algorithm. An interactive Q-learning model was introduced to develop and optimize ATAs, were demonstrated using data from major depressive disorder studies. Additionally, Vincent *et al.* focused on optimizing erythropoiesis-stimulating agent treatments for managing anemia in hemodialysis patients through fitted Q-iteration [69]. It was advised to employ a hybrid approach that combines Q-learning and outcome-weighted learning to determine which ATSs from SMART trials are most effective for treating behavioral disorders such as attention deficit disorder, major depressive disorder, and hyperactivity disorder [70]. Furthermore, a Monte Carlo and value iteration method-based online selection framework is suggested to reduce the typical recovery time for individuals with medical problems [71]. RL approaches have proven valuable in enhancing treatment strategies for various medical conditions [71].

For choosing an HIV therapy, a hybrid strategy combining expert techniques and Bayesian RL is suggested. This strategy accounts for the variety of patient data and adapts to various patient circumstances [72]. Meanwhile, another method utilizes a policy gradient method to improve HIV management ATAs by considering causal factors between anti-HIV medication choices and treatment outcomes.

The study explores three distinct challenges in adaptive medical treatment using RL [69]. It addresses problems with designing electrical stimulation techniques for Parkinson's disease deals with the optimization of electrical stimulation for epilepsy treatment and creates a different model for the response of the cell population to radiation therapy for cancer by formulating fractionation scheduling as an MDP. Additionally, Song *et al.* introduce a novel approach known as penalized Q-learning for optimizing ATAs, evaluating its performance in a clinical trial study focused on depression treatment [73]. These diverse RL methods continue to advance adaptive medical treatment across various healthcare domains.

9.5 Diagnostic Tools in Healthcare and Biomedical Research

Biomarkers are measurable indicators that are used to assess various biological or physiological processes, disease states, or responses to treatments. Biomarkers play a crucial role in disease diagnosis, prognosis, and monitoring, as well as in guiding treatment decisions. They are used across various medical fields, including oncology, cardiology, neurology, and more, to improve patient care and outcomes.

This study acknowledges the significant impact of gut microbiota on obesity but emphasizes the challenge of finding consistent biomarkers across diverse regions and delivering personalized interventions for obesity management. To address this, the study introduces a novel ML framework applied to human stool metagenomes from five different nations. It starts with a comprehensive ensemble feature selection approach to identify the most suitable biomarkers. The gut microbiota is then fine-tuned using deep reinforcement learning to achieve a healthy state. The study highlights the effectiveness of the DDPG algorithm in optimizing biomarker abundance for this purpose [74].

Given its ability to identify alterations in brain activity linked to the severe cognitive and emotional processing impairments that plague this patient population, functional imaging paradigms offer enormous potential as biomarkers for schizophrenia research. To pinpoint the most promising functional imaging biomarkers for long-term memory research, the Cognitive Neuroscience Treatment Research to Enhance Cognition in Schizophrenia (CNTRICS) initiative chose relational encoding and retrieval, item encoding and retrieval, and reinforcement learning as its key long-term memory builders [75].

Maier *et al.* examine the use of model-informed precision dosing and Bayesian data assimilation (DA) in the healthcare industry. RL has been utilized extensively in the healthcare industry but has mainly concentrated on clinical trial design as opposed to the most effective dosage in pharmacokinetics and pharmacodynamics. Model-based RL minimizes the requirement for real-time computing during therapy by teaching optimal actions through simulations in unpredictable contexts. Dosing guided BY RL optimizes dosage decisions by utilizing sophisticated learning techniques, especially the tree-specific upper confidence bound and Monte Carlo tree search. Dosing guided by DA-RL utilizes patient-specific therapeutic drug monitoring data to combine RL and DA to produce a flexible and understandable dosage strategy [76].

Alzheimer's disease is a global health challenge, and there is a need for improved population-based strategies for its detection and management. Current retinal imaging techniques show potential for non-invasive Alzheimer's disease screening by assessing changes in retinal structures linked to brain degeneration. Moreover, the progress in DL and the emergence of DRL suggest a potential synergy with retinal imaging for automated Alzheimer's disease prediction. This review explores DRL's applications in retinal imaging for Alzheimer's studies, discussing its capability for its detection and prognosis. It also addresses challenges and opportunities, such as defining reward functions, standardizing imaging practices, and ensuring data availability for clinical integration [77]. Biomarkers can also include behavioral characteristics of RL agents, such as response times, action sequences, or preferences for certain actions under specific conditions.

9.6 Scope of Deep Reinforcement Learning in Healthcare and Biomedical Applications

The preceding section has provided a comprehensive overview of the historical developments and ongoing advancements in the utilization of RL and DRL within the healthcare and biomedicine domain. It has showcased the diverse range of healthcare applications where RL and DRL have been employed, signifying notable achievements. However, many of these studies have primarily employed existing RL and DRL methods without addressing certain critical challenges, resulting in common limitations. This section delves into the unaddressed challenges within current research, spanning various aspects of RL and DRL, state, action, and reward formulation; learning from global models; and policy evaluation. Furthermore, it addresses theoretical problems that are basic to classic RL and DRL, namely, the credit assignment issue and the exploration–exploitation tradeoff. By addressing these challenges, the potential for further enhancements in the application of RL and DRL to healthcare can be unlocked, promising more effective and robust solutions.

9.6.1 State and Action Space

In the first phase of applying RL and DRL to healthcare, medical data must be gathered, preprocessed, and summarized into manageable state representations that incorporate pertinent confounding factors influencing treatment choices and results. Defining appropriate state representations is complex, including determining the level of detail, trajectory duration, and essential confounders. Deep neural networks and other function approximations that can input the medical data in a raw format, which is frequently used in studies, may not be able to accurately represent changing variables or undetected confounders. Complex representation techniques like causal probabilistic inference networks, which can handle noisy, missing, or censored data and model causal linkages, are required to meet these issues.

Additionally, formulating actions is crucial. While most research focuses on discretizing action spaces, many situations in healthcare require continuous or multidimensional actions. Transitioning to continuous dosing is essential for precision medicine. Progress has been made in achieving continuous control with methods like actor–critic and policy search, especially from the fields of robotic control and DRL. However, selecting actions in large or infinite spaces remains challenging, particularly when considering sample complexity. The capacity to apply RL and DRL to sample- intensive issues in healthcare would be improved by creating effective methods for action selection in high-dimensional environments while managing exploration complexity.

Protein-related tasks often involve high-dimensional state spaces, where each dimension corresponds to specific protein features. Effectively representing these complex states is challenging, as it can overwhelm learning algorithms. Biological data used for protein structures and sequences may be sparse, noisy, or incomplete, hindering accurate state representation due to missing data and measurement errors. Proteins are dynamic, changing in response to factors like temperature, pH, and ligand binding. Capturing this dynamism in state representation is complex, requiring modeling of state transitions over time. Protein structures are multi-scale, ranging from atomic interactions to overall folding. Deciding the right level of detail in state representation is challenging to avoid computational inefficiency or loss of critical information. Designing proteins or predicting their structures often involves discrete decisions like selecting amino acids or modifying structural features. Defining meaningful actions can be intricate, as individual actions may interact in complex ways. Balancing multiple objectives, such as stability, binding affinity, and solubility, in action formulation is complex, requiring consideration of how to combine or prioritize these objectives. Exploration of different structural possibilities is vital, but balancing it with exploiting known strategies is challenging, especially with limited information about action consequences. Incorporating domain-specific biological knowledge into the action space can be difficult, given the need to encode complex biochemical principles and interactions. Evaluating the quality of protein designs or predicted structures based on actions can be computationally intensive and may rely on external simulations or experiments, making timely feedback provision to the learning agent challenging.

9.6.2 Reward

The use of RL in healthcare and bioinformatics relies heavily on numerical reward functions, but defining them precisely is challenging and can lead to subjective biases. These rewards involve thresholds and weights, impacting treatment strategies. To address this, an alternative approach is preference reinforcement learning (PRL), which uses qualitative evaluations, such as ranking functions. This method is more intuitive and can handle complex trade-offs in medical decisions. Additionally, multi-objective reinforcement learning (MORL) is employed due to the inherent conflicts in healthcare choices. However, MORL in clinical applications has mainly focused on static preferences and does not yet handle dynamic patient preferences well. Estimating reward functions directly from expert data, known as inverse reinforcement learning (IRL), is preferable, but it is complex due to the clinical data's uncertainty and complexity, leading to potential biases. Balancing short-term noisy feedback and long-term rewards is challenging. Ignoring short-term feedback can hinder learning critical relationships. A current problem in healthcare and bioinformatics applications is striking the correct balance between immediate gratification and long-term rewards.

In protein-related tasks like design, structure prediction, docking, and drug design, there isn't a single correct answer due to variations in intended function or context. This complexity makes defining a precise reward function challenging, given the potential for multiple valid solutions. These tasks often involve competing objectives, such as optimizing stability, binding affinity, and solubility simultaneously. Balancing these objectives within a single reward function is complex, as improvements in one aspect may trade off against others. Proteins have intricated folding patterns and dynamic behaviors, making it daunting to capture these intricacies in a reward function, especially given the vast conformational possibilities. Assessing the quality of protein designs or structures can be computationally expensive and sparse in terms of feedback. Sparse rewards can slow down the learning process. Subjectivity in evaluating protein quality poses a challenge, with varying expert opinions on what defines a good protein. This subjectivity hampers the creation of a universally accepted reward function. Limited and variable-quality data for training RL agents further complicates matters, as high-quality protein structures or designs may be scarce. Objectives in protein tasks can evolve, requiring adaptations in the reward function to reflect changing goals and insights. Inferring reward functions from expert knowledge or historical data can introduce bias, generalizing new proteins or contexts a non-trivial task. High-dimensional state spaces in protein tasks pose a challenge in defining effective reward functions and can suffer from the curse of dimensionality, hindering learning. To tackle these challenges, researchers use domain knowledge, data-driven methods, and specialized RL techniques. They may also explore alternative reward formulations like ranking-based rewards and multi-objective optimization approaches. Advances in deep reinforcement learning and generative models contribute to more effective solutions in protein related tasks.

9.6.3 Policy

The process of determining a policy's value also referred to as an off-policy evaluation problem, is essential in the healthcare industry. This is because assessing the performance of a policy directly on target populations through experiments is often infeasible due to high costs, treatment risks, or ethical concerns. As a result, before using learned policies in actual clinical situations, it is vital to predict how they could function on historical data. Although RL has produced several noteworthy sampling techniques and strategies for balancing bias and variance for the off-policy estimators, issues such as sparse rewards and significant policy differences between RL models and medical experts can make direct application of these estimators in healthcare settings potentially unreliable. Assessments of treatment policies may be inaccurate or even deceptive due to improper handling of state representations, differences in the weights assigned to sampling-based estimators, and confounding variables in ad hoc measurements. The accuracy of the behavior policy estimation from the data and the assurance that the action probabilities under the approximated behavior policy model appropriately reflect reality is crucial to the quality of off-policy evaluation.

In many practical applications, especially those in bioinformatics or healthcare, obtaining meaningful and frequent rewards to evaluate the performance of a policy can be challenging. This sparsity makes it difficult to provide timely and informative feedback to the learning agent. The state spaces in many applications are high-dimensional, containing many variables or features. Evaluating a policy's performance in such spaces requires efficient methods to explore and assess the quality of actions across a vast and complex state space. Environments can change over time, which means that a policy's effectiveness may vary as conditions evolve. This dynamic nature requires continuous evaluation and adaptation of policies to ensure their relevance and performance. Policy evaluation often involves the use of estimators, and striking the right balance between bias and variance in these estimators can be challenging. Reducing bias may increase variance and *vice versa*, making it difficult to obtain accurate evaluations. When evaluating a policy using data collected by another policy (off-policy evaluation), issues related to policy discrepancy and importance sampling arise. Ensuring that the behavior policy accurately represents the data distribution, and that the importance sampling process is reliable is a complex task.

9.6.4 Model Training

Compared to model-free methods, sample efficiency can increase the efficacy of model-based methods. As the learning process progresses, model-based methods dynamically learn transition and reward functions. The utilization of model-based RL approaches in the healthcare industry is currently understudied in the literature. Numerous model-based RL algorithms currently in use are intended for agents that operate in narrow domains with discrete state spaces. This contrasts with healthcare domains, which typically involve multidimensional, continuous states and actions. There are extra difficulties in modifying model-based techniques to operate effectively in such continuous, large-scale environments. A more intricate issue is the development of exploration strategies tailored for continuous action and state spaces. By creating a finite representation of the system, model learning operations within healthcare systems can be substantially more efficient than existing RL algorithms, enabling intelligent exploration and efficient planning.

9.6.5 Exploration

The act of exploring is a crucial part of RL. The ϵ-greedy technique and many other heuristic-based exploration strategies are used in numerous real-time learning (RL) applications that are currently being used in the healthcare industry. These techniques have shown some success, but they are typically insufficient when dealing with complex dynamics and wide state or action spaces in medical circumstances. They may lead to either large sample complexity or performance that is much below ideal. When only a portion of the state space is visible, this issue becomes more obvious. Naively exploring the entire space in such situations is highly inefficient. This issue is amplified in continuous state and action spaces, as observed in cases like HIV treatment, where the healthy state's basin of attraction is notably smaller compared to the unhealthy state. Traditional exploration methods have shown limitations in achieving performance improvements or generating meaningful treatment strategies, even after prolonged exploration across the entire space. To explore effectively in high-dimensional settings, exploration tactics that can change as learners do or that take advantage of performance measures are therefore urgently needed. Recent years have seen the emergence of several advanced exploration strategies, such as concurrent exploration mechanisms, approaches that ensure exploration complexity in continuous landscapes, and deep reinforcement learning-specific exploration strategies. Such sophisticated investigation techniques must be applied in more difficult medical conditions. This not only reduces sample complexity but also holds the potential to uncover novel and previously undiscovered problem-solving strategies.

9.6.6 Credit Assignment

In reinforcement learning (RL), the credit assignment problem is crucial for determining which actions are responsible for the learning outcome. This challenge is especially pronounced in healthcare due to the varied and sometimes delayed effects of medical treatments. Traditional RL approaches use heuristics like eligibility traces (backward view) or discount factors (forward view) to address this problem. However, these simplified heuristics often struggle to model complex interactions in healthcare. For example, it may be challenging to link certain behaviors such as taking insulin after lunch or working out first thing in the morning to the management of blood sugar levels in individuals with type 1 diabetes, as both may have an impact on blood sugar and lead to hypo-glycemia at a later time. This challenge calls for the modeling of healthcare's time-varying causal links and their incorporation into the learning process. The extensive literature on causal explanation and inference offers a valuable tool for improving causal reasoning within RL algorithms [78]. It can improve learning performance and produce better interpretable and explicable learning strategies when causal hypotheses are generated to explain or predict observed events.

9.7 Conclusions

DRL holds immense promise when it comes to biology and medicine, offering innovative solutions to complex problems. Its capacity to learn optimal strategies from data makes it a valuable tool in various applications, including drug discovery, protein folding, personalized treatment plans, and disease diagnosis. DRL's ability to handle high-dimensional, noisy, and complex biological data allows researchers and clinicians to extract meaningful insights. In precision medicine, it adapts treatments to individual patients, optimizing therapeutic outcomes. Moreover, DRL accelerates drug discovery by expediting the search for compounds with therapeutic potential. While DRL presents robust mathematical foundations for optimal decision-making, applying these solutions directly to healthcare is intricate due to the complexity of medical data. Clinical decision-making has unique features that demand tailored DRL methods to address real healthcare challenges. Challenges persist, including the need for robust reward formulation, complex state and action representations, and ethical considerations in healthcare applications. Ensuring the safety and interpretability of DRL-based models remains a priority. Several underexplored perspectives in the current methodology deserve heightened focus: developing more interpretable

DRL models, enhancing transfer learning capabilities, addressing challenges related to limited data, and Adapting existing DRL methods for omnipresent data in the age of ambient intelligence healthcare systems while assuring in-vivo applications' security, reliability, and effectiveness. The use of DRL in biomedicine and healthcare lies at the nexus of computer science and medicine. Despite receiving comparatively less attention than other research paradigms, including deep learning and conventional machine learning, DRL is gaining traction due to advances in theories and practical needs. DRL holds significant promise in revolutionizing biology and medicine. Despite existing challenges, ongoing advancements in algorithms and computational resources are driving DRL's continued contributions to these critical fields, ultimately improving human health and well-being. The research community will benefit from this survey's systematic grasp of the underpinnings, approaches, difficulties, and novel findings of this growing paradigm.

References

1. Sutton, R.S. and Barto, A.G., Reinforcement Learning : An Introduction, Cambridge, London, The Mit Press, 2018.
2. Poole, D.L. and Mackworth, A.K., Artificial Intelligence: Foundations of Computational Agents, Cambridge, United Kingdom, Cambridge University Press, 2010.
3. Busoniu, L., Babuska, R., De Schutter, B., Ernst, D., Reinforcement Learning and Dynamic Programming Using Function Approximators, Boca Raton, FL, USA, Crc Press, 2017.
4. Sutton, R.S., Maei, H.R., Precup, D., Bhatnagar, S., Silver, D., Szepesv´ari, C., Wiewiora, E., Fast gradient-descent methods for temporal-difference learning with linear function approximation, in: *Proceedings of the 26th annual international conference on machine learning*, pp. 993–1000, 2009.
5. Schaul, T., Horgan, D., Gregor, K., Silver, D., Universal value function approximators, in: *International conference on machine learning*, PMLR, pp. 1312–1320, 2015.
6. Woergoetter, F., and Porr, B., Reinforcement Learning, *Scholarpedia*, 3, 3, 1448, 2008, [online], Available: https://doi.org/10.4249/scholarpedia.1448.
7. Nguyen, A., Yosinski, J., Clune, J., Deep neural networks are easily fooled: High confidence predictions for unrecognizable images, in: *Proceedings of the IEEE conference on computer vision and pattern recognition*, pp. 427–436, 2015.
8. Li, Y., Deep reinforcement learning: An overview, arXiv preprint, arXiv:1701.07274, 2017, [Online], Available: https://doi.org/10.48550/arXiv.1701.07274.
9. Mnih, V., Kavukcuoglu, K., Silver, D., Rusu, A.A., Veness, J., Bellemare, M.G., Graves, A., Riedmiller, M., Fidjeland, A.K., Ostrovski, G. *et al.*, Human-level control through deep reinforcement learning. *Nature*, 518, 7540, 529–533, 2015.
10. Van Hasselt, H., Guez, A., Silver, D., Deep reinforcement learning with double q-learning, in: *Proceedings of the AAAI conference on artificial intelligence*, vol. 30, no. 1, 2016.
11. Hasselt, H., Double q-learning, in: *Advances in neural information processing systems*, vol. 23, 2010.
12. Cooper, S., Khatib, F., Treuille, A., Barbero, J., Lee, J., Beenen, M., Leaver- Fay, A., Baker, D., Popovi´c, Z., Players, F., Predicting protein structures with a multiplayer online game. *Nature*, 466, 756–760, 08 2010, [Online], Available: https://www.nature.com/articles/nature09304.
13. Panou, D.N. and Reczko, M., Deepfoldit–a deep reinforcement learning neural network folding proteins, arXiv preprint, arXiv:2011.03442, 2020, [Online], Available: https://doi.org/10.48550/arXiv.2011.03442.

14. Soltanikazemi, E., Roy, R., Quadir, F., Giri, N., Morehead, A., Cheng, J., Drlcomplex: Reconstruction of protein quaternary structures using deep reinforcement learning, arXiv preprint, arXiv:2205.13594, 05 2022, [Online], Available: https://doi.org/10.48550/arXiv.2205.13594
15. Yan, Y. and Huang, S.-Y., Accurate prediction of inter-protein residue–residue contacts for homo-oligomeric protein complexes. *Briefings Bioinf.*, 22, 5, bbab038, 03 2021, [Online], Available: https://doi.org/10.1093/bib/bbab038.
16. Soltanikazemi, E., Quadir, F., Roy, R.S., Guo, Z., Cheng, J., Distance-based reconstruction of protein quaternary structures from inter-chain contacts. *Proteins: Struct. Funct. Bioinf.*, 90, 3, 720–731, 2022.
17. Aderinwale, T., Christoffer, C., Kihara, D., Rl-mlzerd: Multimeric protein docking using reinforcement learning. *Front. Mol. Biosci.*, 9, 969394, 2022.
18. Inbar, Y., Benyamini, H., Nussinov, R., Wolfson, H.J., Prediction of multimolecular assemblies by multiple docking. *J. Mol. Biol.*, 349, 2, 435–447, 2005.
19. Esquivel-Rodríguez, J., Yang, Y.D., Kihara, D., Multi-lzerd: multiple protein docking for asymmetric complexes. *Proteins: Struct. Funct. Bioinf.*, 80, 7, 1818–1833, 2012.
20. Mirdita, M., Schütze, K., Moriwaki, Y., Heo, L., Ovchinnikov, S., Steinegger, M., Colabfold: making protein folding accessible to all. *Nat. Methods*, 19, 6, 679–682, 2022.
21. Zhou, H. and Skolnick, J., Goap: a generalized orientation-dependent, all-atom statistical potential for protein structure prediction. *Biophys. J.*, 101, 8, 2043–2052, 2011.
22. Zhou, H. and Zhou, Y., Distance-scaled, finite ideal-gas reference state improves structure-derived potentials of mean force for structure selection and stability prediction. *Protein Sci.*, 11, 11, 2714–2726, 2002.
23. Huang, S.-Y. and Zou, X., Statistical mechanics-based method to extract atomic distance-dependent potentials from protein structures. *Proteins: Struct. Funct. Bioinf.*, 79, 9, 2648–2661, 2011.
24. Jose, J., Gupta, K., Alam, U., Jatana, N., Arora, P., Reinforcement learning based approach for ligand pose prediction, bioRxiv preprint, bioRxiv: 2021.04.10.438538, 2021, [Online], Available: https://doi.org/10.1101/2021.04.10.438538.
25. Zhang, Y., Chen, Y., Wang, C., Lo, C.-C., Liu, X., Wu, W., Zhang, J., Prodconn: Protein design using a convolutional neural network. *Proteins*, 88, 819–829, 07 2020, [Online], Available: https://pubmed.ncbi.nlm.nih.gov/31867753/.
26. Wang, C., Chen, Y., Zhang, Y., Li, K., Lin, M., Pan, F., Wu, W. Zhang, J., A Reinforcement Learning Approach for Protein–Ligand Binding Pose Prediction, *BMC Bioinformatics*, 23, 1, 368, 8, 09 2022, [Online], Available, https://doi.org/10.1186/s12859-022-04912-7.
27. Hu, L., Benson, M.L., Smith, R.D., Lerner, M.G., Carlson, H.A., Binding moad (mother of all databases). *Proteins: Struct. Funct. Bioinf.*, 60, 3, 333–340, 2005.
28. Nissink, J.W.M., Murray, C., Hartshorn, M., Verdonk, M.L., Cole, J.C., Taylor, R., A new test set for validating predictions of protein–ligand interaction. *Proteins: Struct. Funct. Bioinf.*, 49, 4, 457–471, 2002.
29. Jose, J., Alam, U., Singh, D., Jatana, N., Arora, P., PandoraRL: DQN and Graph Convolution based ligand pose learning for SARS-COV1 Mprotease, *IEEE International Conference on Bioinformatics and Biomedicine*, IEEE, 380–385, 2022.
30. Chen, Z., Zhang, B., Guo, H., Emani, P., Clancy, T., Jiang, C., Gerstein, M., Ning, X., Cheng, C., Min, M.R., Binding peptide generation for MHC Class I proteins with deep reinforcement learning. *Bioinformatics*, 39, 2, btad055, 01 2023, [Online], Available: https://doi.org/10.1093/bioinformatics/btad055.

31. O'Donnell, T.J., Rubinsteyn, A., Laserson, U., Mhcflurry 2.0: improved pan-allele prediction of mhc class i-presented peptides by incorporating antigen processing. *Cell Syst.*, 11, 1, 42–48, 2020.
32. Mitra, P., Shultis, D., Zhang, Y., Evodesign: de novo protein design based on structural and evolutionary profiles. *Nucleic Acids Res.*, 41, W1, W273–W280, 2013.
33. Mitra, P., Shultis, D., Brender, J.R., Czajka, J., Marsh, D., Gray, F., Cierpicki, T., Zhang, Y., An evolution-based approach to de novo protein design and case study on mycobacterium tuberculosis. *PLoS Comput. Biol.*, 9, 10, e1003298, 2013.
34. Banerjee, A., Levy, Y., Mitra, P., Analyzing change in protein stability associated with single point deletions in a newly defined protein structure database. *J. Proteome Res.*, 18, 3, 1402–1410, 2019.
35. Banerjee, A. and Mitra, P., Estimating the effect of single-point mutations on protein thermodynamic stability and analyzing the mutation landscape of the p53 protein. *J. Chem. Inf. Model.*, 60, 6, 3315–3323, 2020.
36. Shultis, D., Mitra, P., Huang, X., Johnson, J., Khattak, N.A., Gray, F., Piper, C., Czajka, J., Hansen, L., Wan, B. *et al.*, Changing the apoptosis pathway through evolutionary protein design. *J. Mol. Biol.*, 431, 4, 825–841, 2019.
37. Banerjee, A., Kumar, A., Ghosh, K.K., Mitra, P., Estimating change in foldability due to multipoint deletions in protein structures. *J. Chem. Inf. Model.*, 60, 12, 6679–6690, 2020.
38. Wang, Y., Tang, H., Huang, L., Pan, L., Yang, L., Yang, H., Mu, F., Yang, M., Self-play reinforcement learning guides protein engineering. *Nat. Mach. Intell.*, 5, 845–860, 08 2023, [Online], Available: https://www.nature.com/articles/s42256-023-00691-9.
39. Angermueller, C., Dohan, D., Belanger, D., Deshpande, R., Murphy, K., Colwell, L., Model-based reinforcement learning for biological sequence design, in: *International conference on learning representations*, 2019.
40. Schulman, J., Wolski, F., Dhariwal, P., Radford, A., Klimov, O., Proximal policy optimization algorithms, arXiv preprint, arXiv:1707.06347, 2017, [Online], Available: https://doi.org/10.48550/arXiv.1707.06347.
41. Goodfellow, I. J., Mirza, M., Xiao, D., Courville, A., Bengio, Y., An empirical investigation of catastrophic forgetting in gradient-based neural networks, arXiv preprint, arXiv:1312.6211, 2013, [Online], Available: https://doi.org/10.48550/arXiv.1312.6211.
42. Banerjee, A., Pal, A., Pal, D., Mitra, P., Ebolavirus interferon antagonists-protein interaction perspectives to combat pathogenesis. *Briefings Funct. Genomics*, 17, 6, 392–401, 2018.
43. Popova, M., Isayev, O., Tropsha, A., Deep reinforcement learning for de novo drug design. *Sci. Adv.*, 4, 7, eaap7885, 2018.
44. Chuang, L.-Y., Tsai, J.-H., Yang, C.-H., Operon prediction using particle swarm opti- mization and reinforcement learning, in: *2010 International Conference on Technologies and Applications of Artificial Intelligence*, IEEE, pp. 366–372, 2010.
45. Ralha, C.G., Schneider, H.W., Walter, M.E.M.T., Bazzan, A.L., Reinforcement learning method for BioAgents, Eleventh Brazilian Symposium on Neural Networks, *IEEE Xplore*, 109-114, 10 2010, [Online], Available: https://ieeexplore.ieee.org/abstract/document/5715222.
46. Bocicor, M.-I., Czibula, G., Czibula, I.-G., A reinforcement learning approach for solving the fragment assembly problem, in: *2011 13th International Symposium on Symbolic and Numeric Algorithms for Scientific Computing*, pp. 191–198, 2011.
47. Zhu, F., Liu, Q., Zhang, X., Shen, B., Protein-protein interaction network constructing based on text mining and reinforcement learning with application to prostate cancer, in: *2014 IEEE International Conference on Bioinformatics and Biomedicine (BIBM)*, pp. 46–51, 2014.
48. Maiti, S. and Mitra, P., Boosting phosphorylation site prediction with sequence feature-based machine learning. *Proteins: Struct. Funct. Bioinf.*, 88, 2, 284–291, 2020.

49. Sahba, F., Tizhoosh, H.R., Salama, M.M., Application of reinforcement learning for segmentation of transrectal ultrasound images. *BMC Med. Imaging*, 8, 1–10, 2008.
50. Coronato, A., Naeem, M., De Pietro, G., Paragliola, G., Reinforce- ment learning for intelligent healthcare applications: A survey. *Artif. Intell. Med.*, 109, 101964, 2020, [Online], Available: https://www.sciencedirect.com/science/article/pii/S093336572031229X.
51. DiGiovanna, J., Mahmoudi, B., Fortes, J., Principe, J.C., Sanchez, J.C., Coadaptive brain-machine interface via reinforcement learning. *IEEE Trans. Biomed. Eng.*, 56, 1, 54–64, 2008.
52. Mahmoudi, B. and Sanchez, J.C., A symbiotic brain-machine interface through value-based decision making. *PLoS One*, 6, 3, e14760, 2011.
53. Sanchez, J.C., Tarigoppula, A., Choi, J.S., Marsh, B.T., Chhatbar, P.Y., Mahmoudi, B., Francis, J.T., Control of a center-out reaching task using a reinforcement learning brain- machine interface, in: *2011 5th International IEEE/EMBS Conference on Neural Engineering*, IEEE, pp. 525–528, 2011.
54. Mahmoudi, B., Pohlmeyer, E.A., Prins, N.W., Geng, S., Sanchez, J.C., Towards autonomous neuroprosthetic control using hebbian reinforcement learning. *J. Neural Eng.*, 10, 6, 066005, 2013.
55. Wang, F., Xu, K., Zhang, Q.S., Wang, Y.W., Zheng, X.X., A multi-step neural control for motor brain-machine interface by reinforcement learning. *Appl. Mech. Mater.*, 461, 565–569, 2014.
56. Pohlmeyer, E.A., Mahmoudi, B., Geng, S., Prins, N.W., Sanchez, J.C., Using reinforcement learning to provide stable brain-machine interface control despite neural input reorganization. *PLoS One*, 9, 1, e87253, 2014.
57. Wang, F., Wang, Y., Xu, K., Li, H., Liao, Y., Zhang, Q., Zhang, S., Zheng, X., Principe, J.C., Quantized attention-gated kernel reinforcement learning for brain–machine interface decoding. *IEEE Trans. Neural Networks Learn. Syst.*, 28, 4, 873–886, 2015.
58. Lampe, T., Fiederer, L.D., Voelker, M., Knorr, A., Riedmiller, M., Ball, T., A brain-computer interface for high-level remote control of an autonomous, reinforcement-learning-based robotic system for reaching and grasping, in: *Proceedings of the 19th international conference on Intelligent User Interfaces*, pp. 83–88, 2014.
59. Bauer, R. and Gharabaghi, A., Reinforcement learning for adaptive threshold control of restorative brain-computer interfaces: a bayesian simulation. *Front. Neurosci.*, 9, 36, 2015.
60. Murphy, S.A., An experimental design for the development of adaptive treatment strategies. *Stat. Med.*, 24, 10, 1455–1481, 2005.
61. Guez, A., Vincent, R.D., Avoli, M., Pineau, J., Adaptive treatment of epilepsy via batch- mode reinforcement learning, in: *AAAI*, vol. 8, pp. 1671–1678, 2008.
62. Shortreed, S.M., Laber, E., Lizotte, D.J., Stroup, T.S., Pineau, J., Murphy, S.A., Informing sequential clinical decision-making through reinforcement learning: an empirical study. *Mach. Learn.*, 84, 109–136, 2011.
63. Ertefaie, A., Shortreed, S., Chakraborty, B., Q-learning residual analysis: application to the effectiveness of sequences of antipsychotic medications for patients with schizophrenia. *Stat. Med.*, 35, 13, 2221–2234, 2016.
64. Komorowski, M., Celi, L.A., Badawi, O., Gordon, A.C., Faisal, A.A., The artificial in- telligence clinician learns optimal treatment strategies for sepsis in intensive care. *Nat. Med.*, 24, 11, 1716–1720, 2018.
65. Raghu, A., Komorowski, M., Ahmed, I., Celi, L., Szolovits, P., Ghassemi, M., Deep reinforcement learning for sepsis treatment, arXiv preprint, arXiv:1711.09602, 2017, [Online], Available: https://doi.org/10.48550/arXiv.1711.09602.
66. Zhao, Y., Kosorok, M.R., Zeng, D., Reinforcement learning design for cancer clinical trials. *Stat. Med.*, 28, 26, 3294–3315, 2009.

67. Liu, Y., Logan, B., Liu, N., Xu, Z., Tang, J., Wang, Y., Deep reinforcement learning for dynamic treatment regimes on medical registry data, in: *2017 IEEE International Conference on Healthcare Informatics (ICHI)*, IEEE, pp. 380–385, 2017.
68. Jalalimanesh, A., Haghighi, H.S., Ahmadi, A., Soltani, M., Simulation-based optimization of radiotherapy: Agent-based modeling and reinforcement learning. *Math. Comput. Simul.*, 133, 235–248, 2017.
69. Vincent, R., and Pineau, J., Reinforcement Learning in Models of Adaptive Medical Treatment Strategies, McGill University, Montreal, Quebec, Canada, 2014.
70. Liu, Y., Wang, Y., Kosorok, M.R., Zhao, Y., Zeng, D., Robust hybrid learning for estimating personalized dynamic treatment regimens, arXiv preprint, arXiv:1611.02314, 2016, [Online], Available: https://doi.org/10.48550/arXiv.1611.02314.
71. Jaimes, L.G., Llofriu, M., Raij, A., Calma, an algorithm framework for mobile just in time interventions, in: *SoutheastCon 2015*, IEEE, pp. 1–5, 2015.
72. Parbhoo, S., Bogojeska, J., Zazzi, M., Roth, V., Doshi-Velez, F., Combining kernel and model based learning for hiv therapy selection. *AMIA Summits on Translational Science Proceedings*, vol. 2017, p. 239, 2017.
73. Song, R., Wang, W., Zeng, D., Kosorok, M.R., Penalized q-learning for dynamic treatment regimens. *Stat. Sin.*, 25, 3, 901, 2015.
74. Liu, Y., Zhu, J., Wang, H., Lu, W., Lee, Y.K., Zhao, J., Zhang, H., Machine learning framework for gut microbiome biomarkers discovery and modulation analysis in large-scale obese population. *BMC Genomics*, 23, 1, 1–14, 2022.
75. Ragland, J.D., Cohen, N.J., Cools, R., Frank, M.J., Hannula, D.E., Ranganath, C., CNTRICS Imaging Biomarkers Final Task Selection: Long-Term Memory and Reinforcement Learning. *Schizophr. Bull.*, 38, 1, 62–72, 11 2011, [Online], Available: https://doi.org/10.1093/schbul/sbr168.
76. Maier, C., Hartung, N., Kloft, C., Huisinga, W., de Wiljes, J., Reinforcement learning and bayesian data assimilation for model-informed precision dosing in oncology. *CPT Pharmacometrics Syst. Pharmacol.*, 10, 3, 241–254, 2021.
77. Hui, H.Y., Ran, A.R., Dai, J.J., Cheung, C. Y., Deep Reinforcement Learning-Based Retinal Imaging in Alzheimer's Disease: Potential and Perspectives, *Journal of Alzheimer's Disease: JAD*, 94, 1, 39–50, 2023, [Online], Available: pubmed.ncbi.nlm.nih.gov/37212112/, https://doi.org/10.3233/JAD-230055.
78. Greenland, S., Causal analysis in the health sciences. *J. Am. Stat. Assoc.*, 95, 449, 286–289, 2000.

10

Deep Reinforcement Learning in Robotics and Autonomous Systems

Uma Yadav[1]*, Shweta V. Bondre[2] and Bhakti Thakre[3]

[1]*Computer Science and Engineering Department, Shri Ramdeobaba College of Engineering and Management, Nagpur, India*
[2]*Information Technology Department, Shri Ramdeobaba College of Engineering and Management, Nagpur, India*
[3]*Cyber Security Department, St. Vincent Pallotti College of Engineering and Technology, Nagpur, India*

Abstract

Deep reinforcement learning (DRL) has begun as a powerful paradigm for enabling intelligent behavior in robotics and autonomous systems. A potential method for autonomously learning complicated behaviors from cursory sensor data is deep reinforcement learning (RL). Deep RL has shown potential in empowering somatic robots to acquire difficult abilities in the actual world despite a substantial chunk of research focusing on solicitations in audiovisual games and virtual control, which do not link with the restrictions of erudition in existent situations. Actual-world robotics, which is closely related to how individuals learn as personified agents in the actual world—provides an interesting arena for assessing such algorithms. Numerous difficulties arise while learning to see and move in the actual environment; some of these difficulties are simpler to solve than others, and some of these difficulties are frequently overlooked in RL research that solely considers virtual domains. In this chapter, we give many case revisions employing mechanical deep RL. We examine often-recognized deep RL issues and how these issues have been addressed in these studies, building on these case examples. We also give a brief summary of other problems that still need to be solved, numerous of which are specific to the real-world computing environment and are not frequently the subject of conventional RL research. This chapter provides a widespread outline of the solicitation of DRL techniques to address the challenges and complexities inherent in robotic tasks. The chapter also highlights the importance of transfer learning, sim-to-real transfer, and safety considerations in deploying DRL agents in real-world environments. Finally, we outline future directions and the ongoing challenges that researchers and practitioners face in harnessing the potential of DRL for shaping the future of robotics and autonomous systems.

***Keywords*:** Deep reinforcement learning, robotics, autonomous systems, neural networks, reinforcement learning, deep Q-networks, transfer learning, sim-to-real transfer

**Corresponding author*: uma.yadav12@gmail.com

Shubham Mahajan, Pethuru Raj and Amit Kant Pandit (eds.) Deep Reinforcement Learning and Its Industrial Use Cases: AI for Real-World Applications, (207–238) © 2024 Scrivener Publishing LLC

10.1 Introduction

Machine learning and robotics come together in robotic learning. Robotics offers a window into the learning constraints that people and animals face, revealing aspects of intelligence that might not otherwise be apparent to study when we confine ourselves to virtual environments. From the perspective of a machine learning researcher interested in studying intelligence, robotics is an appealing medium to study—for instance, as a result of their activities, robots acquire streams of unprocessed sensory data, and they are virtually unable to obtain extensive quantities of detailed supervision beyond monitoring these sensor inputs. This creates a difficult yet incredibly realistic learning situation. Furthermore, unlike the agents in video games, robots must create their own internal representation of objective success as they do not automatically get a score or reward function that is tailored to their needs.

Robots also lack the ability to easily obtain a groove or reward gathering that is tailored to their requirements, unlike the agents in video games, and must instead create their own inner illustration of objective growth. Using learning-based approaches is intriguing from the standpoint of robotics research since it can allow robots to progress near less controlled settings, handle unfamiliar items, and develop a state representation suited for several jobs. Although it is an intriguing intermediate, there is a big obstacle for a machine learning investigator to move into robotics, and the opposite is also true beyond the robot's price. There are several design considerations when deciding how to configure the procedure and the robot—for instance, RL procedures call for learning from skill that the robot independently gathers itself, providing a variety of options for how to start the learning process, how to stop risky behavior, and how to define the objective or reward. Similar to RL algorithms, machine learning algorithms offer a lot of critical design options and hyperparameters that might be challenging to choose.

RL has become one of the most widely recognized research fields in the field of machine learning since it successfully addresses a number of problems and challenges related to artificial intelligence. It has led to a wide range of notable developments in a number of domains, such as manufacturing, board games, robot control, and autonomous driving. Industry, agriculture, the service industry, healthcare, and aerospace all heavily rely on robotics, which has become one of the most active research fields in the world. Researchers are quite interested in studies on RL algorithms for robotics. While numerous well-known research organizations and businesses like UC Berkeley, OpenAI, DeepMind, and Google Brain have made considerable progress in this area, there are still many obstacles to overcome. How to improve the future employment of an energetic system with historical data is being investigated by RL and optimization control theory. Designing systems that engage in planning, control, and highly structured perception that effectively adapt to changing environmental conditions is the aim. Designing a controller to maximize the system's performance in some indicators is known as optimal control. Value functions and dynamic programming (DP) are frequently used to solve optimum control problems. On the foundation of Hamilton and Jacobi's philosophy, Bellman and colleagues prolonged certain explanations and provided some explanations with a dynamic scheme state and worth function, also known as an optimum return function.

We come to the conclusion that all optimization issues may be categorized as RL problems grounded on the "Markov decision process (MDP)". The use of robots for things like mobility control, path planning, environmental awareness, and behavior decision-making was covered in various reviews from decades ago. Furthermore, great progress has been achieved employing genetic algorithms (GA), fuzzy control algorithms, particle swarm optimization (PSO), 22 neural networks (NNs), ant colony optimization (ACO), and virtual annealing techniques to solve difficult issues across robotic domains. The research described above mostly concentrate on performance enhancement, sample effectiveness, and robot manipulation (such as grasping, handling, and route planning). These techniques frequently become stuck in local optimums and are challenging to converge. Robot vision is experiencing a storm as a result of the emergence of deep learning (DL). It supports the quick advancement of robots in the recognition of interior and outdoor scene, family

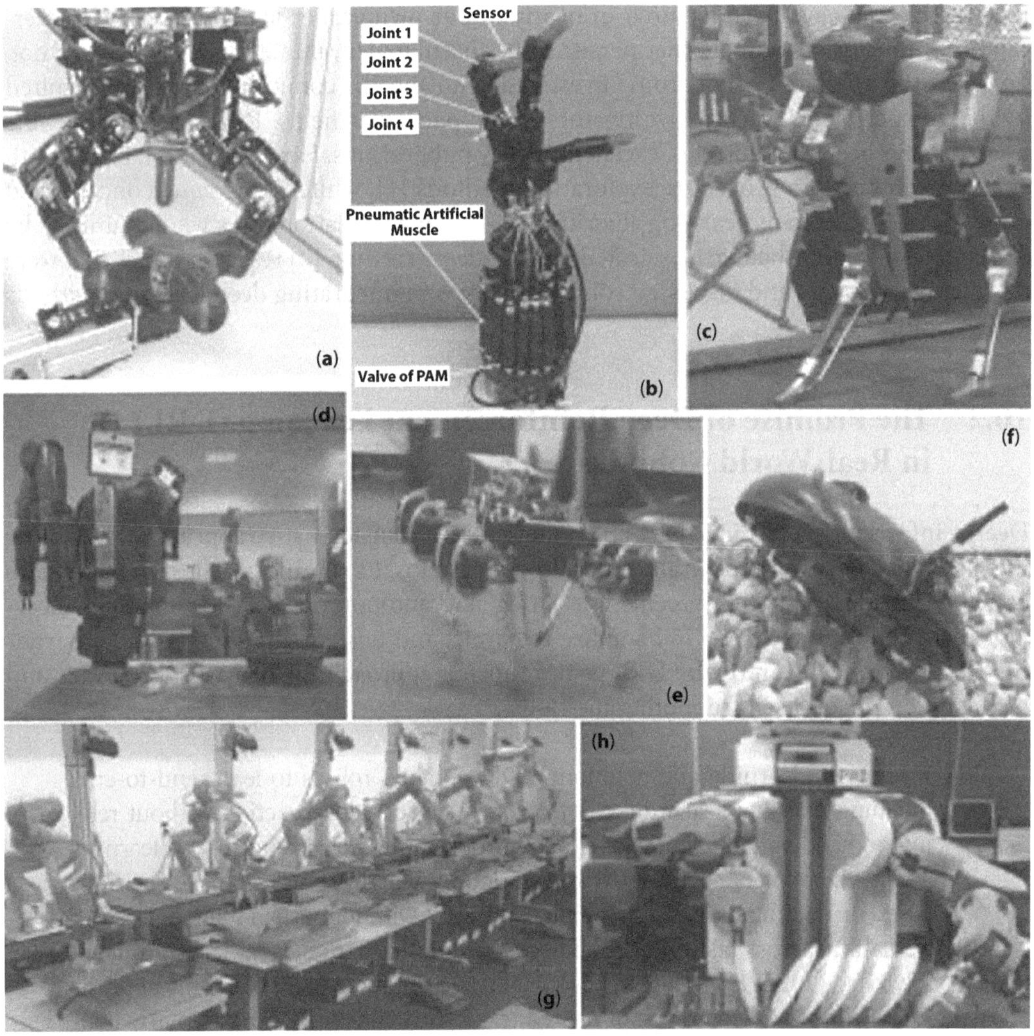

Figure 10.1 Several typical robot investigations established on RL. (a) Rotating a cross-shaped valve with multi-fingered hands [5], (b) shadow dexterous hand [6], (c) Cassie: walking on a treadmill [7], (d) Rethnik robotics baxter [8], (e) a quadruped robot [9], (f) VelociRoACH (a millirobot) [10], (g) seven robots simultaneously perform grasp training [11], and (h) PR2: learning to gently place a dish in a plate rack [12].

and industrial services, and multirobot collaboration. Although DL successfully addresses some issues with target identification, grasping posture, and robot cooperative learning, it is unable to make decisions and manage robots on its own. On the other hand, the benefits of RL in complex robot systems' online flexibility and self-learning capabilities have received a lot of attention. It interacts with the surroundings in a trial-and-error manner until it finds the best control approach. Numerous example robot studies based on RL exist (Figure 10.1). We are aware of no comprehensive study that discusses RL approaches explicitly for robot research, despite numerous replications of prior work and intriguing findings. Future research on robots to tackle general artificial intelligence will be greatly wedged by the merging of AI and RL.

In this editorial, we aim to deliver a high-level indication as to in what way deep RL can be advanced in the context of automation, summarize the traditions in which crucial RL experiments have been lectured in nearly of our own earlier work, and deliver a perception on major encounters that persist to be disentangled, numerous of which are not, however, the emphasis of vigorous investigation in the RL community. We are inspired by these experiments for the investigators in the particular fields. Robotics and machine learning have been the subject of excellent survey publications. For automation, Deisenroth *et al.* concentrated on procedure exploration methods [1], while Kober *et al.* concentrated on RL [2]. More recently, learning algorithms [3] for manipulation tasks were examined by Kroemer *et al.* Sunderhauf *et al.* selected current deep learning [4] research topics that were pertinent to robotics and discussed some difficulties in integrating deep learning methods with robotics.

10.2 The Promise of Deep Reinforcement Learning (DRL) in Real-World Robotics

Deep reinforcement learning (DRL) has emerged as a transformative approach in enabling intelligent behavior in robotics and autonomous systems. Its significance lies in its ability to combine deep learning, reinforcement learning, and autonomous decision-making, allowing robots to adapt and excel in complex, dynamic, and unstructured real-world environments. Here we delve into why DRL is a promising approach for real-world robotics and how it differentiates itself from other methods.

1) End-to-end learning and autonomy: DRL enables robots to learn end-to-end, meaning that they learn directly from raw sensor data to action without relying on handcrafted features or explicit programming. This end-to-end learning approach provides a pathway to true autonomy, where robots can make decisions based on data, adapting to changing conditions without human intervention.
2) Versatility across tasks: DRL's ability to generalize across tasks is a significant advantage. Once a DRL agent has learned in one task, it can potentially apply its knowledge to a wide range of related tasks. This versatility is particularly beneficial in real-world settings, where robots may encounter diverse challenges.
3) Adaptation to real-world dynamics: Unlike purely simulation-based methods, DRL facilitates adaptation to the complexities of the real world. Through reinforcement learning, robots can learn to make decisions in unpredictable,

dynamic, and unstructured environments. They adjust to unforeseen obstacles, variations in lighting, or changes in task requirements.

4) Data-driven learning: DRL leverages the power of data-driven learning. Robots learn from their interactions with the environment, making decisions based on collected data. This approach is highly effective in scenarios where data is abundant, enabling robots to continuously improve their performance.
5) Potential for transfer learning: DRL supports transfer learning, allowing robots to reuse knowledge gained in one context to expedite learning in another. This feature is advantageous in real-world applications, where the ability to apply previous knowledge can significantly reduce training time and resource requirements.
6) Human-level decision-making: DRL has demonstrated the capability to achieve human-level decision-making in a variety of tasks, from playing complex board games to controlling robotic systems. This potential for high-level cognition positions DRL as a frontrunner in developing intelligent robotic agents for real-world scenarios.
7) Comparison to other methods: While traditional control and planning methods have been used in robotics, they often rely on explicit models of the environment, which can be limiting in highly dynamic and unstructured real-world scenarios. DRL, in contrast, learns directly from data and adapts to unforeseen conditions, making it a more promising approach for applications where traditional methods fall short.

In essence, DRL's promise in real-world robotics is rooted in its capacity to learn and adapt autonomously from data, offering unparalleled versatility, adaptability, and potential for transfer learning. It distinguishes itself from other methods by embracing the challenges of real-world complexity and unstructured environments, making it a compelling choice for shaping the future of robotics and autonomous systems.

10.3 Preliminaries

Key Concepts and Terminology

RL is hypothesis-based and goal-oriented. Robots that interact with their surroundings based on RL can learn by making mistakes. Identifying the ideal course of action to maximize long-term advantages is the ultimate objective. Three categories of learning algorithms—RL, supervised learning, and unsupervised learning—are used in connectionist learning. The fact that the learning data are labeled is a hallmark of supervised learning. The model is known, meaning that, before learning, we have already informed it about the type of behavior that is appropriate in each state. In other words, we have a unique teacher who will lead it. It is typically applied to classification and regression issues. On the other hand, RL is used to investigate the properties of data rather than to learn without a label. While it does award a reward, it does not immediately judge whether a condition or action is good or harmful. Data are serialized, feedback is delayed, and there is a correlation between the data. The agent's actions will have an impact on the statistics that follow. An RL system consists of eight essential components, with the exclusion of the agent and environment: action At, state St, reward Rt, value function Vp(st), policy p, state transition probability matrix Pass 0, reward discount factor g, and exploration rate e. The agent takes behaviors based on the fundamental learning process of RL. After arriving in the state St of time t1 and obtaining the reward Rt at time t, according to the present strategy, at state St of time t. Sampling is used to get the observation sequence H, states, actions, and reward. Value function, which may be used to direct robot manipulation, yields the best course of action.

Markov Decision Process

A MDP is a mathematical structure used in the arena of RL and policymaking under indecision. It offers a prescribed method to prototypical and solve consecutive decision-making hitches where an agent interrelates with the situation to achieve a specific goal. Here are the key components and concepts associated with an MDP:

1. States (S): States represent the different situations or configurations that the agent and the environment can be in. These states encapsulate all relevant information about the system at a given time.
2. Actions (A): Actions are the choices or decisions obtainable to the agent. In each form, the agent can select the stroke from the set of possible actions.
3. Transitions (P): The transition function, denoted as P, defines the possibility of transitioning from one form to another when a particular stroke is taken. It is characterized as $P(st'|st, a)$, where st' is the subsequent state, st is the existing state, and a is the action.
4. Rewards (R): Rewards are numerical values that provide immediate feedback to the agent after taking an stroke in a specific state. The reward point, denoted as R, defines the expected reward for moving from one form to another by enchanting a precise action.
5. Policy (π): A policy is an approach that specifies which stroke to take in each form. It maps states to actions and determines the agent's performance in the environment. A policy can be stochastic ($\pi(a|st) = P(a|st)$) or deterministic ($\pi(st) = a$).
6. Discount factor (γ): The discount factor is a parameter that represents the agent's predilection for instantaneous rewards over upcoming rewards. It regulates the weight given to forthcoming rewards in the agent's policymaking process. A higher γ values long-term rewards more.
7. Return (G): The return, signified as G, is the cumulative summation of rewards obtained by the agent when following a particular policy over a sequence of states and actions. It is used as a measure of the policy's quality.
8. Value function (V): The value function, represented as V(st), estimates the anticipated accumulative reward an agent can attain preliminary from a specific state and succeeding a particular policy. It quantifies the desirability of states.

9. Q-value function (Q): The Q-value function, represented as Q(st, a), estimates the probable accumulative reward an agent can accomplish by enchanting a specific action in a specific state and then succeeding a particular policy.
10. Bellman equations: The Bellman equations are recursive associations that direct the value function or Q-value function in relations of future rewards and transitions. They are fundamental to solving MDPs.
11. Policy optimization: The process of discovery the optimum policy, which exploits the estimated return, often involves policy evaluation and policy improvement steps.

MDP serves as a fundamental outline for demonstrating and explaining sequential policymaking problems in various fields, including robotics, economics, game theory, and artificial intelligence. They provide a structured way to formalize decision problems under uncertainty and are a key component of reinforcement learning algorithms used to train intelligent agents.

Modern Reinforcement Learning Systems in Robotics

Numerous RL algorithms are used in robot research. The particular techniques employed in robot learning are determined by the creation of training data. Robots and their environments can interact to provide the information required for robot knowledge, or specialists can supply it. Then, by merging artificial intelligence technology with RL methodologies, a contemporary intellectual robot with independent decision-making and knowledge capabilities is investigated. As a outcome, the robotics associated RL methods of policy-based RL, value-based RL, DRL, model-based RL, inverse RL (IRL) and meta-RL are discussed in this part. Table 10.1 also provides an overview of the advantages and disadvantages of RL techniques.

Table 10.1 List of RL algorithm's advantages and disadvantages.

RL algorithm type	Key features	Strengths	Weaknesses
Value-based RL	Value function (V or Q) estimation	Effective for tasks with discrete action spaces	May struggle with high-dimensional state spaces
	Bellman equation	Stable and well-understood	Can be sample-inefficient
	Policy derivative from value (ex., ε-greedy)	Converges to a global optimum	Prone to overestimation bias (ex., DQN)
		Suitable for deterministic environments	May require large amounts of experience

(*Continued*)

Table 10.1 List of RL algorithm's advantages and disadvantages. (*Continued*)

RL algorithm type	Key features	Strengths	Weaknesses
Policy-based RL	Directly learn a stochastic or deterministic policy	Can handle continuous action spaces	Can be unstable during training
	Policy gradient methods	Well suited for high-dimensional state spaces	Sensitive to hyperparameters
	Exploration strategies (ex., entropy regularization)	Can learn stochastic policies	Sample inefficiency, especially in high dimensions
		Handles environments with noise and uncertainty well	Exploration can be challenging in continuous spaces
Model-based RL	Learn a model of the environment	Improved sample efficiency	Models may have inaccuracies
	Transition and reward models	Useful for real-world applications with risk	Computationally intensive
	Planning and optimization techniques	Can simulate outcomes without real-world interactions	Models may not generalize well
	Model-based policy adaptation	Can plan and optimize more effectively	Requires accurate modeling of environment
Meta-reinforcement learning	Learn to learn (meta-policies)	Enables quick adaptation to new tasks or environments	Complexity in meta-training and architecture design
	Few-shot or one-shot learning	Supports few-shot or one-shot learning	May not perform well in some standard RL tasks
	Transfer learning and generalization	Promotes transfer learning and generalization	Potential for overfitting
	Adaptation to diverse backgrounds	Beneficial for robotic and self-governing systems	Limited practical applications

Value-Based Reinforcement Learning

Value-based reinforcement learning is a subfield of reinforcement learning (RL) in machine learning and artificial intelligence. It focuses on learning a value function that assigns a value to each state or state–action pair in an environment. The primary goal of value-based RL is to find an optimal policy that maximizes the cumulative reward over time. Here are some key concepts and components of value-based reinforcement learning:

1. Value function: state value function (V): This function, denoted as V(s), represents the expected cumulative reward that an agent can achieve starting from a particular state s and following a given policy. It quantifies how good it is to be in a particular state. State–action value function (Q): Also known as the Q-function, denoted as Q(s, a), it represents the expected cumulative reward an agent can achieve by taking action a in state s and then following a specific policy. It helps in evaluating the quality of actions in a given state.
2. Bellman equation: The Bellman equation is a fundamental equation in value-based RL that expresses the relationship between the value function of a state or state–action pair and the values of its neighboring states. It is used for iterative value function updates.

$$\text{Bellman Equation for V: } V(s) = \max[a]\{\Sigma[s',r]\ p(s',r|s,a)\ [r + \gamma V(s')]\}$$

$$\text{Bellman Equation for Q: } Q(s, a) = \Sigma[s',r]\ p(s',r|s,a)\ [r + \gamma \max[a']\ Q(s', a')]$$

3. Exploration vs. exploitation: In value-based RL, agents must balance exploration (trying new actions or states) and exploitation (choosing actions based on the current knowledge to maximize rewards). Exploration is necessary to discover optimal policies.
4. Q-learning: Q-learning is a popular value-based RL algorithm that learns the Q-function iteratively using the Bellman equation. It uses a table to store Q-values for state–action pairs and updates them based on experiences in the environment.
5. Deep Q-networks (DQN): DQN is an extension of Q-learning that uses deep neural networks to approximate the Q-function, making it suitable for high-dimensional state spaces, such as images.
6. Double Q-learning: Double Q-learning is an enhancement to traditional Q-learning that helps mitigate overestimation bias, which can occur in value-based methods.
7. Policy improvement: After learning the value function, the agent can derive an optimal policy by selecting actions that maximize the estimated values. The policy can be derived using a greedy strategy or by adding exploration mechanisms.
8. Discount factor (γ): The discount factor (γ) determines the importance of future rewards compared to immediate rewards. A high γ values future rewards more, while a low γ focuses on immediate rewards.

Value-based reinforcement learning is powerful for solving problems where the agent can learn a policy by estimating the values of states or state–action pairs. However, it may face challenges in environments with high-dimensional state spaces or continuous action spaces, which have led to the development of techniques like deep Q-networks to handle such cases.

Policy-Based Reinforcement Learning

Policy-based reinforcement learning is another fundamental approach to solving problems in the field of reinforcement learning (RL). Unlike value-based RL, which focuses on estimating the value of states or state–action pairs, policy-based RL directly learns a policy, which is a mapping from states to actions. In other words, policy-based RL aims to find the optimal strategy or policy that the agent should follow to maximize its cumulative reward. Here are some key concepts and components of policy-based reinforcement learning:

1. Policy: A policy π is a mapping from states (or state–action pairs) to actions, i.e., $\pi(s) \rightarrow a$. It defines the agent's behavior in the environment.
2. Parameterized Policy: In policy-based RL, the policy is often parameterized by a set of parameters θ. These parameters are adjusted during training to find the best policy. Common parameterized policy representations include neural networks, Gaussian distributions (for continuous action spaces), or discrete probability distributions (for discrete action spaces).
3. Objective function: The objective in policy-based RL is to find the parameters θ that maximize the expected cumulative reward, also known as the return. This is typically expressed as an objective function $J(\theta)$.
   a. $J(\theta) = E[\Sigma\gamma$^t * R_t$|\pi(\theta)]$, where R_t is the reward at time step t, and γ is the discount factor.
4. Policy gradient methods: Policy gradient methods directly optimize the policy by computing gradients of the expected return with respect to the policy parameters. This allows for stochastic policies. The gradient ascent update rule is often used to adjust the policy parameters in the direction that increases expected return: $\Delta\theta \propto \nabla\theta J(\theta)$.
5. Reinforcement learning algorithms: Some popular policy-based RL algorithms include REINFORCE, proximal policy optimization (PPO), trust region policy optimization (TRPO), and actor–critic methods. Actor–critic methods combine policy-based and value-based approaches by maintaining both a policy (actor) and a value function (critic).
6. Exploration strategies: In policy-based RL, exploration is typically handled within the policy. Stochastic policies introduce randomness in action selection, which helps the agent explore the environment.
7. Continuous and discrete action spaces: Policy-based RL can handle both continuous and discrete action spaces. In continuous action spaces, policies may output action parameters (e.g., means and variances) that define probability distributions over actions.
8. Sample efficiency: Policy-based methods can be sample-inefficient, often requiring more samples to converge than value-based methods. Techniques

like entropy regularization can encourage exploration and improve sample efficiency.
9. High-dimensional state spaces: Policy-based RL can be advantageous when dealing with high-dimensional state spaces, such as image inputs, as neural networks can effectively represent complex policies.

Policy-based reinforcement learning is particularly useful in situations where the policy is complex, stochastic, and when the state or action space is high-dimensional or continuous. It has been successfully applied to a wide range of applications, including robotics, natural language processing, and game playing. However, it can also be sensitive to initialization and hyperparameters, requiring careful tuning during training.

Model-Based Reinforcement Learning

Model-based reinforcement learning is an approach to solving problems in reinforcement learning (RL) that involves building and using a model of the environment to make decisions and learn optimal policies. In contrast to model-free RL, where agents directly interact with the environment to learn from trial and error, model-based RL leverages a learned model to simulate the environment and plan actions without the need for extensive real-world interactions. Here are the key components and concepts associated with model-based reinforcement learning:

1. Model of the environment: In model-based RL, the agent constructs a model that approximates the dynamics of the environment. This model can be a representation of how the environment evolves from one state to another after taking actions.
2. Transition model: The transition model, often denoted as $P(s'|s,a)$, estimates the probability distribution of reaching state s' when taking action a in state s. This model is used to simulate the next state's dynamics.
3. Reward model: In addition to the transition model, the agent may also learn a reward model, which approximates the expected reward for transitioning from one state to another.
4. Planning and decision making: With the learned models, model-based RL agents can perform various planning and optimization techniques, such as dynamic programming, Monte Carlo methods, or search algorithms, to find an optimal policy. Common planning algorithms include value iteration and policy iteration.
5. Model learning: Model-based RL requires the agent to learn and update its transition and reward models from interaction with the real environment. This typically involves collecting data by executing actions and observing state transitions and rewards. Techniques like supervised learning, reinforcement learning from real samples, or model-based imitation learning can be used for model learning.
6. Exploration: Model-based RL still requires exploration to improve the accuracy of the learned models. Agents often use exploration strategies, such as

epsilon-greedy, to decide when to exploit the learned models and when to explore the environment.
7. Sample efficiency: One advantage of model-based RL is its potential for improved sample efficiency compared to model-free methods. Since it can plan and simulate outcomes without directly interacting with the environment, it may require fewer samples.
8. Uncertainty handling: Model-based RL often deals with uncertainties in the learned models. Agents may incorporate uncertainty estimates into their decision-making processes to make more informed choices.
9. Hybrid approaches: Some RL algorithms combine both model-based and model-free components—for example, model predictive control (MPC) uses a learned model for planning but directly interacts with the environment for execution.

Model-based reinforcement learning can be particularly useful in scenarios where collecting real-world data is expensive, risky, or time-consuming. However, it also faces challenges such as the need for accurate model estimation and computational complexity, especially when dealing with high-dimensional state spaces or complex environments. Researchers continue to explore techniques to make model-based RL more practical and effective in a wider range of applications.

Deep Reinforcement Learning

DRL is a subfield of reinforcement learning (RL) that combines deep learning techniques with RL algorithms to solve complex decision-making problems, often involving high-dimensional sensory input and action spaces. DRL has gained significant attention and achieved remarkable success in various domains, including robotics, game playing, autonomous vehicles, and natural language processing. Here are key components and concepts associated with deep reinforcement learning:

1. Deep neural networks: DRL leverages deep neural networks, particularly deep artificial neural networks known as deep learning models. These neural networks are used to approximate complex functions, such as value functions or policies, which are central to RL.
2. Value functions:
   a) Deep Q-networks (DQN): DRL extends traditional Q-learning by using deep neural networks to approximate the Q-function, enabling the handling of high-dimensional state spaces.
   b) Deep deterministic policy gradient (DDPG): DDPG combines value-based and policy-based methods to learn a deterministic policy and a Q-function, making it suitable for continuous action spaces.
   c) Twin delayed deep deterministic policy gradient (TD3): TD3 is an improvement over DDPG that addresses overestimation bias in the Q-function.
   d) Soft actor–critic (SAC): SAC is another DRL algorithm that uses entropy regularization to encourage exploration and improve learning stability.

3. Policy networks:
   a) TRPO: TRPO is a policy-based DRL algorithm that optimizes a parameterized policy using trust region methods.
   b) PPO: PPO is a policy optimization method that aims to strike a balance between sample efficiency and ease of implementation.
4. Actor–critic architectures:
   a) Actor–critic methods combine value-based and policy-based components. The actor represents the policy, and the critic estimates the value function.
   b) Asynchronous advantage actor–critic (A3C) and advantage actor–critic (A2C) are examples of actor–critic DRL algorithms.
5. Experience replay: Experience replay is a technique borrowed from DQN that stores and samples past experiences to break temporal correlations in the data and stabilize learning.
6. Exploration strategies: DRL algorithms use various exploration strategies, such as epsilon-greedy policies, noisy networks, or stochastic policies, to encourage exploration in the environment.
7. Transfer learning: Transfer learning techniques, including fine-tuning pre-trained models (e.g., convolutional neural networks for vision tasks), can be applied in DRL to accelerate learning in new tasks.
8. High-dimensional state and action spaces: DRL is well-suited for problems with high-dimensional sensory inputs, such as images or text, as deep neural networks can effectively represent complex mappings.
9. Challenges: DRL often requires a large number of samples, making it sample-inefficient compared to traditional RL in some cases. Hyperparameter tuning and instability during training can be challenges in DRL. Exploration in continuous action spaces can be tricky, and handling partial observability may require advanced techniques.

DRL has demonstrated impressive capabilities, including achieving superhuman performance in games like Go, Dota 2, and video games. It has also been applied in robotics for tasks like autonomous driving and manipulation. However, it remains an active area of research with ongoing efforts to improve its sample efficiency, stability, and generalization capabilities.

Inverse Reinforcement Learning

Inverse reinforcement learning (IRL) is a subbranch of reinforcement learning (RL) that deals with the problem of learning a reward function or cost function from observed behavior data. In traditional RL, the goal is to treasure the optimal policy (a charting from states to actions) that exploits a known reward function. In contrast, IRL aims to infer the underlying reward function from observed demonstrations or trajectories of an expert, and then use this learned reward function for decision-making or policy optimization. Here are the key concepts and components associated with inverse reinforcement learning:

1. Demonstrations or trajectories: In IRL, you typically have admittance to a usual of demonstrations or trajectories generated by an expert in the environment. These trajectories consist of sequences of state–action pairs that reflect the expert's behavior.
2. Reward function: The reward function in IRL is typically represented as a function R(s, a), which allocates a scalar rate to each state–action duos. This reward function encodes the desirability or cost associated with being in a particular form and pleasing a precise action.
3. Optimal policy recovery: The main aim of IRL is to recover the reward function from the expert's data and then use this reward function to compute an optimum or near-optimal plan. This recovered policy should ideally mimic the expert's behavior.
4. Assumption of rationality: IRL typically assumes that the expert's behavior is rational, meaning it is driven by an underlying reward function. The objective is to infer this fundamental reward function that explains the observed behavior.
5. Maximum likelihood IRL: One common approach to IRL is to formulate it as a maximum likelihood estimation problem. The objective is to find the reward function that exploits the possibility of the observed trajectories under the learned policy.
6. Generative models: IRL can be approached using generative models, such as generative adversarial networks (GANs) or maximum entropy models, to model the reward function and generate plausible trajectories.
7. Application areas: IRL has been pragmatic in countless fields, including robotics, self-directed driving, and game design. It can be used to learn from human demonstrations and adapt RL agents to real-world tasks.
8. Challenges: IRL can be the exciting tricky because the true reward function is often not directly observable. It depends on the supposition that the expert's comportment is optimum with esteem to some underlying reward, which may not always hold.
9. Inverse optimal control (IOC): IOC is a closely related concept to IRL and is sometimes used interchangeably. IOC often focuses on finding the reward function that explains expert behavior in control and optimization problems.

Inverse reinforcement learning is a valuable technique for transferring human expertise into RL systems and enabling agents to perform tasks without explicit hand-crafted reward functions. It has practical applications in areas where experts can provide demonstrations, but specifying a reward function may be challenging or subjective.

Meta-Reinforcement Learning

Meta-reinforcement learning (meta-RL) is a subfield of reinforcement learning (RL) that emphases on developing agents capable of quickly adapting to new tasks or environments. Unlike traditional RL, where agents learn a fixed policy for a specific task, meta-RL trains agents to learn how to learn. In other words, they learn a set of skills or update their policies efficiently based on past experiences to adapt to new, unseen tasks or environments. Here are the key concepts and components associated with meta-reinforcement learning:

1. Meta-training and meta-testing: Meta-RL typically involves two stages: meta-training and meta-testing. During meta-training, agents are exposed to a variety of training errands. In the meta-testing phase, they are assessed on new, unseen tasks to assess their adaptation capabilities.
2. Meta-policies: Agents in meta-RL often learn meta-policies, which are policies that govern the learning process itself. These meta-policies determine how the agent should adapt its behavior based on the current task or context.
3. Task distribution: In meta-RL, there is typically an underlying distribution of tasks. During meta-training, agents learn to generalize from a set of training tasks sampled from this distribution.
4. Fast adaptation: Meta-RL emphasizes the capability to quickly familiarize to novel errands or environments. Agents are expected to update their policies or meta-policies with relatively few samples or interactions.
5. Gradient-based methods: Many meta-RL algorithms employ gradient-based optimization techniques, such as gradient descent, to adapt to new tasks. They learn to modernize policy considerations in a manner that exploits expected rewards on new tasks.
6. Few-shot or one-shot learning: In some cases, meta-RL aims for few-shot or one-shot learning, where agents must adapt to novel errands with only a minor amount of instances or interactions.
7. Transfer learning and generalization: Meta-RL promotes transfer learning and generalization across tasks. Agents should be able to leverage knowledge from past tasks to perform well on new tasks.
8. Algorithm variants: There are various meta-RL algorithms and approaches, including model-agnostic meta-learning (MAML), Reptile, and various recurrent neural network (RNN) or meta-learner architectures.
9. Applications: Meta-RL has applications in robotics, autonomous systems, natural language processing, and other domains where agents need to adapt to different conditions or tasks.
10. OpenAI Gym Meta-World: OpenAI Gym Meta-World is a popular benchmark environment for evaluating meta-RL algorithms. It provides a diverse set of robotic manipulation tasks for testing adaptation and generalization capabilities.

Meta-reinforcement learning is the thrilling area of investigation with the probable to suggestively enhance the adaptability and versatility of autonomous agents. It has the promise of enabling agents to acquire new skills and perform tasks efficiently in the extensive variety of real-life situations. Nevertheless, it also poses challenges related to sample efficiency, stability, and scalability, which researchers continue to address.

10.4 Enhancing RL for Real-World Robotics

The importance of transfer learning, sim-to-real transmission, and safety considerations in deploying DRL agents in real-world environments cannot be overstated. These three components are critical for the successful integration of DRL into practical robotic applications:

Transfer learning: Transfer learning is a pivotal procedure in the deployment of RL agents in real-world robotics. It enables RL agents to leverage knowledge gained in one domain or task to accelerate learning in a new, related domain or task. This dramatically reduces the time and resources required for training in real-world settings. Moreover, transfer learning enhances the generalization capability of RL agents, allowing them to adapt their learned policies to a broader range of scenarios. It also makes RL agents more data-efficient, as they require fewer interactions with the environment to achieve competent performance. This adaptability and efficiency are essential for practical real-world applications where conditions may change unpredictably, and data collection can be costly or resource-intensive. In essence, transfer learning empowers RL-driven robots to learn efficiently, generalize effectively, and excel in various domains, making it a cornerstone of real-world robotic applications.

- Generalization capability: Transfer learning enhances the generalization capability of RL agents. Instead of being specialized for a single task or environment, agents can adapt their learned policies to a broader range of scenarios. This adaptability is crucial for real-world applications where conditions may change unpredictably. Example: Consider a robot that learns to grasp objects in a controlled environment. Transfer learning enables it to adapt quickly to different tasks, such as picking up items in a warehouse, handling tools in manufacturing, or assisting in household chores.
- Efficient learning: Transfer learning is essential for speeding up the learning process for RL agents. In real-world applications, training an RL agent from scratch can be time-consuming and data-intensive. Transfer learning allows agents to start with pre-trained knowledge, significantly reducing training time and data requirements.
- Data efficiency: Real-world data collection can be expensive, risky, or impractical. Transfer learning makes RL agents more data-efficient. They require fewer communications with the surroundings to accomplish competent enactment, which is particularly valuable in settings where data collection is resource-intensive.
- Real-world applicability: Many real-world problems involve tasks that are related to or share common elements with previously learned tasks. Transfer learning allows knowledge gained in one context to be applied to another, making RL practical for the extensive variety of applications, from robotic control to autonomous driving.

Sim-to-real transfer: Sim-to-real transmission is a pivotal technique in the deployment of RL agents in real-world robotics. It enables RL agents to undergo initial training and policy refinement in virtual environments, which are safe, controlled, and cost-effective. This reduces the risk associated with real-world experimentation and minimizes resource-intensive data collection. Simulations can closely mimic real-world dynamics, but differences inevitably exist. Sim-to-real transfer addresses these domain gaps by adapting policies learned in replication to real-life conditions. Techniques such as domain adaptation facilitate the transition, ensuring that learning in simulation effectively translates to reality. In essence, sim-to-real transfer offers a bridge between the virtual and physical worlds, allowing RL-driven robots to learn, adapt, and excel in complex real-world environments while mitigating risks and reducing costs.

- Safe and controlled training: Training RL agents in simulations provides a safe and controlled environment. Agents can explore and learn without the risk of causing physical damage or harm. This is especially critical when experimenting with novel algorithms or in high-stakes applications like autonomous vehicles.
- Cost reduction: Real-world data collection and experimentation can be prohibitively expensive. Sim-to-real transfer reduces costs by using virtual environments, which are more affordable and accessible. This enables extensive experimentation and iteration.
- Domain adaptation: Simulations can mimic real-world dynamics, but there are inevitable differences. Sim-to-real transfer addresses these domain gaps by adapting policies learned in simulation to real-world conditions. Techniques such as domain adaptation help bridge the gap between the two domains, ensuring that learning in simulation effectively translates to reality.
- Risk mitigation: Training in simulation allows for risk mitigation. RL agents can undergo initial training and policy refinement in the safety of simulations, reducing the chances of catastrophic failures when deployed in the real world. This iterative process enhances the safety of RL-driven systems.
- Bridging the reality gap: Sim-to-real transfer is instrumental in bridging the gap between simulated and real-world environments. It ensures that policies learned in controlled simulations can effectively translate into practical, unstructured settings, reducing risks and costs associated with real-world experimentation. Example: Training a drone in a simulated environment to fly in diverse real-world conditions, like urban landscapes or challenging weather, exemplifies the importance of sim-to-real transfer.

Safety considerations: Safety considerations are paramount in the deployment of RL agents in real-world robotics. Ensuring the physical safety of robots, humans, and the environment is a fundamental concern. This involves implementing safety mechanisms, constraints, and fail-safes to prevent dangerous actions and collisions. Moreover, ethical behavior is crucial, with RL agents expected to adhere to ethical guidelines in their decision-making processes, particularly in applications involving human interaction. Building and maintaining human trust in robots is equally important, necessitating transparency and predictability in robot actions. Regulatory compliance adds another layer of complexity, as adhering to safety and ethical standards, along with relevant regulations and guidelines, ensures that robots operate responsibly within legal boundaries. These considerations collectively create a framework that not only enhances safety but also fosters trust and ethical behavior in RL-driven robots, making them reliable partners in various real-world scenarios.

- Physical safety: Ensuring the physical safety of robots and their interactions with the environment and humans is of utmost importance. Safety mechanisms, constraints, and fail-safes are incorporated into RL-driven robots to prevent dangerous actions or collisions.
- Ethical behavior: Safety considerations extend beyond physical harm to ethical behavior. RL agents need to adhere to ethical guidelines and principles in their decision-making processes. This is particularly crucial in applications involving human interaction, such as healthcare or autonomous vehicles.
- Human trust: Building and maintaining trust in human–robot interactions is essential. Safety considerations encompass the transparency and predictability of robot actions. Humans should be able to understand and anticipate robot behavior, fostering trust in collaborative and assistive settings.
- Regulatory compliance: Deploying RL-driven robots often involves adhering to safety and ethical standards, regulations, and guidelines. Compliance ensures that robots operate within legal boundaries, protecting both users and the public.
- Fundamental for responsibility: Safety considerations are foundational for the responsible deployment of DRL in robotics. Ensuring physical safety, ethical behavior, and regulatory compliance is not just a requirement but a prerequisite for human trust and ethical use of autonomous systems. Example: In applications like medical robotics, ensuring that robotic surgeons prioritize patient safety, ethical decision-making, and compliance with healthcare regulations is paramount for responsible use.

In summary, transfer learning, sim-to-real transfer, and safety considerations are pivotal components for the responsible and effective deployment of DRL agents in real-world environments. They enable robots to learn efficiently, adapt to dynamic conditions, and operate safely, making DRL a practical and transformative technology across a wide variety of claims, from healthcare to manufacturing and beyond.

10.5 Reinforcement Learning for Various Robotic Applications

Reinforcement learning (RL) has gained prominence in the field of robotics for its ability to enable autonomous decision-making and control in a wide range of applications. Here are various robotic applications where reinforcement learning is used:

1. Robot arm manipulation: pick-and-place: In pick-and-place tasks, robots are trained to grasp objects from one location and place them in another, often with precision. RL can be applied to teach robotic arms the optimal strategies for grasping and placing objects [2]. DRL procedures, such as deep deterministic policy gradients (DDPG), TRPO, or PPO, have been used for training robot arms in pick-and-place tasks [13]. Figure 10.2 illustrates a robotic arm execution for pick-and-place with the help of RL.

Assembly: Robotic assembly chores involve joining multiple parts together to create a final product. RL techniques can optimize the assembly process by learning how to manipulate parts and perform assembly steps efficiently [14]. Researchers have employed RL

Figure 10.2 Pick-and-place execution by robotic arm.

Figure 10.3 Robotic assembly scenario using RL to control the assembly process.

algorithms, including PPO and TRPO, for training robots in assembly tasks. These algorithms enable robots to learn assembly strategies from scratch [15]. Figure 10.3 shows a robotic assembly scenario where RL is used to control the assembly process.

2. Autonomous navigation

Self-driving cars: RL is applied to enable autonomous vehicles to make real-time decisions while navigating roads. These decisions include lane-keeping, overtaking, intersection negotiation, and handling complex traffic scenarios [16]. DRL systems, such as DDPG and PPO, have been used to train RL policies for self-driving cars. These policies take sensory inputs (e.g., camera images, LiDAR data) and produce control commands (steering, acceleration, braking) [17]. Figure 10.4 illustrates a self-driving car navigating a complex urban environment using RL-based control.

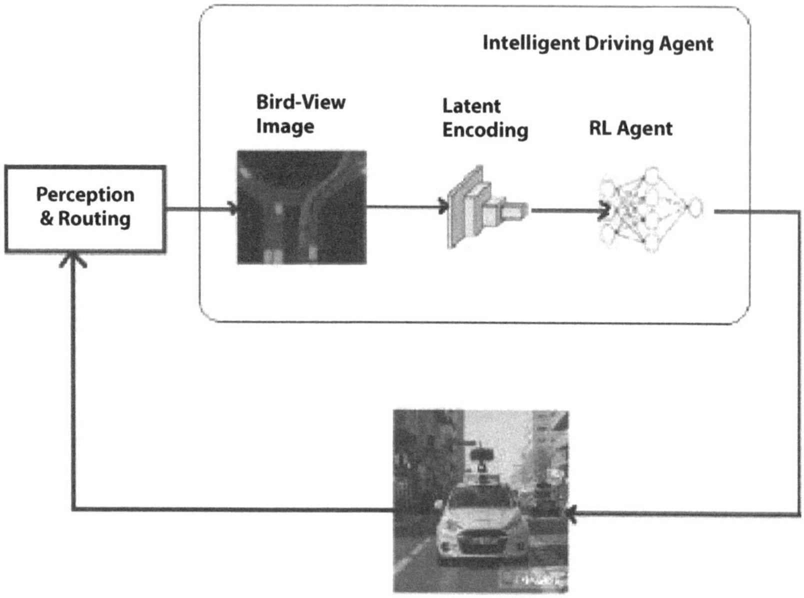

Figure 10.4 Self-driving car navigating a complex urban environment using RL-based control.

Unmanned aerial vehicles (UAVs): UAVs, including drones, rely on RL to perform tasks like navigation, obstacle avoidance, surveillance, and data collection. RL algorithms enable UAVs to adapt to varying environmental conditions [18]. TRPO and PPO are commonly used DRL systems for training UAV control policies [19]. Figure 10.5 shows a UAV using RL for autonomous navigation and obstacle avoidance in a complex outdoor environment.

3. Legged locomotion: Legged robots, such as quadrupeds and bipeds, utilize RL techniques to achieve stable and agile locomotion across challenging terrains. These robots need to maintain balance and adapt to dynamic environments [20]. RL procedures, including TRPO and PPO, have been instrumental in training legged robots to perform tasks like

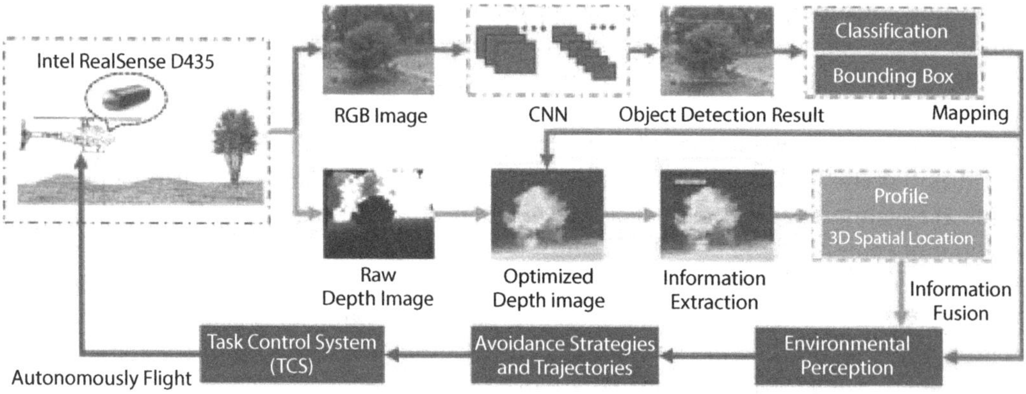

Figure 10.5 UAV using RL for autonomous navigation and obstacle avoidance.

Figure 10.6 Quadrupedal robot employing RL to navigate uneven terrain and maintain stability.

walking, running, climbing, and jumping [21]. Figure 10.6 showcases a quadrupedal robot employing RL to navigate uneven terrain and maintain stability.

4. Humanoid robotics:
Walking: Humanoid robots often use RL to learn and adapt their walking patterns. RL algorithms enable robots to adjust their gaits and balance based on sensor feedback to move gracefully on various surfaces [22]. Gesture recognition: Humanoid robots can employ RL to recognize and respond to human gestures effectively. This involves training the robot to interpret gestures using sensory data, such as cameras and depth sensors [23]. Human–robot interaction: RL plays a pivotal role in improving human–robot interaction. By learning from human feedback and demonstrations, robots can become more responsive and adapt to the preferences and needs of users [24]. Figure 10.7 illustrates a humanoid robot walking structure.

5. Human–robot collaboration and assistance
Healthcare: RL-driven robots are increasingly used in healthcare settings—for instance, robotic exoskeletons use RL to assist with rehabilitation exercises for patients recovering from injuries or surgeries [25]. Elderly care: Robots with RL-based algorithms can provide support and companionship to elderly individuals. They can assist with daily activities and monitor vital signs [26]. Dangerous environments: In high-risk environments like nuclear reactors or disaster zones, RL-driven robots can be deployed for tasks such as inspection, maintenance, or searching for survivors. These robots can adapt to changing conditions and make autonomous decisions [27]. Figure 10.8 shows a robotic exoskeleton for rehabilitation.

Figure 10.7 Humanoid robot walking.

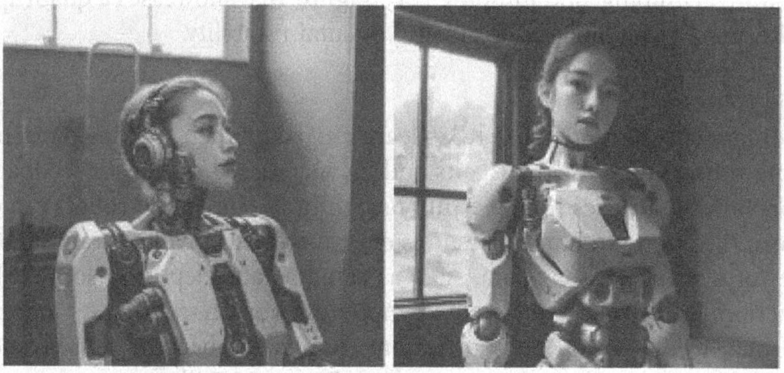

Figure 10.8 Robotic exoskeleton for rehabilitation.

Warehouse automation: Navigation: RL is applied to optimize the navigation of autonomous mobile robots in large warehouses. These robots use RL algorithms to plan efficient routes and avoid collisions [28]. Goods picking: RL-driven robots are used for automated picking and sorting of goods in warehouses. These robots can adapt to changing inventory and optimize their picking strategies [29]. Efficiency: By employing RL-based control, warehouse automation systems can increase the efficiency of order fulfillment processes, reducing human labor and operational costs. Figure 10.9 illustrates how a robot is doing the warehouse automation task.

Figure 10.9 Warehouse automation.

6. Sim-to-real transfer: Sim-to-real transmission in RL for robotics is a vital technique employed to facilitate the seamless transition of trained RL policies from virtual environments to real-world robotic platforms. This approach addresses the substantial gap that often exists between controlled simulations and the complexities of real-world scenarios. The primary motivation behind sim-to-real transfer is the necessity for safe and cost-effective training environments, which simulations provide. These virtual environments accurately replicate a robot's dynamics and interactions using sophisticated 3D or physics-based simulators like MuJoCo, Gazebo, or Unity. Within these virtual environments, RL agents are trained using established algorithms like PPO or TRPO. Figure 10.10 shows the sim-to-real transmission.

These agents learn policies that map sensor observations to actions while optimizing rewards within the context of virtual tasks. However, the "reality gap" emerges as a significant challenge. This term refers to the disparities among the replication surroundings and the actual world, encompassing differences in astronomy, sensor noise, lighting conditions, and unaccounted-for variables. Consequently, RL policies that excel in simulations often fail to perform optimally in the real world due to this gap. To address this challenge, sim-to-real Transfer involves the transfer of learned policies from simulation to the physical robot.

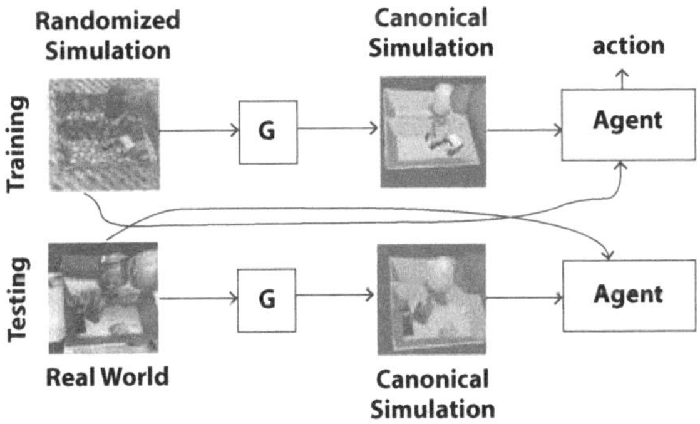

Figure 10.10 Sim-to-real transmission.

Nevertheless, this transition necessitates fine-tuning and adaptation to account for the real-world dynamics and uncertainties. Various techniques, including domain adaptation, reinforcement learning from human feedback (RLHF), and safety-aware RL, are employed in the real-world phase to ensure that the robot can adapt effectively to unpredictable real-world conditions and enhance its performance [29–35].

7. Robotics in agriculture: Agriculture has seen significant advancements with the integration of reinforcement learning (RL) in robotics. RL algorithms empower agricultural robots to perform a diverse range of tasks, from precision planting to crop monitoring and harvesting. These robots can make data-driven decisions based on environmental conditions and sensor inputs, ultimately enhancing agricultural productivity while reducing labor costs [36]. One crucial task that RL-driven robots excel at is planting and seeding. They can precisely plant seeds, optimizing spacing and depth to maximize crop growth. Moreover, these robots can monitor crop health using sensors like cameras and multispectral imaging, allowing for early pest detection and efficient resource allocation. In the realm of weeding, autonomous robots leverage RL to identify and remove weeds without causing harm to crops, minimizing the need for herbicides [37]. During harvesting, RL algorithms enable robots to pick crops efficiently and accurately, reducing the demand for human labor. In addition to crop-related tasks, RL-driven robots play a vital role in optimizing agricultural processes such as irrigation and fertilization. By considering real-time data on soil moisture, weather forecasts, and nutrient levels, these robots can make informed decisions to optimize resource usage. Furthermore, they contribute to data collection efforts by gathering information on soil quality, temperature, humidity, and more, enabling farmers to make data-driven decisions [38]. Figure 10.11 illustrates the application of robotics in agriculture and farming.

These are just a few examples of how RL is transforming various domains of robotics by enabling robots to learn and adapt to their environments. As research in RL endures to improvement, we can presume uniform further refined and accomplished robotic structures in the forthcoming.

Figure 10.11 Robotics in agriculture using RL.

10.6 Problems Faced in RL for Robotics

In addition to the applications and advancements in reinforcement learning (RL) for robotics, there are several challenges and problems specific to real-world robotics environments that require attention:

1. Sensor noise and uncertainty: Real-world devices, like LIDAR, and cameras are susceptible to noise and uncertainty. Developing RL algorithms that can effectively handle noisy sensor data and make robust decisions is an ongoing challenge.
2. Safety assurance: Ensuring the safety of RL-driven robots in unpredictable environments is paramount. Developing methodologies and frameworks for guaranteeing safety in RL policies, particularly in situations where human interaction is involved, remains a significant challenge.
3. Sample competence: Many RL procedures require a huge amount of collaborations with the surroundings, which can be time-consuming and impracticable for real-life robots. Improving sample efficiency is crucial for practical deployment.
4. Adaptation to dynamic environments: Real-world environments are dynamic, with objects moving, people interacting, and conditions changing. Developing RL algorithms that can adapt to these dynamic environments in real-time is a challenging problem.
5. Resource constraints: Real-world robots often have limited computational resources and power constraints. Optimizing RL algorithms to operate efficiently under these constraints is a pressing concern.
6. Long-term planning: Many real-world tasks require long-term planning and decision-making. Developing RL algorithms that can handle long time horizons and consider delayed consequences is a complex problem.
7. Real-time decision-making: Real-world robots must make decisions quickly to respond to changing conditions or unexpected events. Balancing the need for real-time decision-making with the complexity of RL algorithms is a challenge.
8. Human–robot interaction: Integrating RL-driven robots into human-centric environments, such as healthcare or public spaces, poses unique challenges in terms of safety, ethics, and user acceptance.
9. Ethical and legal considerations: Deploying RL-driven robots raises ethical and legal questions, such as liability in case of accidents or ensuring that robots behave ethically in sensitive contexts.
10. Generalization across environments: RL agents often struggle to generalize their learned behaviors from one environment to another. Improving the adaptability and generalization capabilities of RL algorithms is essential.
11. Robust perception: Real-world perception systems must handle variations in lighting, weather, and object appearance. Developing robust perception algorithms that can reliably interpret sensor data is a vital challenge.

12. Interpretable and explainable AI: As robots operate alongside humans, there is a growing need for interpretable and explainable AI models to enhance trust and facilitate human–robot collaboration.

These challenges are crucial to address for the successful deployment of RL-driven robots in real-world applications. Solving these problems requires interdisciplinary collaboration among researchers in robotics, machine learning, computer vision, ethics, and policy-making, among others. Overcoming these challenges will pave the way for more capable, adaptable, and safe robotic systems in diverse real-world scenarios.

10.7 RL in Robotics: Trends and Challenges

In the rapidly evolving field of robotics, the integration of reinforcement learning (RL) has opened new frontiers for autonomous decision-making and adaptability. As robots venture into complex real-world environments, it becomes crucial to examine emerging trends, potential research directions, and the lingering challenges that demand innovative solutions. This section offers a wide-ranging outline of the landscape, offering insights into what lies ahead for RL in robotics.

Emerging trends: The arena of RL in robotics is witnessing a transformative shift characterized by several key trends. These trends include the incorporation of multi-modal sensing, the imperative for real-time learning and adaptation, the rise of collaborative human–robot interaction, the emphasis on ethical AI, and the pursuit of transfer learning across diverse domains. These developments set the stage for exciting advancements and novel challenges in the application of RL to real-world robotic systems.

- Multi-modal sensing: As robots become more sophisticated, they are equipped with a variety of sensors that capture different types of data. Vision, audio, tactile, and proprioceptive sensors are integrated to enable robots to perceive and interact with their environment in a more holistic manner. The trend is to develop algorithms that can effectively fuse and interpret data from these multi-modal sensors.
- Real-time learning and adaptation: Real-world environments are dynamic and unpredictable. To navigate and operate effectively, robots need to learn and adapt in real-time. Emerging research focuses on developing RL algorithms that can update policies and strategies on the fly, allowing robots to respond to changing conditions and unexpected events dynamically.
- Human–AI collaboration: The future of robotics involves humans and AI working collaboratively. This trend emphasizes safety, trust, and natural interaction between humans and robots. It includes research into human–robot interfaces, shared decision-making, and ethical considerations in human–robot collaboration.
- Ethical AI: Ethical considerations are gaining prominence in RL research. Ensuring that RL agents behave ethically and responsibly is crucial, especially in applications where robots interact with humans. Researchers are exploring value-aligned RL and ethical frameworks to guide robot behavior.

- Transfer learning across domains: A major challenge in RL is making learned policies transferable across different domains and environments. Researchers are working on techniques to improve the adaptability of RL agents, especially when transferring knowledge from virtual environments to real-world scenarios.

Potential research directions: As the landscape of reinforcement learning (RL) for robotics continues to evolve, researchers are charting exciting pathways for future exploration. These research directions encompass improving sample efficiency, addressing long-term planning challenges, enhancing robustness to uncertainty, striving for interpretable and explainable AI, and optimizing resource efficiency. These directions promise to reshape how RL-equipped robots interact with and adapt to dynamic real-world environments, marking significant progress in the field.

- Sample effectiveness: RL procedures often need the enormous amount of samples for training, which can be impractical in real-world settings. Research is focused on developing more sample-efficient algorithms, possibly incorporating techniques like meta-learning or improved exploration strategies.
- Long-term planning: Many real-world tasks require robots to make decisions with long time horizons. Future research may involve combining RL with other planning techniques to handle extended time frames effectively.
- Robustness to uncertainty: Real-world environments are inherently uncertain, with sensor noise and unpredictable events. Research aims to enhance the robustness of RL agents to handle uncertainty and adapt to unforeseen challenges.
- Interpretable and explainable AI: Understanding and trusting RL-driven robot decision-making is essential. Researchers are working on making RL policies more interpretable and explainable, which is particularly critical in applications where human collaboration or oversight is required.
- Resource efficiency: In resource-constrained environments or robotic platforms, computational efficiency is crucial. Future research will focus on developing lightweight RL algorithms suitable for constrained hardware.

Open challenges: Despite remarkable progress, the integration of reinforcement learning (RL) into robotics faces persistent and complex challenges. These challenges encompass sample complexity reduction, the intricate task of sim-to-real transfer, the development of generalization capabilities across diverse tasks, enabling continuous learning in dynamic environments, ensuring safety at all levels of operation, orchestrating effective human–robot collaboration, establishing ethical and legal frameworks, and enhancing perception systems' robustness. Navigating these challenges is essential to unleash the full potential of RL-driven robots in real-world applications. In the dynamic landscape of robotics and autonomous systems, the complexities of deploying deep reinforcement learning (DRL) agents become particularly evident. Real-world applications introduce a multitude of challenges that demand thorough examination.

- Sample complexity: Reducing the sample complexity of RL algorithms to make them practical for real-world deployment is a persistent challenge. This involves improving algorithms for efficient learning. Example: In the realm of autonomous vehicles, the challenge of dealing with adverse weather conditions, such as heavy rain or snow, can severely affect sensor accuracy and hinder the vehicle's ability to navigate safely.
- Sim-to-real transmission: Adapting RL strategies from imitation to realism, while accounting for discrepancies in dynamics and sensor fidelity, remains a complex and ongoing challenge. Example: In industrial automation, the precise control of robotic arms for tasks like assembling delicate components or surgical procedures necessitates mitigating issues like sensor noise and ensuring minimal latency.
- Generalization across tasks: Developing RL agents that can generalize their learned policies across diverse tasks and environments is a multifaceted problem that requires innovative solutions. Example: Consider training a drone in a simulator to fly indoors with no wind, and then deploying it outdoors in a real-world environment with varying wind conditions. The domain shift presents a significant challenge.
- Continuous learning: Ensuring that RL agents can continually learn and adapt over extended periods, even in the face of changing conditions and requirements, is a challenge in long-term deployments. Example: Training a robot to perform complex tasks in the real world often requires an extensive number of trials, which can be resource-intensive—for example, training a robot to perform advanced dexterous manipulation in a cluttered environment may demand thousands of real-world trials.
- Safety assurance: Ensuring the safety of RL-driven robots in both physical and ethical terms is an ever-present and evolving challenge. Research will need to address safety at multiple levels, including perception, action, and decision-making. Example: In applications like autonomous vehicles, there are ethical dilemmas—for instance, should a self-driving car prioritize the safety of its occupants over that of pedestrians? Striking the right ethical balance is challenging.
- Human–robot collaboration: Effectively designing and implementing human–robot collaboration, including shared decision-making and intuitive interfaces, requires interdisciplinary research spanning psychology, human factors, and AI. Example: In medical robotics, ensuring safe and effective interaction between a surgical robot and a human surgeon is crucial. Designing intuitive interfaces for remote robotic surgery or addressing challenges in telerobotics is a complex task.
- Ethical and legal frameworks: Establishing ethical and legal frameworks for the responsible deployment of RL-driven robots in sensitive and regulated domains is an ongoing societal challenge. Example: In the aerospace industry, deploying autonomous aircraft involves navigating a web of regulatory considerations, including airworthiness certification and airspace regulations, which are essential to ensure safety and compliance.

- Robust perception: Developing perception systems that can reliably interpret complex and noisy sensor data in real-world scenarios is a critical challenge, particularly for tasks like autonomous driving and healthcare. Example: A mobile robot navigating a factory floor must adapt to dynamically changing conditions, such as the appearance of new obstacles or variations in lighting. Ensuring reliable adaptability in real-time is a significant challenge.

These emerging trends, research directions, and open challenges collectively shape the future of RL for robotics. Addressing these issues will pave the way for more capable, adaptable, and responsible robotic systems that can thrive in diverse real-world applications.

10.8 Conclusion

In summary, this chapter has taken a comprehensive dive into the realm of DRL within the framework of robotics and autonomous systems. We have explored not only the current state of the field but also the potential trajectories it may take in the future. Throughout our discussion, we have highlighted several emerging trends that are reshaping the landscape, such as the integration of multi-modal sensing, the emphasis on real-time learning and adaptation, the growth of collaborative human–robot interaction, the imperative of ethical AI, and the pursuit of transfer learning across domains. These trends collectively reflect the dynamic nature of the field, promising innovative and impactful applications.

Moreover, we have ventured into the potential research directions that researchers and practitioners may choose to pursue. These directions include tackling the challenge of sample efficiency to make RL more practical in resource-intensive environments, addressing the complexities of long-term planning in dynamic settings, enhancing the robustness of RL agents to uncertainty and unexpected events, striving for interpretable and explainable AI to foster trust and collaboration, and optimizing resource efficiency for resource-constrained robotic platforms. These research avenues represent opportunities to elevate the capabilities of RL-driven robots, ensuring they adapt and excel in increasingly complex real-world scenarios.

However, it is crucial to acknowledge that alongside these opportunities, we face persistent and multifaceted challenges. These challenges encompass reducing the sample complexity of RL algorithms to make them more feasible for real-life deployment, connecting the gap among recreation and realism in sim-to-real transfer, developing RL agents that can generalize effectively across a wide array of tasks and environments, enabling continuous learning in the face of changing conditions, ensuring safety at all levels of robot operation, fostering effective human–robot collaboration, establishing ethical and legal frameworks for responsible robot behavior, and enhancing the robustness of perception systems in real-world scenarios. Navigating these challenges is not only essential but also central to unleashing the full potential of RL-driven robots in practical applications.

As we look forward, the integration of DRL into robotics and autonomous systems promises to reshape industries, revolutionize how we interact with technology, and address complex real-world problems. Researchers and practitioners in this field have an exciting journey ahead, marked by opportunities for innovation and the fulfillment of long-anticipated visions. By addressing these challenges head-on and collaborating across disciplines, we can collectively shape a future where robotics and autonomous systems play transformative roles in various domains, enhancing our daily lives and pushing the boundaries of what is possible.

References

1. Deisenroth, M.P., A Survey on Policy Search for Robotics. *Found. Trends Rob.*, 2, 1–2, 1–142, 2011, doi: 10.1561/2300000021.
2. Kober, J., Bagnell, J.A., Peters, J., Reinforcement learning in robotics: A survey. *Int. J. Rob. Res.*, 32, 11, 1238–1274, 2013, doi: 10.1177/0278364913495721.
3. Kroemer, O., Niekum, S., Konidaris, G., A review of robot learning for manipulation: Challenges, representations, and algorithms. *J. Mach. Learn. Res.*, 22, 1–82, 2021.
4. Sünderhauf, N. *et al.*, The limits and potentials of deep learning for robotics. *Int. J. Rob. Res.*, 37, 4–5, 405–420, 2018, doi: 10.1177/0278364918770733.
5. Zhu, H., Gupta, A., Rajeswaran, A., Levine, S., Kumar, V., Dexterous manipulation with deep reinforcement learning: Efficient, general, and low-cost, in: *Proceedings - IEEE International Conference on Robotics and Automation*, vol. 2019-May, 2019, doi: 10.1109/ICRA.2019.8794102.
6. Cui, Y., Matsubara, T., Sugimoto, K., Pneumatic artificial muscle-driven robot control using local update reinforcement learning. *Adv. Rob.*, 31, 8, 397–412, 2017, doi: 10.1080/01691864.2016.1274680.
7. Liu, X., Liu, Y., Chen, Y., Reinforcement learning in multiple-UAV networks: Deployment and movement design. *IEEE Trans. Veh. Technol.*, 68, 8, 8036–8049, 2019, doi: 10.1109/TVT.2019.2922849.
8. Polydoros, A.S. and Nalpantidis, L., Survey of Model-Based Reinforcement Learning: Applications on Robotics. *J. Intell. Rob. Syst.*, 86, 2, 153–173, 2017, doi: 10.1007/s10846-017-0468-y.
9. Tan, J. *et al.*, Sim-to-Real: Learning Agile Locomotion For Quadruped Robots, Robotics: Science and Systems, MIT Press Journals, 2018, doi: 10.15607/RSS.2018.XIV.010.
10. Nagabandi, A. *et al.*, Learning Image-Conditioned Dynamics Models for Control of Underactuated Legged Millirobots, *IEEE/RSJ International Conference on Intelligent Robots and Systems (IROS)*, Madrid, Spain, 2018, pp. 4606–4613, 2018, doi: 10.1109/IROS.2018.8594193.
11. Kalashnikov, D. *et al.*, Scalable Deep Reinforcement Learning for Vision-Based Robotic Manipulation, in: *Proceedings of Machine Learning Research*, vol. 87, 2018.
12. Finn, C., Levine, S., Abbeel, P., Guided cost learning: Deep inverse optimal control via policy optimization, in: *33rd International Conference on Machine Learning, ICML 2016*, vol. 1, 2016.
13. Lillicrap, T.P. *et al.*, Continuous control with deep reinforcement learning, *4th International Conference on Learning Representations, ICLR 2016 - Conference Track Proceedings*, pp. 1–14, 2016.
14. Karnati, N., Kent, B., Engeberg, E.D., Backdrivable periodic finger joint synergies: Human observations applied to a dexterous robotic hand, *2011 IEEE International Conference on Robotics and Biomimetics, ROBIO* 2011, 1122–1127, 2011, doi: 10.1109/ROBIO.2011.6181438.
15. Learning dexterity, *Int. J. Robotics Res.*, 39, 1, 3–20, 2020, https://openai.com/research/learning-dexterity (accessed Oct. 16, 2023).
16. Pan, Y. *et al.*, Agile Autonomous Driving using End-to-End Deep Imitation Learning, *Robotics: Science and Systems*, MIT Press Journals, pp. 1–10, 2018, doi: 10.15607/RSS.2018.XIV.056.
17. Ackerman, E., Drive.ai Brings Deep Learning to Self-Driving Cars. *IEEE Spectr.*, 2016.

18. long Liang, X., kai Yao, Z., wen Ge, Y., yong Yao, J., Reinforcement learning based adaptive control for uncertain mechanical systems with asymptotic tracking. *Def. Technol.*, 34, April 2024, 19–28, 2023, doi: 10.1016/j.dt.2023.05.016.
19. Mnih, V. *et al.*, Asynchronous methods for deep reinforcement learning, in: *33rd International Conference on Machine Learning, ICML 2016*, vol. 4, 2016.
20. Narikiyo, T., Matsumoto, T., Ugurlu, B., Kawanishi, M., Compliant Locomotion Control for a Quadruped Robot with Damper Coefficients Assigned by Reinforcement Learning. *J. Rob. Soc. Jpn.*, 35, 5, 414–423, 2017, doi: 10.7210/jrsj.35.414.
21. Robine, J., Uelwer, T., Harmeling, S., Smaller World Models for Reinforcement Learning. *Neural Process. Lett.*, 11397–11427, 2023, doi: 10.1007/s11063-023-11381-3.
22. Portnoy, S. *et al.*, Chapter 8 - Multimodal Localization for Embedded Systems: A Survey. *Comput. Hum. Behav.*, 25, 1, 24–32, 2017.
23. Li, G., Whiteson, S., Knox, W.B., Hung, H., Social interaction for efficient agent learning from human reward. *Auton. Agents Multi-Agent Syst.*, 32, 1, 1–25, 2018, doi: 10.1007/s10458-017-9374-8.
24. Senft, E., Lemaignan, S., Baxter, P.E., Bartlett, M., Belpaeme, T., Teaching robots social autonomy from *in situ* human guidance. *Sci. Rob.*, 4, 35, 1–13, 2019, doi: 10.1126/scirobotics.aat1186.
25. Rubio, F., Valero, F., Llopis-Albert, C., A review of mobile robots: Concepts, methods, theoretical framework, and applications. *Int. J. Adv. Rob. Syst.*, 16, 2, 1–22, 2019, doi: 10.1177/1729881419839596.
26. Broadbent, E. *et al.*, Using robots at home to support patients with chronic obstructive pulmonary disease: Pilot randomized controlled trial. *J. Med. Internet Res.*, 20, 2, e45, 2018, doi: 10.2196/jmir.8640.
27. Li, H., Zhang, Q., Chen, Y., Zhao, D., SIM: A Scenario IMagination Based Deep Reinforcement Learning Method for Outdoor Transportation Environment Exploration, *IEEE 11th Data Driven Control and Learning Systems Conference (DDCLS)*, pp. 636–642, Chengdu, China, 2022, doi: 10.1109/DDCLS55054.2022.9858473.
28. Göçer, Ö. *et al.*, A Ship in a Box. *Appl. Mech. Mater.*, 29, 2, 2014.
29. Tobin, J., Fong, R., Ray, A., Schneider, J., Zaremba, W., Abbeel, P., Domain randomization for transferring deep neural networks from simulation to the real world, in: *IEEE International Conference on Intelligent Robots and Systems*, vol. 2017-September, 2017, doi: 10.1109/IROS.2017.8202133.
30. Rajeswaran, A. *et al.*, Learning Complex Dexterous Manipulation with Deep Reinforcement Learning and Demonstrations, *Proceedings of Robotics: Science and Systems*, 2018, doi: 10.15607/RSS.2018.XIV.049.
31. Chebotar, Y. *et al.*, Closing the sim-to-real Loop: Adapting simulation randomization with real world experience, in: *Proceedings - IEEE International Conference on Robotics and Automation*, vol. 2019-May, 2019, doi: 10.1109/ICRA.2019.8793789.
32. Sadeghi, F. and Levine, S., CAD2RL: Real single-image flight without a single real image, in: *Robotics: Science and Systems*, vol. 13, 2017, doi: 10.15607/rss.2017.xiii.034.
33. Coady, P., Learning Artificial Intelligence. *Learn. AI*, 2017.
34. Todorov, E., Erez, T., Tassa, Y., MuJoCo: A physics engine for model-based control, *IEEE/RSJ International Conference on Intelligent Robots and Systems*, Vilamoura-Algarve, Portugal, 2012, pp. 5026-5033, doi: 10.1109/IROS.2012.6386109.
35. Zhang, F., Leitner, J., Milford, M., Corke, P., Modular deep Q networks for sim-to-real transfer of visuo-motor policies, in: *Australasian Conference on Robotics and Automation, ACRA*, vol. 2017-December, 2017.
36. Hemming, S., de Zwart, F., Elings, A., Petropoulou, A., Righini, I., Cherry tomato production in intelligent greenhouses-sensors and ai for control of climate, irrigation, crop yield, and quality. *Sensors (Switzerland)*, 20, 22, 1–30, 2020, doi: 10.3390/s20226430.

37. Hashemi-Beni, L., Gebrehiwot, A., Karimoddini, A., Shahbazi, A., Dorbu, F., Deep Convolutional Neural Networks for Weeds and Crops Discrimination From UAS Imagery. *Front. Remote Sens.*, 3, 1–12, 2022, doi: 10.3389/frsen.2022.755939.
38. Din, A., Ismail, M.Y., Shah, B., Babar, M., Ali, F., Baig, S.U., A deep reinforcement learning-based multi-agent area coverage control for smart agriculture. *Comput. Electr. Eng.*, 101, 2022, doi: 10.1016/j.compeleceng.2022.108089.

11

Diabetic Retinopathy Detection and Classification Using Deep Reinforcement Learning

H.R. Manjunatha* **and P. Sathish**

Department of Computer Applications, SOSS, CMR University, Bangalore, India

Abstract

Diabetic retinopathy (DR), a complication, is most usually connected to diabetes. These patients have constructive vision loss because it damages the retina. The majority of blindness worldwide is caused by this condition, and early identification can save individuals from vision loss. If the abnormalities are not detected in the early stage then there is no treatment for restoring the eyesight. Therefore, the only ways to resolve this irreversible situation is to diagnose and treat the disease early. Even if the existing DR datasets have been examined using a number of deep learning and machine learning approaches, better outputs have not been achieved in terms of preprocessing accuracy optimization of the classification feature extraction processes. Early and accurate detection of DR can prevent visual impairment in most DR cases. Even with today's technology, it is quite challenging to identify DR early because symptoms do not show up until the very end of the disease. Smart medicine could take advantage from the application of deep reinforcement learning models. As a result, DR classification requires an automated process that may be carried out using deep reinforcement learning (deep RL). This paper explains that deep reinforcement learning is used to identify and categorize DR. Compared to earlier approaches, this system would achieve a very low false positive rate and more than 95% performance in terms of sensitivity, accuracy, F1 score, precision, and specificity.

Keywords: Detection and classification, deep reinforcement learning, diabetic retinopathy

11.1 Introduction

This condition of diabetes is further divided down into two categories: diabetes mellitus and diabetes insipidus. Low levels of the hormone insulin secreted by the pancreas result in diabetes mellitus. The blood glucose levels could change as a result of these circumstances. Diabetes mellitus is the condition in which the blood glucose levels are variable. The leading cause of diabetic retinopathy is diabetes mellitus [1]. Blood vessel damage results from persistently increased blood glucose levels. Diabetic retinopathy is the term used to describe the situation if the blood vessel damage is close to the retina. Light rays from the eye's outer

**Corresponding author*: manjunatha.hr@cmr.edu.in

Shubham Mahajan, Pethuru Raj and Amit Kant Pandit (eds.) Deep Reinforcement Learning and Its Industrial Use Cases: AI for Real-World Applications, (239–264) © 2024 Scrivener Publishing LLC

surface are concentrated in the retina, which is its inner region, where messages are transmitted to the brain.

The retina in the human eye is a thin tissue layer. The retina is in charge of vision; it takes in light and transforms it into certain neural-based signals, which are then sent to the brain for visual identification. Diabetic retinopathy (DR) is the term for the condition when the retina of the eye got harmed by diabetes. Here the words "retino" and "pathy" both refer to the retina [5].

Diabetic retinopathy (DR) ultimately damages the retina's blood vessels, resulting in the loss of vision (an inner part of the eye tissue is sensitive to light), that develops more slowly when people with diabetes have less control over their blood sugar. Hyperglycemia is a condition that the body gets as the blood sugar levels increase, which harms the retinal pericytes. These retinal pericytes are essential for controlling the blood flow. Damage to the pericytes may prevent cells from adequately metabolizing glucose. Diabetic retinopathy is the name given to this early stage of the disease. Only a microscope can be used to find these qualities. Large molecules like proteins and lipids can flow into and out of the arteries when capillary or vascular permeability is increased [15]. Blurred vision, difficulties with vision floaters or dots when trying to see properly at night, and a black or empty patch in the middle of the vision are all signs of DR.

The eye is the smallest part when compared with other body organs, but its structure is incredibly complex. In particular, the eye's retina is complicated. Loss of vision might result if there is retinopathy. Therefore, it is critical to distinguish between a normal retina and a retina afflicted by diabetes. Since this complex issue must first start to affect the small blood vessels in the retina before being identified, this diabetic retinopathy frequently goes unnoticed until it develops to a point when one's ability to see is threatened. This little blood channel generates blood flow, and retinal fluid results in the formation of characteristics. When the condition progresses to the following stage, the development of new blood vessels allows oxygen to pass between the retina and impairs vision [8].

Since diabetic retinopathy can result in blindness and visual loss in patients, it is regarded as a fatal eye ailment. The blood vessels in the retina suffer considerable harm which results from extremely high blood sugar levels. The macula swells or thickens as a result of fluid leakage from the blood vessels in the eye, which prevents the passage of blood. On the retina, abnormal new blood vessel development does occur sometimes. The mentioned disorders can all result in irreversible visual loss. At first, diabetic retinopathy does not cause symptoms, but over time it might deteriorate and cause visual loss.

Early diagnosis could help to preserve their vision. In the first phases of diabetic retinopathy, symptoms may not be present. Reading or seeing faraway items may become challenging. The following signs and symptoms appear when the illness worsens or spreads: They have cloudy vision, increased floating spots, fluctuating vision, poor night vision, or more floating spots. Having trouble recognizing colors due to color vision impairment, patches of darkness or emptiness in the field of vision, such as shadows produced by in-eye specks, and complete loss of eyesight are indications of vision loss [9].

Diabetic retinopathy can appear in two different types of people with the disease: non-proliferative diabetic retinopathy (NPDR) and proliferative diabetic retinopathy (PDR). Microaneurysms, cotton wool spots, and hemorrhages are all included in NPDR, while retinal neo-vascularization, or abnormal blood vessel development, is included in PDR.

Non-proliferative retinopathy of the iris or retina is defined by microaneurysms, cotton wool regions, and hemorrhages. Proliferative retinopathy is defined by neo-vascularization.

The early stage of disease in retina is NPDR (non proliferative diabetic retinopathy) which causes tiny red spots. The abnormal pouching of blood arteries in these tiny spots might be microaneurysms, while the tiny spots themselves could be hemorrhage. Exudates, a term for fluid and fatty material that leaks from these blood vessels, can occur when the lining of the blood vessels maintains a significant damage. While PDR is an advanced stage of DR, it causes the creation of abnormal blood vessels in the retina, which is significant loss during this stage; non-proliferative is the milder type—it is largely symptomless. For research purposes, these characteristics can be rated clinically as normal, mild, moderate, or severe diabetic retinopathy by opthalmoscopy or fundus imaging. In addition, the DR is divided into four grades: normal stage represents level 0, mild stage represents level 1, moderate stage represents level 2, severe stage represents level 3, and proliferative stage represents level 4.

Vision blurring is a symptom of proliferative DR, the ultimate and most serious stage of DR. When DR reaches the third stage, it turns proliferative and can be quite severe, which leads to irreversible loss of the whole eyesight. In PDR, in the retina, abnormal blood vessels that already exist are created. Additional disadvantages of PDR include the retina's separation from the post-operative eye and the possibility that newly developed blood vessels may rupture, and hemorrhage inside the retina is causing the patient permanent blindness. When DR is detected early, it can be treated, since when it gets too bad, it is more challenging to treat. There are several forms of diabetic retinopathy, and each form contains certain fundamental components. These foundational elements include: (i) microaneurysm, (ii) exudates, (iii) cotton wool spots, (iv) loss of blood capillaries, (v) hemorrhage, and (v) over-creation of blood capillaries.

In order to execute appropriate therapy and avoid vision loss, regular DR screening is crucial. Blood pressure and controlling the blood sugar can be used to delay the onset of DR, and vision loss can be minimized late through photocoagulation or intravitreal injection. Regular screening is necessary for diabetic patients to monitor the progression of diabetic retinopathy among its four severity stages; this includes proliferative and normal. The most dangerous stage of diabetic retinopathy is the proliferative stage since it enhances the risk of blood loss, which might result in irreversible vision loss. In the early stages, DR may not show any symptoms or just cause mild vision issues; if treatment is not performed or no treatment is received, visual loss may later occur. To identify DR, a number of physical tests are utilized, including tonometry, pupil dilation test, and ocular activity test. The issue with these tests is that they take more time, which has an impact on patients and may result in total blindness. Additionally, it is very challenging to diagnose a disease in its early stages through testing. In order to treat the disease at an earlier stage, it is necessary to develop a novel diagnosis procedure [17].

However, there is a shortage of skilled ophthalmologists who can evaluate fundus images to identify DR, which occasionally leads to misdiagnosis. Additionally, in local regions with a higher concentration of diabetes patients and a lengthy investigative process, there is a shortage of educated physicians. As a result, using an automated diagnosis infrastructure in place of manual diagnosis can effectively save time and money [12].

Medical image analysis is a technology that has gained popularity over the last 10 years for creating visual images of inside body components in medical research and therapeutic intervention. Images are the subject of medical image analysis, which involves tasks including image capture, storage, transmission, and presentation. Digital images may also be used for low-cost processing, quick, immediate quality assessment, low-cost duplication, and adaptive alteration. According to the medical region, it was challenging to identify the stages of the retinal disease. The noise factors were released while using retinal images; therefore, the outcome was not clear. Medical image diagnosis was made simpler by image processing [4].

The images were typically easy to examine and provided a thorough explanation of each bodily part. The most developed tangible organ in the human body, the retina lines are the eye's inner layer; this is a remarkable type of photographed tissue. The incident light signal is converted to a neural signal, which the mind then modifies with the use of visual devices and concentrates. The focal retinal connector, vein, and their branches are made up in part of retinal veins. Any changes in the location of these retinal veins are used to identify a few anomalies, such as high blood sugar levels, diabetic retinopathy, and suddenly increased circulatory strain [13].

As a result, the techniques for evaluating images to find out the amount of DR diagnostic models are available. Additionally, the computer vision-based methodologies are divided into two categories: hands-on engineering and end-to-end learning. End-to-end learning produces the best categorization since the hidden rich characteristics are automatically learnt. The existing approaches, such as SIFT (Scale Invariant Feature Transform), HOG (Histogram of Oriented Gradients) Gabor Filters, LBP (Local Binary Patterns) etc., which are not encoded, the size and rotation differences and illumination are removed by manual engineering procedures. Additionally, the moderate level of DR cannot be identified by any of the approaches, despite the fact that several end-to-end learning and hands-on engineering techniques are used to find DR in the Kaggle dataset. Early diagnosis is mostly dependent on the discovery of the mild stage. The procedures mentioned earlier are commonly found to be inaccurate since they cannot be understood [7].

Computer scientists and engineers can significantly contribute to the identification of DR with modern machine/deep learning models and powerful computing resources [24]. Utilizing new technology is one of the most prevalent strategies for increasing performance in this region, specifically machine learning and deep learning. Numerous studies are investigating techniques to discover the DR faster in the research process [6].

Ophthalmologists are experts in the field who examine human retinas. They take images of the fundus with a high-resolution digital fundus camera. For the purpose of picture annotation and any future treatment needs, the fundus identifies different types of DR retinal defects. Early detection of DR is difficult to utilize due to the demanding physical diagnosis procedure and the lack of appropriate therapeutic options. As a result, such a medical condition requires sophisticated tools for accurate diagnosis and treatment. To overcome the challenges of labor-intensive DR identification, researchers have proposed intelligent expert systems that use deep learning (DL) for the analysis and in-depth evaluation of DR features using the fundus images [19].

Intelligent systems are more efficient in terms of time, feature extraction, fault identification, and early diagnosis and treatment compared to conventional methods. These intelligent systems identify depression as mild non-proliferative DR (NPDR), mild proliferative DR (PDR), early proliferative DR (PDR), etc., by using fundus images as their input, which are subsequently enhanced and analyzed for the extraction of important properties [3].

Among "machine learning (ML)," deep learning (DL) is a subset that enables machines to learn on their own and improve performance through practice without using human involvement. By absorbing environmental information, this resembles a human brain functions, and DL utilizes the neural network concept found in a brain, this propagates the training approach through learning. The development of computer programs that can utilize data to learn for themselves and are trained to perform with high accuracy on data that the model is unfamiliar with is unlabeled is at the basis of deep learning [2].

Learning begins with observation, in this case of information and instructions. When analyzing the information and making future decisions, the system then searches out patterns that are similar to those during training. The main goal is to enable the systems to learn autonomously, without interference or support from humans, and to adapt their behavior as a result. Networks that can learn from unsupervised training, unlabeled, or unstructured data are incredibly successful built using deep learning [16]. Pupil dilating examinations, vision tests, optical coherence tomography, and other physical procedures are available to diagnose diabetic retinopathy. However, they take a lot of time, and the patients must deal with a lot of pain [25].

The field of machine learning called deep reinforcement learning (deep RL) combines reinforcement learning (RL) with deep learning. In agent learning, a computational agent that learns by making errors is a problem. Deep learning (DL) allows agents to form opinions based on unstructured input data, hence removing the necessity for human state space engineering. Deep RL algorithms can select which plan of action is better achieve an objective (such as maximising the game score) by analysing extremely large inputs, like each pixel displayed on a video game screen. Deep reinforcement learning's numerous applications in computer vision, robotics, natural language processing, finance, education, healthcare, and video games. Additionally, reinforcement learning for the analysis of medical images has advanced quickly in recent years. DRL is used to automate decision-making within these ongoing treatment regimes. For a wide range of computer vision tasks, including image categorization, object and face detection, captioning, disease prediction, and detection, deep reinforcement learning is a useful technique.

Therefore, this analysis presents the identification and classification of diabetic retinopathy using deep reinforcement learning. The text's remaining sections are arranged as follows: In section 11.2, the literature survey is explained. The diabetic Retinopathy detection and classification using Deep Reinforcement Learning is presented in this section 11.3. In Section 11.4 evaluated the outcome analysis. The conclusion is covered in section 11.5.

11.2 Literature Survey

There have been several investigations and studies on the automated diagnosis of diabetic retinopathy utilising various advanced technologies and progressively precise exactness processes is developments were established. To create the current work, extensive research and analysis were done on the works listed below.

K. Suwarna Gothane, Raju Srujan, Nuthanakanti Bhaskar, and Divya G. *et al.*, [10] explains the use of deep learning to detect diabetic retinopathy. Currently, studies show that neural networks have a strong potential for therapeutic applications, grayscale fundus images and to identify DR, the Convolutional neural network (CNN) model is applied.

There are five steps for DR detection. There are four age groups with the following classifications: Mild, Proliferative, Moderate, Severe and DR. The primary benefit of CNN is that it automatically recognizes all significant features and the unique features of each class. CNN has a high computational efficiency.

Shah, S.S. and Bakal, J.W *et. al.*,[11] describes DL to detect the DR. In this analysis, gray scale fundus images are used and CNN is used to detect the DR. Here, the detection of DR is done is various stages such as Mild, No DR, severe, Proliferative and Mild classified on different age groups. The main advantage of using CNN is automatic detection of all the important features and learns each class peculiar features.

Vikramathithan A C, Bhaskar S, Pooja P, Navya M S, Rakshith K B *et. al.* [14] explains that image processing may be used to find Diabetic retinopathy (DR). The authors suggest that an efficient and reliable approach for analysing retinal images is provided by combining morphological procedures with geometric relationships relating to certain properties. The development of this system was predicated on the monitoring of multiple parameters in order to realize and create effective techniques or algorithms for the detection of DR. Automated DR grading provides benefits, but it is not disadvantages. This can be due to abnormalities in retinal images or the use of neural networks. Even though, there is a lot of world research and they are now conducting research, but it will take some time before we find the most precise method of detecting diabetic retinopathy.

Ramasamy Lakshmana Kumar, Padinjappurathu Shynu Gopalan, Seifedine Kadry and Robertas Damaševičius *et al.*, [18] Textural and ridgelet properties of retinal images are combined with a sequential minimum optimization classifier to detect diabetic retinopathy and textural aspects of retinal images. With this method, a model for DR diagnostics is developed. They first separate and combine the ophthalmoscopic characteristics from the retinal images using textural gray-level data such co-occurrence, run-length matrix, and Ridgelet Transform coefficients. Diabetic retinopathy is categorized using Sequential Minimal Optimization (SMO) based on retinal characteristics. The results of the studies show the quality and efficacy of the publicly available retinal image datasets, which are utilized for performance analysis.

Zhuang Ai, Xuan Huang, Yuan Fan, Jing Feng, Fanxin Zeng and Yaping Lu *et al.*, [20] combining deep ensemble learning and attention methods to find diabetic retinopathy DR-IIXRN is discussed in this paper. A complete detection model is composed of Inception V3, Inception ResNet V2, ResNeXt101, NASNetLarge, and Xception. They adjusted the network model for each base classifier utilizing fine-tuning, transfer learning, and multiple techniques to increase DR detection's efficiency. In order to decide which group the images belonged to (mild, normal, moderate, severe, or proliferative DR), a weighted voting technique is employed. The trained network model was fine-tuned by the authors, the efficiency of the algorithm in diagnosing diabetic retina was further validated by the use of hospital data and actual test samples in the hospital. The recommended method has higher AUC (Area Under the Curve), accuracy, and recall rates from the conventional single network model detection algorithm to 95, 92, and 92%, respectively, according to the experiments, demonstrating its adaptability and correctness.

Enas M.F. El Houby *et al.*, [21] explains the way to classify the stages of diabetic retinopathy using transfer learning. The identification of DR and the categorization of its phases were performed in this study using a Convolutional neural network (CNN) based on colored retinal fundus images. CNN offers an automated diagnostic and can recognize complex structures on the retina. To use the previously learned parameters in the detection, Transfer learning (TL) was used to apply the pre-trained VGG (Visual Geometry Group)

-16 CNN model. The results obtained from carrying out multiple studies designed with various severity groups are encouraging. For 5-class, 4-class, 3-class and 2- classes, the best-achieved accuracies are 86.5, 80.5, 63.5, and 73.7, respectively.

Kahlil Muchtar, Ramzi Adriman, Novi Maulina *et al.*, [22] explains that deep learning methods utilizing texture features are performed in the binary categorization of DA. A procedure for classifying and identifying DRs is explained. This method has been divided into two primary phases: The first step extracts texture characteristics using Local binary patterns (LBP), and advanced deep learning techniques for detection and classification issues are thoroughly examined in the second phase. Deep learning methods include ResNet, DenseNet, and DetNet. ResNet, DenseNet, and DetNet all produced positive early results for accuracy. Additionally, they assess the way each detection setting performs.

Yasuyuki Ishikaw, Ayaka Sugeno, Toshio Ohshima, Rieko Muramatsu *et al.*, [23] demonstrates the basic image processing and transfer learning techniques can be used to find lesions and evaluate the extent of diabetic retinopathy. EfficientNet-B3, a freshly created convolutional neural network, using the Kaggle training data through the 2019 Asia pacific tele-ophthalmology society (APTOS), which includes synthetic noise were utilized to develop a severity evaluation system for DR. After the dataset's blurred and duplicate images were removed using a numerical threshold, the trained model's specificity and sensitivity scores for recognizing DR retinas increased. Through the use of convolutional neural network visual explanation techniques, it was shown that EfficientNets-B3 may be used to assess both the particular retinal regions and the degree of DR. By using an experimentally lesion extraction is done using a threshold value that was calculated for the enhanced retinal images. Identifying red lesions and removing blood vessels at the same time, there was no that the soft and firm exudates from the red and white lesions had also been eliminated. With generally accurate results, using images from the DIARETDB1 database, the areas of the detected lesions were further validated with ground truth.

S. Saket Kajol Gupta, Chaturvedi, Vaishali Ninawe, S.Prakash Prasad *et al.*, [28] explains the way to grade diabetic retinopathy automatically using deep convolutional neural networks. Regarding the pre-trained DenseNet121 network, the training dataset was the APTOS 2019 (Asia Pacific Tele-Ophthalmology Society 2019 Blindness Detection) dataset., which received a number of modifications. The suggested approach obtained high accuracy and beat other advanced networks in early-stage identification. Accuracy, recall, and f1-score were also provided for our network, and quadratic weighted kappa values of 87%, 86%, 91.96%, and 86%, respectively. This design is extremely accurate, simple, and effective in terms of computing time and space.

Lourdhu Regina Suganthi S., Sneha U K, Shwetha S *et al.*, [29] explains that machine learning techniques are used to categorize diabetic retinopathy. Using domain knowledge, different elements are retrieved from the eye image to reveal various illness pattern features. In general, machine learning approaches used for automatic categorization are inflexible. The classifier is trained for binary classification using the pre-processed eye image data set, to determine if the patient's eye is infected or healthy. The model has been assessed using several metrics, including F-Score, Precision, and Recall. Machine learning algorithms are used to quantify and categorize the illness severity into several groups. The performance of the Decision Tree classifier outperforms the SVM classifier and the Random Forest approach, and according on the confusion matrix results.

S Ambaji. Jadhav, B Pushpa. Patil, Sunil Biradar *et al.*, [30] provides an enhanced rider optimisation technique using deep learning that uses optimum feature selection to identify diabetic retinopathy. Four categories are outlined by the deep belief network (DBN)-based categorization method for images: normal, early, moderate, or severe phases. An improved meta-heuristic technique is the modified gear and steering-based rider (MGS-ROA) Optimisation Method., adjusts the DBN's weight by choosing the best characteristics. Lastly, the recommended model's greater stability is demonstrated through efficient performance, comparison analysis dependability over current models. When compared to the current classifiers, the suggested model's performance is evaluated. Based on the results, the proposed MGS-ROA-DBN outperforms SVM (Support Vector Machine) by 17.1%, KNN (K-Nearest Neighbour) by 32.2%, and NN (Neural Network) by 30.1% in terms of accuracy.

R. Subhashini, T.N.R. Nithin, U.M.S. Koushik *et al.*, [31] explains that image processing GUI (Graphical User Interface) is used to find diabetic retinopathy. An image processing technique-based graphical user interface is being developed to find out if diabetic retinopathy affects the fundus/retinal image that the patient has access to. The graphical user interface will show the effect, the severity, and the action that needs to be taken by the patient or user. This substantially reduces the time, it takes to process a disease diagnosis, and as a backup to confirm or support in identifying the condition, the ophthalmologists may also utilize this graphical user interface.

BK Tripathy, Kaushick Parui *et al.*, [32] explains the use of image processing to detect diabetic retinopathy. This method's main goal is to use image processing techniques to find Diabetic retinopathy (DR) eye condition. MATLAB (Matrix Laboratory), the tool utilized in this approach is an applications that is frequently used for image processing. In order to identify DR, using a medical image of the retinal fundus of the human eye, this study provides a method for separating blood vessels. This approach makes use of the open CV (computer vision) framework to develop the CLAHE (contrast limited Adaptive histogram equalization) method. The results show that the DR is observable in the fundus is consider of the injured eye whereas it is not found in the fundus image of the healthy vision.

Naveen R., Maruthi Shankar B., Sivakumar S. A., Keerthana Priyaa A. *et al.*, [33] offers Image Processing for Diabetic Retinopathy Detection. The outcome demonstrated that the DR (Diabetic Retinopathy) in the injured fundus images, diseases was seen, since in the healthy fundus image, DR was unaffected. The final accuracy evaluation of the method is computed using the attributes of Specificity (SP), Accuracy (ACC), and Sensitivity (SE). These parameters for DR detection are determined using training images from the High resolution Fundus (HRF) images collection.

N. Yalçin, S. Alver and N. Uluhatun *et al.*, [34] explains the use of deep learning to classify retinal pictures with the purpose of detecting diabetic retinopathy early days. There are two phases in this method. Pre-treatments were carried out in the first step to eliminate retinal pictures and standardise their size from various data sets. Convolutional Neural Networks, a deep learning method, were used in the second stage of data classification. The ability of this study's deep learning network to automatically construct itself in a very short amount of time sets it apart from other approaches that use the Central Processing Unit (CPU) and Graphics Processing Unit (GPU) during the training phase.

Dutta Suvajit, CS Bonthala Manideep, Syed Muzamil Basha, D. Ronnie Caytiles and Iyengar N. Ch. S. N. *et al.*, [35] uses deep learning algorithms to provide categorization of photos of diabetic retinopathy. After evaluating the models using CPU-trained neural networks, this model was created using three different backpropagation techniques: Deep neural networks (DNN),

NN, and convolutional neural networks (CNN). One hidden layer helps deep learning models perform better than NN. Deep learning models can be used to classify features such as blood vessels, fluid drop, exudates, hemorrhages, and microaneurysms. The weights that indicate the patient's eye's severity level will be calculated by the model. The primary obstacle facing this analysis is the precise determination of each feature class thresholds. Using a weighted fuzzy C-means technique, the target class thresholds were identified. The algorithm will be useful in classifying images of diabetic retinopathy according to their severity.

Ö. Deperlıoğlu and U. Köse *et al.*, [36] explains that image processing and a convolutional neural network are used to diagnose diabetic retinopathy. The HSV (Hue Saturation Value) transform algorithm and histogram equalisation techniques were applied in a practical way. The picture of the retinal fundus was then subjected to a Gaussian low-pass filter. Using a convolutional neural network, categorization was done after image processing. To determine the effectiveness of the presented method, through the Kaggle Diabetic Retinopathy Detection database, and this 400 retinal fundus pictures were utilized. For each step of the image processing process, classification work has been done in experiments. The categorization analysis carried out following image processing. For each stage, average values were found after 20 experiments are completed. The gathered data show as good and effective this method is in identifying diabetic retinopathy from retinal fundus images.

Shruti Nair Sisodia Dilip Singh, and Pooja Khobragade *et al.*, [37] Feature extraction and preprocessing of diabetic retinal fundus images to enable early detection of diabetic retinopathy. In this method, before performing scaling, histogram equalization, and image enhancement, the green channel is taken out of the original retinal fundus images. Additionally, from pre-processed images, fourteen features are extracted for quantitative analysis. The diabetic retinopathy dataset from Kaggle is used for the tests, the evaluation of the results takes into consideration the mean value and standard deviation for the acquired features. With a mean difference of 1029.7, exudate area was the parameter that received the highest ranking in the analysis. The outcome was described to its total absence in images of diabetics with normal blood sugar levels and its simultaneous appearance in photographs of all three types of diabetic retinopathy is mild, normal, and severe.

Sai Sudha, G., Praveena, M., SandhyaRani, G., Harish, T.N.S.K., Charisma, A., Asish, A *et al.*, [26] desecribes a DR detection and classification model using Deep Learning. The main aim of this work is to make the detection process as easy as possible while using CNN classifier. This model made the DR detection is easy for doctors, however, appropriate testing and validation are needed before doing the treatment direct on a patient.

Skariah, S.M. and George, J. *et al.*, [27] presents A Novel Study on Automated Diabetic Retinopathy: Detection and Grading. In this analysis, various ML, ensemble learning, transfer learning and DL techniques are compared to detect and grade the DR. From the results, it was observed that, ML approaches do not produce any overhead but the grading and identification consumes time and tedious. The transfer learning and ensemble learning supports the DR identification and grading to large extent. The DL models achieved better performance for grading and detection.

Chandrakumar, T. and Kathirvel, R *et al.*, [38] describes Classifying Diabetic Retinopathy using Deep Learning Architecture. DCNN (Deep Convolutional Neural Network) is used and it provided better performance through spatial analysis. The DCNN is one of the complex architectures because it is inferred from the visual perspects of human. This DCNN architecture was deployed with dropout later models and showed better performance.

Priya, R. and Aruna, P *et al.*, [39] describes Diagnosis of Diabetic Retinopathy Using Machine Learning Techniques. In this model, SVM, PNN (probabilistic neural network) and Bayes are used and their performances are compared. The retina features are extracted to identify the amount of disease spread in the retina. Image processing models are used to extract features like NPDR, blood vessels, PDR exudates from raw images. Compared to PNN and Bayes classifier, SVM obtained better results.

Venkatesan, R., Chandakkar, P., Li, B., Li, H.K *et al.*, [40] presents classification of diabetic retinopathy images using multi-class multiple-instance learning based on color correlogram features. The main concern is automatic classification of DR images with two important clinical findings namely MA (Micro-Aneurysm) and NV (Neo-Vascularization). A Modified color auto-correlation feature is tuned spectrally with less dimensionality towards the DR images. The results are promising than other models.

11.3 Diabetic Retinopathy Detection and Classification

In this analysis, diabetic retinopathy detection and classification using deep reinforcement learning is presented. The most important diagnostic method for the majority of retinal-based disorders is the recognition and categorization of diabetic retinopathy. Retinopathy is a malignant illness that may easily affect both men and women and is classified. According to these issues, the affected individuals may lose their eyesight capacity at a young age. The most crucial component is to recognize and categorize the retinal illness at an earlier stage because tracking the damaged parts is disadvantaged. Deep reinforcement learning is utilized to identify and categorize the DR in order to get around these problems. The proposed technique's block diagram is shown in Figure 11.1.

A collection of FUNDUS retinal images with pixels larger than 2000x3000 has been developed. The dataset, which is available for free on the website kaggle.com, has 35,000 images for testing and 15,000 images for training. This dataset consists of a sizable collection of retinal pictures in high resolution that was captured under various imaging circumstances. Each subject has a left and a right field. Images are identified by left or right and a subject id (e.g., 1_left.jpeg indicates the left eye of patient number 1). A doctor used a scale ranging from 0 to 4 to determine whether diabetic retinopathy was present in each photo: All of the following are zero: no, mild, moderate, severe, and proliferative.

Thirty Five Thousand One Hundred Twenty Six images in all make up this dataset. The validation set, test set, and training set are each given a sample size of 21,074, 7,026, and 7,026, respectively, following a 3:1:1 creation of training, validation, and test sets from the data source. The images in the dataset were taken using various camera types and configurations, which might have changed that the left and right seem to the eye. The retina is shown anatomically in multiple positions, with the optic nerve on the right and the macula on the left for the right eye. Others are shown upside-down, or using a microscope condensing lens, much like they would during an actual eye exam.

If a picture is inverted, there are often 2 techniques to detect: The macula, the tiny, black core spot is inverted if it is positioned slightly above the midline of the optic nerve. If the macula lies below the midline of the optic nerve, it is not inverted. Any line that runs down the side of an image (such as a square, triangle, or circle) is not reversed. In the absence of one, it will be reversed. Like with any real-world data gathering, there will be noise in

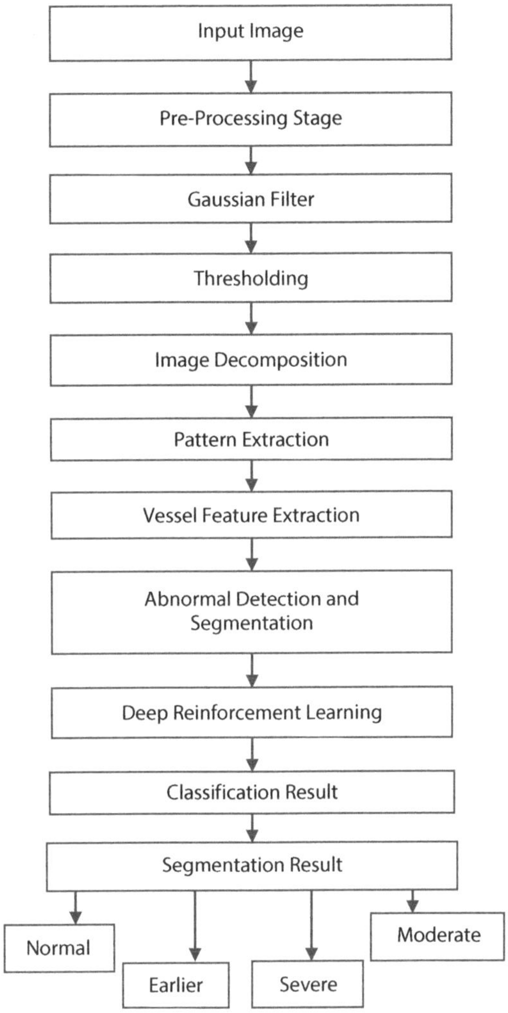

Figure 11.1 Block diagram of diabetic retinopathy detection and classification using deep reinforcement learning.

the labels as well as the images. Images could contain artefacts, be underexposed or overexposed, or be out of focus. Implementing strong algorithms that work in reducing noise and variability is the primary goal of this competition. To reduce the noise, images must be pre-processed. The images from the gathered Kaggle dataset are initially set up for pre-processing and standardization. Data from the bigger database must be removed, divided into different subsets for different purposes, and then combined in order to execute the pre-processing procedure. Additionally, a pre-processing technique that transforms a big dataset into a sensible arrangement is used for the goal of removing undesired cases. In this case, the real-world dataset is invariably fragmented, and sending such information through a model could impact the entire model.

Here, the Laplacian notion of a Gaussian filter is used with a double-factor filtering strategy to increase the image rate during preprocessing, and a pad array is built using the binary state of the image. Every pixel in the area of interest is sent through the filter, which is constructed as an odd-sized symmetric kernel (a DIP version of a matrix), in order to obtain the desired result.

Apply Gaussian filtering to get rid of noise and detail. Apply Gaussian filtering to get rid of noise and detail. As a point-spread function, the Gaussian filter makes use of the 2D distribution. Convoluting the picture with the 2D Gaussian distribution function allows. Uneven low pass filtering is achieved through the Gaussian filter. The kernel coefficients decrease as one moves away from the kernel's center. The kernel coefficients decrease with increasing distance from the kernel core. A bigger peak (more blurring) results from larger values of σ. For the filter to remain Gaussian, the kernel size must grow as rises σ. The value of σ depends on the Gaussian kernel coefficients. Coefficients must be near zero at the mask's edge. There is no directional bias in the kernel, which is rotationally symmetric. The separability of the Gaussian kernel enables quick computing. Due to the separability of the Gaussian kernel, calculation is quick. It is possible that Gaussian filters don't maintain image brightness.

In one dimension, the Gaussian function is:

$$G(x) = \frac{1}{\sqrt{2\pi\sigma^2}} e^{-\frac{x^2}{2\sigma^2}} \tag{11.1}$$

When the distribution's standard deviation is σ. This is predicted that the distribution's mean is zero. They have to utilize the two-dimensional Gaussian function while working with photographs. This equation is simply the union of two 1D (1-Dimensional) Gaussian functions, one for every possible direction.

$$G(x, y) = \frac{1}{\sqrt{2\pi\sigma^2}} e^{-\frac{x^2+y^2}{2\sigma^2}} \tag{11.2}$$

The 2D Gaussian function is sampled for the Gaussian kernel coefficients. Where the distribution standard deviation is σ. It is assumed that the distribution has a mean of 0. To save continuous Gaussian functions as discrete pixels, they must discretize them initially. Using 2D distributions are implemented as a point-spread function by the Gaussian filter to operate. Convoluting the picture with the 2D Gaussian distribution function allows. The Gaussian function has to be approximated discretely. This may need the use of an infinitely large convolution kernel since the Gaussian distribution does not include zeros everywhere.

Images that were captured in poor light and have a lot of noise can be have that noise reduced through gaussian blur. A non-uniform low pass filter is the Gaussian filter. As one goes away from the kernel's center, Kernel coefficients are decreasing. More weight is placed on central pixels than on outside ones. Wider peaks (more blurring) are produced by bigger values. The segmentation disc region is visible in the filtered image's threshold. Grey threshold is used to remove the background.

When blood glucose levels are out of balance, diabetic retinopathy occurs when veins in the retina alter with lesion defects. The vessels expand and completely swell the liquid. On the exterior of the retina, abnormal blood vessels can appear in a number of situations. The mean value of an image with a normal and grey level is used as the basis for thresholding.

$$Th = \frac{1}{2}(M_{normal} + M_{gray}) \quad (11.3)$$

The term "image decomposition" refers to the division and separation of an image's constituent parts. In the domains of computer graphics and image processing, image decomposition which divides a given input picture into structure and texture images are has been applied in a number of ways. DWT (Discrete Wavelet Transform) is utilized in this instance to decompose images. One of the most often used techniques for image processing is DWT. This DWT transform technique divides the image into four parts, and these wavelet coefficients precisely simulate the characteristics of the human visual system.

The image being processed is divided into a number of bands with different scales, especially low-high, high-low, and high-high, with the exception of the low-low band, which is incorrect and randomly laid off. The two included input images were combined utilizing the fusion procedure while taking the wavelet into consideration I1(x, y) and I2(x, y) are modified by x, the denotation can be described, and then, it is determined the inverse wavelet change x-1. The result of the reimplemented image is I(x, y).

$$I(x,y) = \omega - 1\Big(\varphi\Big(\omega\big(I1(x,y)\big), \omega\big(I2(x,y)\big)\Big)\Big) \quad (11.4)$$

A shape pattern develops when a series of shapes is repeated more than twice. Identifying a form pattern requires the order in which forms are repeated, and to finish it, the must first examine the most recent form that is known. then adding the following shape in the sequence. This system extracts the recurring patterns. An actual thing or an abstract idea both make up a pattern. A description of an animal would be a pattern when discussing the many types of animals. When discussing different ball types, a pattern is used to describe each ball. Football, cricket, table tennis and other balls may fall within the classification of pattern ball in this scenario. The class of the pattern must be ascertained for a new pattern. In order to classify patterns, choosing the qualities and an essential stage is learning how to express them. A good representation is one that makes advantage of distinguishing characteristics and also reduces the computing load in pattern categorization.

For features that are texture-oriented, repeated patterns are extracted. An essential method for finding and recognizing objects is pattern extraction. In order to accurately identify the lesion site in the retinal fundus, this analysis introduces an improved feature analysis for segmentation and detection. The vessels feature is examined during the segmentation procedure in order to acquire the extraction phase of the retina by correcting the optical disc for abnormal detection.

The procedure of obtaining a binary vessel map using the extraction of retinal blood vessels labels using retinal background pixels as logic 0 (black) and retinal blood vessels as logic 1 (white), or the different. Four local characteristics are retrieved during the feature extraction stage: the FAZ (Foveal Avascular Zone) region, the fractal dimension, the blood vessel density, and the skeletal vessel density.

The percentage of a specific region that is made up of blood vessels is known as vessels density. The percentage of vessel area with blood flow over the whole area measured was used to determine the vessel density. Binarized and skeletonized images show blood vessels as white pixels, hence the vessel density was determined from these images by dividing the proportion of white pixels to all other pixels in a given region. An established terminology known as VSD (Vessel Skeletal Density) is frequently used to calculate capillary density. The complexity of the branching patterns in the vasculature is measured by the retinal Fractal dimension (FD). A possible biomarker for the identification of numerous disorders, including diabetes and hypertension, has been identified as FD. A more complicated and thicker retinal vasculature is indicated by higher noninteger values between 1 and 2. The complexity of the retinal vascular tree may be measured using a concept called retinal vascular fractal dimension (D_F).

The Foveal avascular zone (FAZ) is a region that lacks retinal blood vessels within the center of the human retina, or the fovea centralis. It is believed that the absence of blood vessels and the surrounding inner retinal tissue improve the fovea pit's optical quality by minimising light scattering. The foveal avascular zone refers to this foveal avascular area.

The task of segmenting vessels in retinal images is known as retinal image segmentation. Segmenting retinal vessels is one of the significant developments in retinal image processing. For automated vessels analysis, segmenting retinal blood vessels is a crucial initial step. Advanced analysis may be carried out after blood vessel segmentation looking at its diagnostic and prognostic values for eye illnesses including arteriosclerosis and hypertension. In ophthalmology, segmentation of blood vessels is a key part of color retinal fundus image processing for tracking and identifying retinal diseases such as diabetic retinopathy, arteriosclerotic, and age-related macular degeneration.

Using a technique called grey-level thresholding, which assumes that the images has a bimodal histogram, This image has been segmented and cleaned up of blood vessels. Consequently, An easy way to differentiate an object from the background is to compare the image with the threshold TL (Threshold Limit) value, which separates the image into two modes. The binary image that results from the threshold process is shown in Eq. In this case, the objects are represented by pixels with intensities of 1, whereas the backdrop is represented by pixels with intensities of 0.

$$In_{BVR} = \begin{Bmatrix} 1\ if Im(x,y) > TL \\ 0\ if Im(x,y) \leq TL \end{Bmatrix} \tag{11.5}$$

After the blood vessels are removed, the image is represented as In_{BVR} and is then put through abnormality segmentation.

Deep Reinforcement Learning is the outcome of combining Deep Learning and Reinforcement Learning. In robotics, deep reinforcement learning is frequently used. The robot must do successive activities, which are required. Agents can be used in manufacturing, healthcare, and industrial automation because they can learn to deal with constantly changing surroundings. The subfield of artificial intelligence (AI) and machine learning is called reinforcement learning (RL). The Learning Method aims to maximize a cumulative reward signal by learning from interactions with an environment. The idea of trial and error is a key component of reinforcement learning. An reinforcement learning agent

(RL) performs a sequence of tasks in a dynamic environment and receives feedback in the form of rewards and penalties, represented by a reward function, in order to learn from experience. As the AI agent gains experience, it should be able to maximize certain goals that are presented as cumulative rewards. Finding the most effective method of action, or a mapping between states and actions that maximizes predicted cumulative reward, is the agent's goal. The idea behind reinforcement learning is that an agent would learn from its surroundings by interacting with it and earning rewards for actions.

Reinforcement learning has the potential to handle a wide range of computer vision problems, such as object recognition, face detection, captioning, image classification, and more. Reinforcement learning is an essential component in interactive perception, where perception and interaction with the environment would be advantageous to one another. This covers operations like object segmentation, estimate of articulation models, object dynamics, estimation of haptic properties, object identification or classification, estimation of object position, estimation of grasp planning, and learning of manipulation skills.

Deep Reinforcement Learning (RL) instructs an autonomous agent to interact with a given environment by using deep neural networks to accomplish predetermined goals. Deep reinforcement learning applies information from the past to new data. Deep reinforcement learning eliminates the requirement to predefine the environment by accepting raw data, such as pixels, as input, enabling the model to be applied in a wide range of cases. Deep reinforcement learning algorithms can be created in a way that makes them general and permits the use of the same model for many tasks by utilizing this layer of abstraction. The utilization of deep learning technology in reinforcement learning offers the potential for generalization and efficient performance on unknown inputs. Neural networks that have been trained for image identification, for instance, are capable of recognizing the presence of a bird in an image even though they have never seen the specific bird or image.

Deep RL models may be used to a number of applications since there is less need to predefine the environment raw data, such as pixels, can be used as input. Deep reinforcement learning algorithms can be created in a way that makes them general and allows the application of the same model to both detection and classification problems with the help of this abstraction layer. Traditional RL algorithms are unable to solve high-dimensional MDP states, such as images from a camera or the raw sensor stream from a robot, which are present in many real-world decision-making issues. By expressing the policy $\pi(a/s)$ or other learnt functions as a neural network and creating specialized algorithms that work well in this environment, deep reinforcement learning methods use deep learning to solve such MDPs (Markov Decision Process). DRL is a perfect fit for problems that require sequential decision-making – that is, a series of decisions that all affect one another. DR detection and classification is one of the most challenging tasks and it is essential to classify them accurately. Here DRL is used to detect and classify the Diabetic Retinopathy. The main objective is to detect and classify different stages of DR very effectively. For this purpose DRL is used. Deep Reinforcement Learning (DRL) is used to ensure accurate results at a low cost. The algorithm stages are shown in Figure 11.2:

Algorithm Steps:

1. Get the Dataset
2. Pre-process the Collected data
3. Train, validate and Test the data

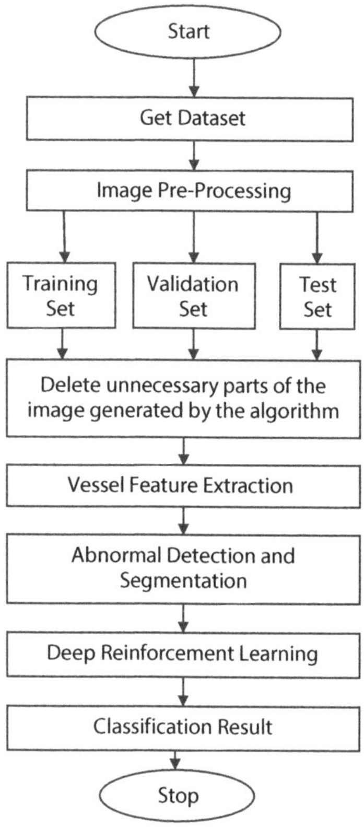

Figure 11.2 Algorithm.

4. Delete the unnecessary features
5. Vessel feature extraction
6. Abnormal Detection and Segmentation
7. Classification using DRL

In this approach, Diabetic retinopathy detection dataset from Kaggle is used. The dataset image contains noise. To remove the noise, Image processing is used. The data is then split into three categories, including training data, validation data, and test data. It is possible for certain pixels outside the "eyeball" area must be transformed by the random-angle rotation to have pixel values greater than 7. As a result, extraneous features are eliminated. Create a circle for each upsampled image, with the image's center serving as the circle's center and its radius equal to half the image's side length. Then, put pixels with values of 0 outside the circle. The unnecessary areas of sampling images are removed and up sampling images are obtained with circular eye areas. The feature extraction technique makes use of the energy and homogeneity determination factors.

Finally, the DRL classifier detected and classified the Diabetic retinopathy in to four different types Normal, Moderate, Earlier and Severe which are as follows: In the early phases of diabetic retinopathy, the retinal blood vessel walls weaken. Small bulges that sometimes leak or oozing fluid and blood into the retina protrude from the vessel walls. The retina's tissues might enlarge, resulting in white spots.

At a moderate level, the small blood vessels continue to grow, obstructing the retina's access to blood and preventing proper nourishment. The disease's more severe form is Proliferative diabetic retinopathy (PDR). Circulation issues are now denying oxygen to the retina. The creation of fine blood vessels is caused by the gel-like fluid that fills the vitreous, the retina, and the vitreous gel.

In this way, different DR stages are classified. If the disease is detected and after being properly categorized, a diagnosis is given. As a result, the loss of vision is reduced greatly. Accuracy, precision, sensitivity, specificity, and f1-score are used to validate the performance of the provided technique using the confusion matrix characteristics of false positive, true positive, false negative, and true negative. This is a description of the confusion matrix's parameters:

True positive (TP): A TP result indicates that the model classified the DR illness properly.
True negative (TN): A TN result indicates that the model accurately identified DR is not present.
False positive (FP): When the model improperly categorizes the DR, the result is a false positive.
False negative (FN): False negative is a result that occurs when the model misclassified the existence of DR.

Accuracy: The total number of predictions, amount of correct forecasts. Using the equation shown below, accuracy for binary classification can also be assessed in terms of positives and negatives:

$$Accuracy = \frac{TP + TN}{TP + FN + FP + TN} \times 100 \tag{11.6}$$

Precision: The quality of a model's successfully detected instances. By dividing the entire number of accurate positive predicts by the total number of real positives, precision is computed.

$$Precision = \frac{TP}{TP + FP} \times 100 \tag{11.7}$$

Sensitivity: Recall or true positive rate (TPR) is other names for it. It is the ratio of accurately identified positive instances to all positive instances used.

$$Sensitivity = \frac{TP}{TP + FN} \times 100 \tag{11.8}$$

Specificity: Specificity is the algorithm's or model's capacity to recognise a real negative for each category that is made available.

$$Specificity = \frac{TN}{TN + FP} \times 100 \tag{11.9}$$

F1 score: An assessment metric that evaluates a model's accuracy is the F1-score. The accuracy and recall scores of a model are combined. Across the whole dataset, the accuracy statistic records the number of times a model made a correct prediction.

$$F1 - score = 2 \times \frac{Precision \times Sensitivity}{Precision + Sensitivity} \times 100 \tag{11.10}$$

False Positive Ratio (FPR): The possibility of incorrectly categorizing a certain test is represented by the false positive ratio, sometimes referred to as fall-out or false alarm ratio. It is written as

$$FPR = \frac{FP}{FP + TN} \times 100 \tag{11.11}$$

11.4 Result Analysis

Diabetic retinopathy detection and classification using deep reinforcement learning is implemented in this section. Here 35,216 images from the diabetic retinopathy dataset Kaggle are utilized. The DR is found and classified using deep reinforcement learning. This method recognizes the DR and categorizes it into four different categories, including normal, earlier, moderate, and severe. Sensitivity, precision, accuracy, specificity, and F1-score are used to evaluate the way the presented technique works.

The evaluation of performance metrics is shown in Table 11.1.

Comparisons are made between the performance of the presented approach and that of artificial neural networks and decision trees. The DRL classifier has better performance than ANN classifier. Figure 11.3 shows the sensitivity comparative graph. The x-axis in Figure 11.3 shows several classifiers, while the y-axis shows performance results expressed as percentages.

Table 11.1 Performance evaluation.

Metrics/methods	**DT**	**ANN**	**DRL**
Accuracy (%)	91.24	94.23	98.45
Sensitivity (%)	89.43	91.34	97.64
Specificity (%)	90.49	92.45	97.23
Precision (%)	88.24	93.53	98.23
F1-score	89.31	92.47	97.46
FPR (%)	13.43	10.23	2.1

Compared to DT and ANN, presented DRL classifier has high sensitivity. Figure 11.4 shows the specificity performance comparison.

In Figure 11.4, the x-axis indicates different classifiers whereas y-axis indicates performance values. The DRL classifier has high specificity than ANN and DT classifiers. Figure 11.5 shows the precision comparison graph.

In Figure 11.5, the x-axis shows classifiers, whereas the y-axis shows performance values in terms of percentage. The high precision value is achieved by DRL classifier. Figure 11.6 shows the accuracy comparison.

In Figure 11.6, y-axis indicates the performance values and x-axis indicates the classifiers. The DRL classifier achieved high accuracy than DT and ANN classifier.

The Figure 11.7 shows the F1-score comparative graph where x-axis indicates the deep learning classifiers and y-axis indicates performance values. Presented DRL model has achieved better F1-score than DT and ANN classifiers. Figure 11.8 shows the FPR comparison.

The DRL classifier has very less FPR than DT and ANN. Figure 11.9 shows the accuracy of different DR stages.

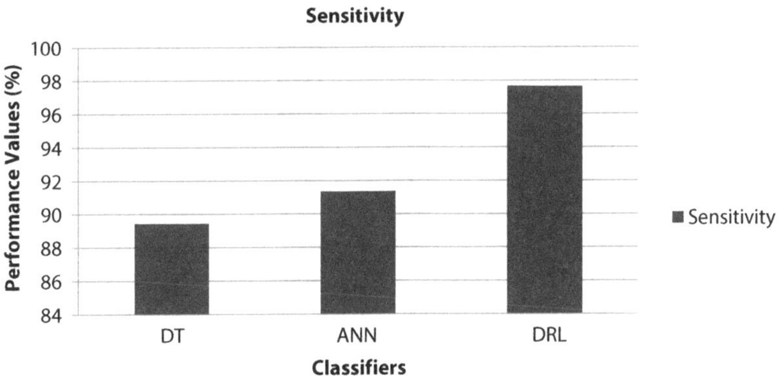

Figure 11.3 Sensitivity comparison.

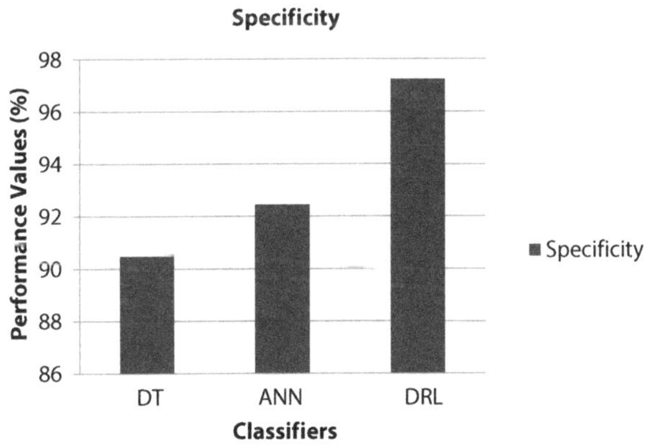

Figure 11.4 Specificity performance.

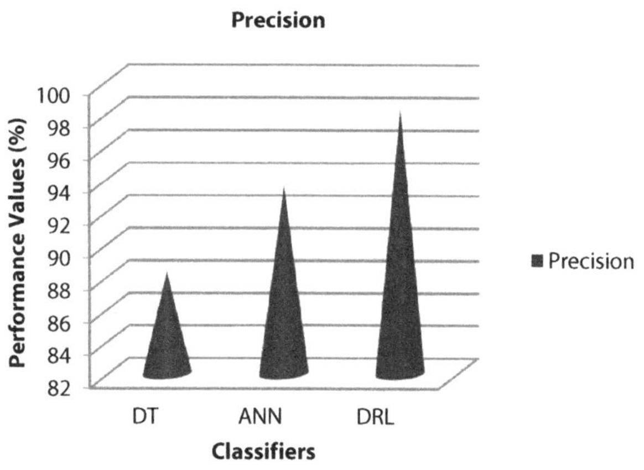

Figure 11.5 Comparative graph for precision.

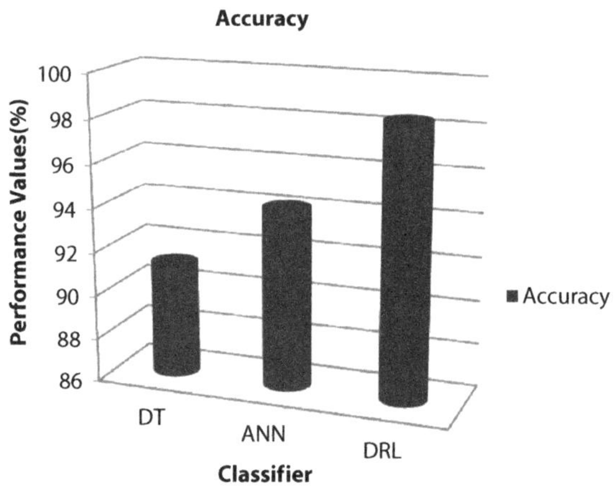

Figure 11.6 Accuracy comparative graph.

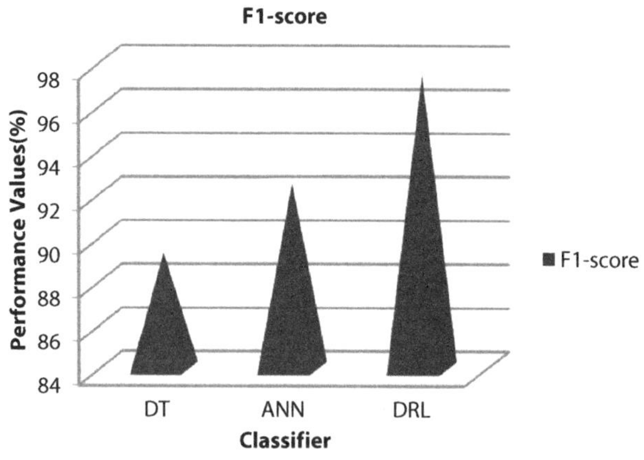

Figure 11.7 F1-score comparative graph.

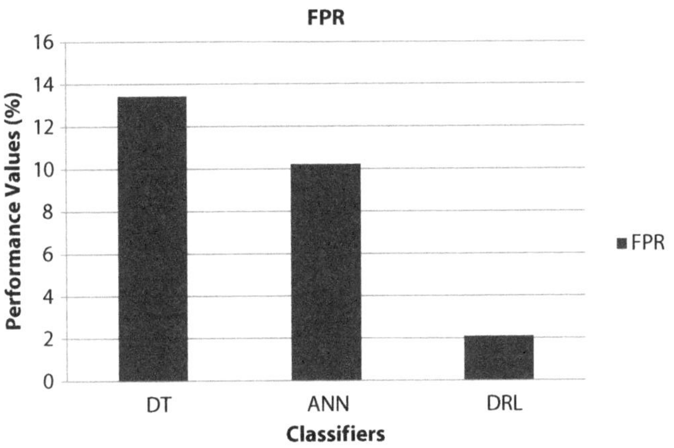

Figure 11.8 False positive rate comparison.

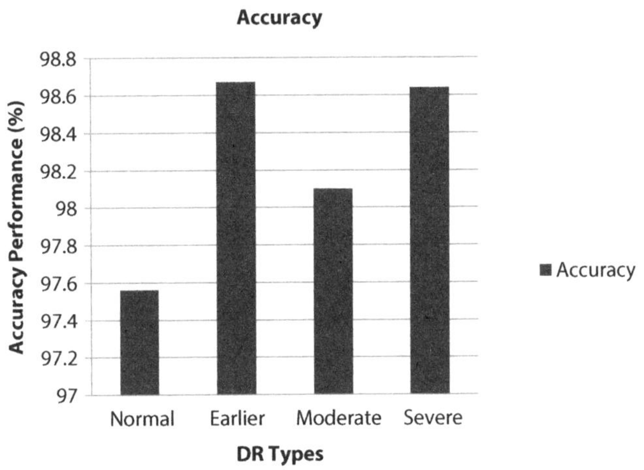

Figure 11.9 Accuracy comparison of DR stages.

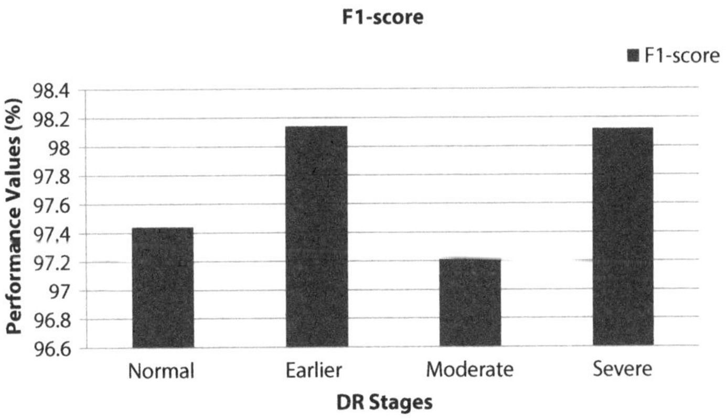

Figure 11.10 F1-score comparison for different DR stages.

In Figure 11.9, y-axis displays accuracy performance while the x-axis different forms of diabetic retinopathy. From the results, it is clear that earlier severe stage have high accuracy than earlier moderate stages. Figure 11.10 shows the F1-score of DR stages.

In Figure 11.10 y-axis displays the F1-score performances while the x-axis different forms of diabetic retinopathy. From the results, it is clear that earlier severe stage have high f1-score than earlier moderate stages. Therefore, this approach has better and high results in terms of accuracy, precision, F1-score, specificity and sensitivity.

11.5 Conclusion

The diagnosis and categorization of retinopathy plays a major part in recognising the abnormalities in the eye, which helps healthcare center's are providing better treatment. This paper describes the identification and categorization of diabetic retinopathy using deep reinforcement learning. The diabetic retinopathy detection dataset from Kaggle is used in this analysis to train, validate, and test the classifier. Extraction of blood vascular features during preprocessing, segmentation of variations, and classification were all steps in this DR detection and classification procedure. The segmentation of retinal vessels makes it possible to determine the impact of eye pathogen infection. DR detection and classification is done for four stages: normal, earlier DR, moderate DR, and severe DR. The diabetic retinopathy illness is detected and classified using deep reinforcement learning (DRL). Lesion on the retina caused by diabetic retinopathy results in loss of vision, but it can be prevented through the use of the suggested strategy. Precision, accuracy, sensitivity, f1-score, specificity, and FPR are utilized to verify the effectiveness of the presented strategy. Compared to previous classifiers, the presented DRL has high sensitivity, specificity, F1-score, FPR, accuracy, and precision. The system for extracting and segmenting retinal vessels using deep reinforcement learning was shown to perform well, according to an extensive examination of the results.

References

1. Bajwa, A., Nosheen, N., Talpur, K.I., Akram, S., A Prospective Study on Diabetic Retinopathy Detection Based on Modify Convolutional Neural Network Using Fundus Images at Sindh Institute of Ophthalmology & Visual Sciences. *Diagnostics (Basel)*, 13, 3, 393, 2023 Jan 20, doi: 10.3390/diagnostics13030393.
2. Mohanty, C., Mahapatra, S., Acharya, B., Kokkoras, F., Gerogiannis, V.C., Karamitsos, I., Kanavos, A., Using Deep Learning Architectures for Detection and Classification of Diabetic Retinopathy. *Sensors*, 23, 5726, 2023, https://doi.org/10.3390/s23125726.
3. Das, D., Biswas, S.K., Bandyopadhyay, S., Detection of Diabetic Retinopathy using Convolutional Neural Networks for Feature Extraction and Classification (DRFEC). *Multimed. Tools Appl.*, 2023, 82, 29943–30001, 2022, Springer, https://doi.org/10.1007/s11042-022-14165-4.
4. Maithili, K., Latha, Y.M., Gangopadhyay, A., Varghese, I.K., Sundaram, A., Pandey, C., Kiran, A., Optimized CNN Model for Diabetic Retinopathy Detection and Classification. *Int. J. Intell. Syst. Appl. Eng., IJISAE*, 11, 7s, 317–331, 2023, doi: https://ijisae.org/index.php/IJISAE/article/view/2957, ISSN:2147-6799.

5. Rajendran, S.K., Mathew, D., Rajendiran, B., Kandasamy, V., Diabetic Retinopathy Detection Using Deep Learning Techniques. *AIP Conf. Proc.*, 2790, 020106, 2023, https://doi.org/10.1063/5.0152424.
6. Pradeep, J., Jeffery, N.E., Saranraj, M., Hussain, J.N., Enhanced Recognition system for Diabetic Retinopathy using Machine Learning with Deep Learning Approach. *J. Popul. Ther. Clin. Pharmacol.*, 30, 11, e452–e466, 12 May 2023, DOI: 10.47750/jptcp.2023.30.11.047.
7. Ryu, G., Lee, K., Park, D., Kim, I., Park, S.H., Sagong, M., A Deep Learning Algorithm for Classifying Diabetic Retinopathy Using Optical Coherence Tomography Angiography. *Transl. Vis. Sci. Technol.*, 11, 2, 39, 2022, https://doi.org/10.1167/tvst.11.2.39.
8. Gundluru, N., Rajput, D.S., Lakshmanna, K., Kaluri, R., Shorfuzzaman, M., Uddin, M., Khan, M.A.R., Enhancement of Detection of Diabetic Retinopathy Using Harris Hawks Optimization with Deep Learning Model. *Hindawi Comput. Intell. Neurosci.*, 2022, Article ID 8512469, 13, https://doi.org/10.1155/2022/8512469.
9. Yasashvini, R., Vergin Raja Sarobin, M., Panjanathan, R., Graceline Jasmine, S., Jani Anbarasi, L., Diabetic Retinopathy Classification Using CNN and Hybrid Deep Convolutional Neural Networks. *Symmetry*, 14, 1932, MDPI Journal, 2022, https://doi.org/10.3390/sym14091932.
10. Suwarna G., Raju, K.S., Bhaskar, N., Divya, G., Diabetic Retinopathy Detection Using Deep Learning, in: *Data Engineering and Intelligent Computing*, 2022, https://doi.org/10.1007/978-981-19-1559-8_39.
11. Shah, S.S. and Bakal, J.W., Deep Learning for the Detection of Diabetic Retinopathy. *Int. J. Res. Appl. Sci. Eng. Technol. (IJRASET)*, 10, VII, 1201–1213, July 2022, doi: 10.22214/ijraset.2022.45175. ISSN: 2321-9653.
12. Mohanarathinam, A., Manikandababu, C.S., Prakash, N.B., Hemalakshmi, G.R., Subramaniam, K., Diabetic Retinopathy Detection and Classification using Hybrid Multiclass SVM Classifier and Deeplearning Techniques. *Math. Stat. Eng. Appl.*, 71, 3, 891–903, 2022, doi: https://doi.org/10.17762/msea.v71i3.251, ISSN: 2326-9865.
13. Bakhshi, A., Hajizadeh, K., Tanhayi, M.R., Diabetic retinopathy diagnosis using image processing methods. *Adv. Obes. Weight Manage. Control*, 12, 5, 132–134, 2022, DOI: 10.15406/aowmc.2022.12.00375.
14. Vikramathithan, A.C., Pooja, P., Bhaskar, M.S., Navya, S., Rakshith, K.B., Identification of Diabetic Retinopathy (DR) using Image Processing. *J. Phys. Conf. Ser.*, 2161, 012051, 2022, doi: 10.1088/1742-6596/2161/1/012051.
15. Goe, S., Gupta, S., Panwar, A., Kumar, S., Verma, M., Bourouis, S., Ullah, M.A., Deep Learning Approach for Stages of Severity Classification in Diabetic Retinopathy Using Color Fundus Retinal Images. *Hindawi Math. Probl. Eng.*, 2021, Article ID 7627566, 8, https://doi.org/10.1155/2021/7627566.
16. Harshitha, C., Asha, A., Pushkala, J.L.S., Anogini, R.N.S., Karthikeyan, Predicting the Stages of Diabetic Retinopathy using Deep Learning. *2021 6th International Conference on Inventive Computation Technologies (ICICT)*, IEEE, 2021, 978-1-7281-8501-9/21, DOI: 10.1109/ICICT50816.2021.9358801.
17. Oh, K., Kang, H.M., Leem, D., Lee, H., Seo, K.Y., Yoon, S., "Early detection of diabetic retinopathy based on deep learning and ultrawidefield fundus images. *Sci. Rep.*, 11, 1897, 2021, https://doi.org/10.1038/s41598-021-81539-3.
18. Ramasamy, L.K., Padinjappurathu, S.G., Kadry, S., Damaševičius, R., Detection of diabetic retinopathy using a fusion of textural and ridgelet features of retinal images and sequential minimal optimization classifier. *PeerJ Comput. Sci.*, 7, e456, 2021, DOI 10.7717/peerj-cs.456.
19. Dai, L., Wu, L., Li, H., Cai, C., Wu, Q., Kong, H., Liu, R., Wang, X., Hou, X., Liu, Y., Long, X., Wen, Y., Lu, L., Shen, Y., Chen, Y., Shen, D., Yang, X., Zou, H., Sheng, B., Jia, W., A deep learning

system for detecting diabetic retinopathy across the disease spectrum. *Nat. Commun.*, 12, 3242, 2021, https://doi.org/10.1038/s41467-021-23458-5.

20. Ai, Z., Huang, X., Fan1, Y., Feng, J., Zeng, F., Lu, Y., DR-IIXRN: Detection Algorithm of Diabetic Retinopathy Based on Deep Ensemble Learning and Attention Mechanism. *Front. Neuroinf.*, 15, Article 778552, 1–16, December 2021, doi: 10.3389/fninf.2021.778552.
21. El Houby, E.M.F., Using transfer learning for diabetic retinopathy stage classification. *Appl. Comput. Inf.*, 1–16, 2021, DOI 10.1108/ACI-07-2021-0191, e-ISSN: 2210-8327.
22. Adriman, R., Muchtar, K., Maulina, N., Performance Evaluation of Binary Classification of Diabetic Retinopathy through Deep Learning Techniques using Texture Feature. *Proc. Comput. Sci.*, 179, 88–94, Elsevier, 2021, doi: 10.1016/j.procs.2021.12.012.
23. Sugeno, A., Ishikaw, Y., Ohshima, T., Muramatsu, R., Simple methods for the lesion detection and severity grading of diabetic retinopathy by image processing and transfer learning. *Comput. Biol. Med.*, 137, 104795, October 2021, doi: 10.1016/j.compbiomed.2021.104795.
24. Priyadharsini, C., Jeyapriya, J., Kannan, R.J., Kumar, G.B., Sakriki, T.P., Classification of Diabetic Retinopathy using Residual Neural Network. *Mater. Sci. Eng.*, 925, 012033, 2020, doi: 10.1088/1757-899X/925/1/01/2023.
25. Revathy, R., Nithya, B.S., Reshma, J.J., Ragendhu, S.S., Sumithra, M.D., Diabetic Retinopathy Detection using Machine Learning. *IJRET*, 09, 06, 122–126, (June 2020), 2020, DOI : 10.17577/IJERTV9IS060170, ISSN(Online) : 2278-0181.
26. Sai Sudha, G., Praveena, M., SandhyaRani, G., Harish, T.N.S.K., Charisma, A., Asish, A., Classification and Detection Of Diabetic Retinopathy Using Deep Learning. *Int. J. Sci. Technol. Res.*, 9, 04, 3186–3192, April 2020, ISSN 2277-8616.
27. Skariah, S.M. and George, J., A Novel Study on Automated Diabetic Retinopathy: Detection and Grading. *Int. J. Eng. Res. Technol. (IJERT)*, 9, 10, October-2020, doi: 10.17577/IJERTV9IS100095. ISSN: 2278-0181,IJERTV9IS100095.
28. Chaturvedi, S.S., Gupta, K., Ninawe, V., Prasad, P.S., Automated Diabetic Retinopathy Grading using Deep Convolutional Neural Network, IEEE, 12, 2020, arXiv>eess>arXiv:2004.06334, https://doi.org/10.48550/arXiv.2004.06334.
29. Suganthi, S.R.L., Sneha, U.K., Shwetha, S., Diabetic Retinopathy Classification Using Machine Learning Techniques. *Int. J. Eng. Trends Technol. (IJETT)*, 68, 1, 51–56, 2020, doi: 10.14445/22315381/IJETT-V68I1P207, ISSN: 2231-5381.
30. Jadhav, A.S., Patil, P.B., Biradar, S., Optimal feature selectionbased diabetic retinopathy detection using improved rider optimization algorithm enabled with deep learning. *Evol. Intell.*, Springer, 14, 1431–1448, 2020, https://doi.org/10.1007/s12065-020-00400-0.
31. Subhashini, R., Nithin, T.N.R., Koushik, U.M.S., Diabetic Retinopathy Detection using Image Processing (GUI). *Int. J. Recent Technol. Eng. (IJRTE)*, 8, 2S3, 538–542, July 2019, DOI: 10.35940/ijrte.B1097.0782S319, ISSN: 2277-3878.
32. Tripathy, B.K. and Parui, K., Diabetic Retinopathy Detection using Image Processing. *Int. J. Eng. Adv. Technol.*, Special Issue, 8, 6, 937–941, 2019, doi: 10.35940/ijeat.f11790.0886s19.
33. Naveen, R., Sivakumar, S.A., Shankar, B.M., Priyaa, A.K., Diabetic Retinopathy Detection using Image Processing. *Int. J. Eng. Adv. Technol. (IJEAT)*, 8, 6S, 937–941, August 2019, doi: DOI:10.35940/ijeat.F1179.0886S19, ISSN: 2249 - 8958.
34. Yalçin, N., Alver, S., Uluhatun, N., Classification of retinal images with deep learning for early detection of diabetic retinopathy disease. *2018 26th Signal Processing and Communications Applications Conference (SIU)*, Izmir, Turkey, pp. 1–4, 2018, doi: 10.1109/SIU.2018.8404369.
35. Dutta, S., Manideep, B.C.S., Basha, S.M., Caytiles., R.D., Iyengar, N.C.S.N., Classification of Diabetic Retinopathy Images by Using Deep Learning Models. *Int. J. Grid Distrib. Comput.*, 11, 1, 89–106, 2018, http://dx.doi.org/10.14257/ijgdc.2018.11.1.09.

36. Deperlıoğlu, Ö. and Köse, U., Diagnosis of Diabetic Retinopathy by Using Image Processing and Convolutional Neural Network. *2018 2nd International Symposium on Multidisciplinary Studies and Innovative Technologies (ISMSIT)*, Ankara, Turkey, pp. 1–5, 2018, doi: 10.1109/ISMSIT.2018.8567055.
37. Sisodia, D.S., Nair, S., Khobragade, P., Diabetic Retinal Fundus Images: Preprocessing and Feature Extraction For Early Detection of Diabetic Retinopathy. *Biomed. Pharm. J.*, 10, 2, 615–626, 2017, doi: http://dx.doi.org/10.13005/bpj/1148.
38. Chandrakumar, T. and Kathirvel, R., Classifying Diabetic Retinopathy using Deep Learning Architecture. *Int. J. Eng. Res. Technol. (IJERT)*, 5, 06, 19–24, June-2016, ISSN: 2278-0181, IJERTV5IS060055.
39. Priya, R. and Aruna, P., Diagnosis Of Diabetic Retinopathy Using Machine Learning Techniques. *ICTACT J. Soft Comput.*, 03, 04, 563–575, July 2013, DOI: 10.21917/ijsc.2013.0083, ISSN: 2229-6956(ONLINE).
40. Venkatesan, R., Chandakkar, P., Li, B., Li, H.K., Classification of diabetic retinopathy images using multi-class multiple-instance learning based on color correlogram features. *2012 Annual International Conference of the IEEE Engineering in Medicine and Biology Society*, San Diego, CA, USA, pp. 1462–1465, 2012, doi: 10.1109/EMBC.2012.6346216.Top of Form.

12

Early Brain Stroke Detection Based on Optimized Cuckoo Search Using LSTM-Gated Multi-Perceptron Neural Network

Anita Venaik[1], Asha A.[2], Dhiyanesh B.[3]*, Kiruthiga G.[4], Shakkeera L.[5] and Vinodkumar Jacob[6]

[1]*IT, Amity Business School, Amity University, Noida, India*
[2]*ECE, Rajalakshmi Engineering College, Chennai, Tamil Nadu, India*
[3]*SCE, SRM Institute of Science and Technology, Vadapalani Campus, Chennai, Tamil Nadu, India*
[4]*IT, Karpagam College of Engineering, Coimbatore, Tamil Nadu, India*
[5]*SoCSE, Presidency University, Bengaluru, Karnataka, India*
[6]*ECE, M.A. College of Engineering, Kothamangalam, Kerala, India*

Abstract

A brain stroke is the deadliest attack that leads to sudden death and affects human nature. It is challenging to analyze the symptoms in the initial phases of the disease, as each patient may have different priorities. The effects of a brain stroke can be prevented by receiving immediate treatment since it is the second leading cause of death worldwide. Early stroke detection can control and reduce severity and mortality. Causes of symptoms may include headaches, high blood pressure, vomiting, paralysis, rheumatoid arthritis, and lethargy. Efficient data analysis is necessary to examine feature dependencies and diagnose each function based on data. An efficient method identifies risk variables using machine learning (ML) algorithms. The critical issue is the presence of feature dimensions that support disease classification. Recognition and classification accuracy decrease as dimensionality and false positive rates increase. Weighted adjustments are made for disease incidence rates to expand feature ranges, resulting in imprecise categories with a high bias. To solve this issue, we must develop an automatic feature selection and classification system to analyze disease features. This will accurately detect brain strokes at an early stage. In this research, we developed an optimized ML system based on Decision Tree Optimized Cuckoo Search (DTOCS) feature selection using a Long Short-Term Memory gated Multilayer Perceptron Neural Network (LSTM-gated MLPNN). Finally, the classifier identifies patients' risks earlier and takes necessary action. Initially, we collected the dataset from Kaggle for brain stroke detection. Secondly, the data was transformed and normalized to be processed using the actual medical margin. In addition, the Genetic Feature Sequence Algorithm (GFSA) estimates the brain impact normalization score. Fitness evaluation calculates the disease-prone factor (DPF). The DTOCS algorithm then selects the acute feature to ease the dimensionality ratio. Furthermore, the selected parts of the analytical assessment of the LSTM-gated MLPNN can be observed by obtaining the maximum values of the processing logical units based on the proposed method. To improve the brain stroke identification accuracy, the LSTM finds the most

**Corresponding author*: dhiyanu87@gmail.com

Shubham Mahajan, Pethuru Raj and Amit Kant Pandit (eds.) Deep Reinforcement Learning and Its Industrial Use Cases: AI for Real-World Applications, (265–300) © 2024 Scrivener Publishing LLC

significant feature weights supported for the risk identification class to improve the classification accuracy. It groups the actual feature thresholds by category. It creates a class for each threshold and reduces irrelevant feature biases. This increases the real disease impact factor in the MLPNN classifier. Classifier performance can be evaluated using classifier performance. The proposed model achieves 95.30% accuracy, outperforming other methods.

Keywords: Feature selection, classification, classifier, neural network, data transformation, fitness function, machine learning, brain stroke

12.1 Introduction

Among the top causes of death in the modern world, strokes pose a significant threat to public health. Cerebrovascular accident (CVA) is one of the most common accidents in the world today. Moreover, a medical emergency like a stroke requires immediate medical attention. The World Health Organization (WHO) classifies vascular insufficiency as a disease that can lead to stroke and severe cumulative damage. This results in brain stroke-like paralysis, severe morbidity, and loss of consciousness. Strokes can be ischemic or hemorrhagic in nature, depending on their underlying cause. Inappropriately, with the world population growing, the number of deaths and morbidity from this disease is expected to increase. However, many tools use the Cox proportional hazards model to predict that strokes can be prevented by early treatment and a positive prognosis. This is a traditional method, so high-dimensional data can make it difficult to predict strokes effectively. Early detection and adequate organization are critical for preventing brain damage and other complications in different parts of the body from becoming serious. WHO estimates that one person dies every 4 to 5 min, and 15 million people suffer from strokes worldwide annually [1, 2].

Health is essential to everyone's life, and a record system is needed to track disease data and their relationships. Furthermore, patient case summaries, clinic medical records, and other manually maintained records can be implemented to track multiple pieces of information. Several risk factors associated with stroke have been identified by examining affected individuals. Moreover, many studies use these risk factors to predict stroke [3].

Strokes damage the brain and cause death. Moreover, people die within a few years or become permanently disabled when circulation problems such as brain failure or stroke occur. Worldwide, strokes are the second leading cause of death after ischemic heart disease. Brain strokes are also connected to metabolic disorders, obesity, high blood pressure, and depression. Furthermore, ischemic stroke is classified into two categories: cerebral infarction, stroke. In addition, these are caused by the constriction of blood vessels in the ischemic stroke brain. There are several disadvantages of manual diagnosis, including its time-consuming nature and the need for multiple specialists. Artificial intelligence (AI) technology plays an essential role in medical diagnosis by reducing mortality, time, and costs due to medical errors [4].

Furthermore, these cerebrovascular diseases contribute to increased morbidity that is incurable. There are also a number of other factors that contribute to a stroke, such as weight gain, excessive smoking and drinking, sedentary living, and stress. Individuals who have suffered a stroke may experience seizures, which can cause communication difficulties, sudden muscle weakness, and memory impairment. In ischemic strokes, blood clots block the brain blood flow, causing reduced blood flow and death. After an ischemic stroke, patients may experience stroke bleeding, which can lead to severe complications. Stiffness in other parts of the body can cause a stroke if a blood vessel ruptures and flows into nearby brain tissue. Vascular conditions, including concussions, high blood pressure, aneurysms, arterial malformations, and bleeding disorders, typically result in hemorrhagic strokes [5].

In most cases, brain strokes are identified by severe disease features, which make the disease more dangerous at the affected stage. The main problem is early risk prediction and identification, which is important to avoid, so analyzing the significant feature margins based on disease findings is essential to early diagnosis and treatment. Using deep learning to select features and classify stroke risks, the research contributes to early stroke identification. LSTM-gated multi-perceptron neural networks are used for the classification of disease-related symptoms by reducing feature dimensionality. The outcome findings prove that the proposed system attains accuracy from the collected brain stroke dataset from Kaggle. To process the data using the actual medical margins, we preprocessed the data through transformation and normalization. In addition, the brain damage score was calculated and normalized by the GFSA method. The TPF method can also be calculated by incorporating fitness evaluation. Furthermore, the DTOCS method can be applied to select critical features for dimension reduction. Moreover, we developed a method to quickly analyze disease characteristics and accurately detect stroke using the proposed LSTM-gated MLPNN classification. The performance of this LSTM method can be enhanced by estimating the variance of clinical edge feature metrics. This is done by grouping them by category according to their accurate feature range. The MLPNN method generates classes for each threshold, reducing the bias weight of irrelevant features and thus increasing the actual disease impact coefficient in the classifier. Therefore, the proposed method can be evaluated for improved accuracy related to other processes using the classifier's performance in accuracy, sensitivity, error rate, FPR, and FNR.

The section discusses the framework for improving brain stroke accuracy through pre-processing, normalization score fitness evaluation, feature selection, and classification, as shown in Figure 12.1.

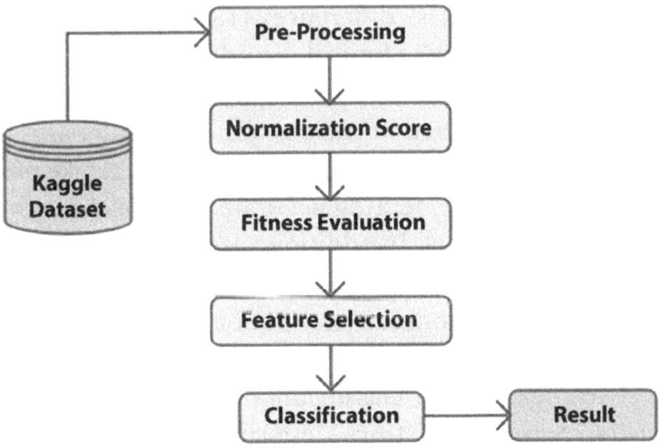

Figure 12.1 Architecture diagram for brain stroke detection.

12.2 Literature Survey

The novel suggests that ML techniques can be proposed and explored in various ways, including generalization ability and prediction accuracy metrics. However, it is a much larger global problem with significant health and economic implications [6]. The novel approach to spatio-spectral characterization of spatial neglect (S.N.), enabling EEG based on BCI to better detect S.N., can be carried out with EEG responses to visual stimulation of contralateral and contralateral lesions despite the significant global problems mentioned above [7]. The novel approach suggests that left and right hemiplegia can be simulated by introducing the SLR method and manipulating the segmentation function. A comparison between the two pipelines [8] will allow us to demonstrate the effectiveness of the designed SLR method and how it compares to the original method. In addition, the ECG and bio signals composed of three electrodes on the index finger at a selection rate of 1,000 Hz/s can be accurately interpreted using the photoplethysmography (PPG) method [9]. The Thrombolysis in Cerebral Infarction (TICI) method score can be defined on a rough ordinal scale based on visual examination. It is common practice to use it for assessing technical outcomes after endovascular therapy (EVT) [10].

In addition, optical coherence tomography (OCT) methods have been studied and tested *in vivo* for acute injury in ischemic stroke [11]. The novel approach presents interventions using brain–computer interfaces (BCIs) that can be customized to fit patients' BCI symptoms. BCI-based adaptive functional electrical stimulation (FES) treatment has clinical implications [12]. Nevertheless, a brain temperature model separated into gray and white matter can be proposed, considering spatially resolved permeation [13]. The effect of TDCS on stroke patients can be investigated at different stages by analyzing EEG data and brain activity networks [14]. The brain–machine interfaces (BMI) model uses radiofrequency signals with minimal risk to describe its paradigm [15].

The proposed Adaptive Clustering Distorted Born Iterative Method (AC-DBIM) can be compared to three conventional methods: stroke head models and digital head models for shape and location reconstruction and contrast source inversion (SLCSI) [16]. Stroke patients underwent various studies that illustrated the novel effects of walking performance, dual cognition, and motor tasks on brain function [17]. It has been demonstrated on several occasions that it is possible to predict cerebral palsy using ML methods based on incomplete physiological data and class differences [18]. Five ML algorithms compute the possibility of stroke based on an entity's physical disorder and medical reports [19]. Medical fields can save time and improve accuracy by using deep learning (DL) models. Moreover, 113 analyzed papers could be published in various academic databases [20].

The novel approach discusses the development of automated stroke prediction algorithms that enable early intervention. Implementing accurate and effective forecasting systems is becoming increasingly important [21]. In addition, stroke is tracked through lesion detection and segmentation to gain critical insights based on specific criteria [20]. The objective is to classify advanced stroke ML techniques into four collections based on their utility and comparability [22]. Furthermore, they described a decision tree learning (D.T.) method that helps recognize stroke features for classification [23]. The novelty is accomplished by considering the decay of connected, distinct scattering operators into single values. When designing, we rigorously consider dynamics and signal-to-noise ratios, depending on the measurement system [24].

A recent study suggests that the entire S-matrix of the antenna array can be measured using a prototype, and the phantom stroke can be adjusted to position the phantom in various predefined positions [25]. The novel approach suggested a two-stage technique for actual grouping and localization of hemorrhagic or ischemic strokes based on microwave scattering data collected throughout the human head [26]. Furthermore, these researchers have proposed preliminary experiments to validate the MWT method using the DBIM-TwIST algorithm for stroke detection and classification [27]. The health monitoring system for real-time predicting stroke precursors in elderly individuals during regular walking. However, real-time models that predict health illnesses and diseases while using various medical services are gaining more attention [28]. The proposed method allows for the development of stroke prediction systems based on real-time bio signals and an AI approach to power diagnosis [29].

Furthermore, these systems can be instigated according to the Oxford Shire Community Stroke Program (SCSP) system to categorize strokes into three groupings [30], suggesting a fuzzy-level segmentation algorithm for detecting ischemic stroke lesions while addressing challenges posed by lesion size and shape differences [31]. To address these issues, a classification learning model can be developed for two-dimensional datasets using hybrid feature selection [32]. In addition, an improved gate strategy for trial initiation is implemented to select an appropriate or inappropriate trial start time point [33]. Moreover, these may require using a gelatin-based head model that mimics brain tissues and targets to simulate stroke types like hemorrhagic or ischemic [34].

Improving computer capabilities for patient health information requires security and reliability [35]. The novel antenna design features a 4 × 4 radiating patch and an open U-shaped slot extended symmetrically using serial and corporate transmission lines [36]. Based on the LBE paradigm, the proposed multi-step method is well suited to effective clinical diagnosis at various stages and under time constraints [37]. The novel approach provides detailed results from an experimental handheld system that uses MWT for early stroke detection and monitoring [38]. The novel approach suggests that these markers could be utilized to develop different ML models for predicting stroke probability [39].

12.2.1 Problem Statement

- Blood flow problems to the brain, as well as their failure to flow, can have serious consequences on the individual's health, including disability and even death in some cases.
- A stroke is a severe medical condition in which an artery in the brain bursts, causing damage to the brain that is permanent and that ultimately results in incapacitation due to the permanent damage that is caused by that ruptured artery.
- There are a variety of vascular and neurologic events that can cause the brain tissue to be injured in an ischemic stroke.
- In the present day, MRI tests are both expensive and time-consuming, which make the diagnosis and prediction of strokes quite challenging.
- Since stroke lesions usually appear in shapes, sizes, and locations that are complex, it can be difficult to detect them automatically.

12.3 Proposed Methodology

This proposed LSTM-gated MLPNN method offers the ability to quickly analyze disease features and employ the classification to accurately diagnose brain strokes, so we first collect brain stroke data from the Kaggle dataset to determine its accuracy. Then, the data are transformed and normalized to process the data using the actual clinical margin. Moreover, they manipulate the GFSA method to calculate a normalized brain impact score. Fitness evaluations determine the DPF method. Furthermore, using DTOCS, the critical feature can be selected to reduce the aspect ratio.

Furthermore, the LSTM-gated MLPNN proposed method evaluates variations in feature measurement between medical stumps and groups the actual feature thresholds by category. In addition, LSTM can be implemented with gate units to obtain the most accurate values of the MLPNN and observe the selected parts of the analytical evaluation by processing the logical units. In Figure 12.2, the architectural diagram of the LSTM gate MLPNN method illustrates how gate units can help implement LSTM and achieve the optimal value of MLPNN. By processing the logic units, this approach further allows the detection of selected parts of the analytical evaluation.

12.3.1 Dataset Collection

Strokes result from insufficient blood flow to the brain, resulting in the death of brain cells. Strokes can be classified as either ischemic or hemorrhagic in nature. Ischemic strokes occur when blood flow is reduced, while hemorrhagic strokes occur when bleeding occurs. Regardless of which type of stroke a person suffers, it will cause the brain to stop functioning

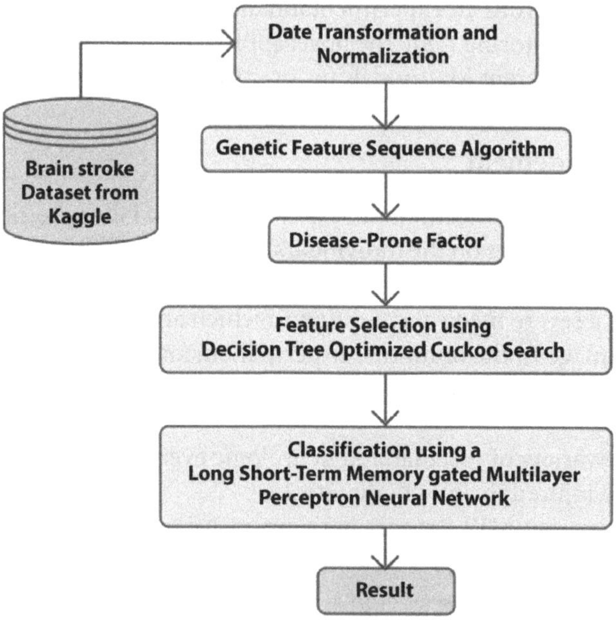

Figure 12.2 Architecture diagram for the proposed LSTM-gated MLPNN method.

properly. Strokes are categorized by symptoms, such as the loss of movement or sensation on one side of the body. These symptoms include difficulty speaking or comprehending, dizziness, and blindness, all symptoms of strokes. These symptoms typically appear after a stroke occurs. There are two types of transient ischemic attacks (TIAs); there are transient ischemic attacks that last from 1 to 2 h and transient mini-strokes.

In addition to the obvious symptoms of a stroke, a hemorrhagic stroke may also result in severe headaches. Long-term complications include pneumonia and bladder control loss. High blood pressure greatly increases stroke risk. Atrial fibrillation is a condition in which the heart does not maintain its rhythm correctly because of hyperlipidemia, smoking, obesity, diabetes, and a history of a TIA. Blocked blood vessels often lead to strokes, although other less common causes exist. A hemorrhagic stroke can cause bleeding directly into the brain or the space between the meninges. If it ruptures, a brain aneurysm can occur. Also, strokes can be corrected surgically. Treatment for lost function is better addressed under the category of brain damage and is recommended as stroke rehabilitation. Figure 12.3 illustrates the brain stroke data set architecture. To prevent brain strokes, it is essential to reduce risk factors. This can involve surgery to widen the brain arteries for those with carotid artery stenosis and warfarin for those with atrial fibrillation. In cases where a brain stroke is analyzed within 3 to 4 1/2 hours, blood clot-dissolving medications can be used.

12.3.2 Preprocessing

The data is transformed and normalized for processing using the actual medical margin in this preprocessing section. Pre-processing is an essential step in optimizing and utilizing data effectively. Additionally, accuracy, operability, and reliability are characteristics of data quality. Data transformation and normalization are crucial stages in data preprocessing. Data transmission can detect missing terms and correct errors, even in noisy data. Data integration refers to merging data from various sources to create an interconnected and unified dataset. Data reduction describes the process of reducing detailed data. Moreover, redundant data is one of the biggest problems in pre-processing-type data integration. As previously mentioned, redundancy is a critical issue in data integration. There are several reasons redundancy can happen, such as attribute naming, consistency, and the source of the attributes. Data transformation involves converting data into a format appropriate for

gender	age	hypertension	heart_disease	ever_married	work_type	Residence_type	avg_glucose_level	bmi	smoking_status	stroke
Male	67	0	1	Yes	Private	Urban	228.69	36.6	formerly smoked	1
Male	80	0	1	Yes	Private	Rural	105.92	32.5	never smoked	1
Female	49	0	0	Yes	Private	Urban	171.23	34.4	smokes	1
Female	79	1	0	Yes	Self-employed	Rural	174.12	24	never smoked	1
Male	81	0	0	Yes	Private	Urban	186.21	29	formerly smoked	1
Male	74	1	1	Yes	Private	Rural	70.09	27.4	never smoked	1
Female	69	0	0	No	Private	Urban	94.39	22.8	never smoked	1
Female	78	0	0	Yes	Private	Urban	58.57	24.2	Unknown	1
Female	81	1	0	Yes	Private	Rural	80.43	29.7	never smoked	1
Female	61	0	1	Yes	Govt_job	Rural	120.46	36.8	smokes	1
Female	54	0	0	Yes	Private	Urban	104.51	27.3	smokes	1
Female	79	0	1	Yes	Private	Urban	214.09	28.2	never smoked	1
Female	50	1	0	Yes	Self-employed	Rural	167.41	30.9	never smoked	1
Male	64	0	1	Yes	Private	Urban	191.61	37.5	smokes	1
Male	75	1	0	Yes	Private	Urban	221.29	25.8	smokes	1
Female	60	0	0	No	Private	Urban	89.22	37.8	never smoked	1
Female	71	0	0	Yes	Govt_job	Rural	193.94	22.4	smokes	1
Female	52	1	0	Yes	Self-employed	Urban	233.29	48.9	never smoked	1
Female	79	0	0	Yes	Self-employed	Urban	228.7	26.6	never smoked	1
Male	82	0	1	Yes	Private	Rural	208.3	32.5	Unknown	1
Male	71	0	0	Yes	Private	Urban	102.87	27.2	formerly smoked	1

Figure 12.3 Brain stroke dataset.

mining purposes. Furthermore, data normalization can be pre-processed and predefined to extract parameters such as patient symptoms, risk variables, stroke classification, and type prediction. The energy of each channel in the frequency dimension is equalized by normalizing every matrix element. Furthermore, it has been demonstrated that transmitted data has a wide dynamic range of scattering parameters, enabling complex scattering matrix models for a specific frequency. A complex logarithmic function can be applied to decrease normalized variance values. Normalization techniques can also be utilized to prepare the data. Numeric column values can be rescaled to a fixed scale of 0 to 1. In addition to this, data normalization is useful when a dataset with a mean of 0 and a standard deviation of 1 needs to be adjusted. ML modeling often requires normalization of the data before it can be used in an analysis. There is no harm in normalizing a dataset if the values in its numerical columns are converted into a constant ratio, which keeps the information intact and does not distort the range of standards in the dataset. Therefore, preprocessing makes sense in terms of balancing data with actual clinical margins to improve accuracy.

In accordance with Equation 12.1, you would have to calculate the means and standard deviations for the data normalized to the brain mask. Let us assume a-data, μ_{zs} -determine mean, σ_{zs} -standard deviation, μ -Mu, σ – Sigma, and u-normalized data.

$$a_{ZScore}(u) = \frac{a(u) - \mu_{zs}}{\sigma_{zs}} \tag{12.1}$$

The white matter (WM) mask can be calculated using fuzzy c-means normalization, as shown in Equation 12.2. Let us assume, fdm-fuzzy c-means, d-cluster,

$$\mathrm{a}_{\mathrm{fdm}}(\mathrm{u}) = \frac{\mathrm{d}_1 .\, \mathrm{a}(\mathrm{u})}{\mu_{\mathrm{fdm}}} \tag{12.2}$$

Calculate the indices corresponding to the W.M. locations in the data as shown in Equation 12.3, where Z-white means

$$\mu_{\mathrm{fdm}} = \frac{1}{|\mathrm{z}|} \sum\nolimits_{\mathrm{z} \in \mathrm{Z}} \mathrm{a}(\mathrm{z}) \tag{12.3}$$

A Gaussian mixture model (GMM)-based normalization represented in Equation 12.4 can calculate the mixture of distributions brain mask the intensity histogram in the brain mask. Let us assume μ_{GMM} - component of mixture mean.

$$\mathrm{a}_{\mathrm{GMM}}(\mathrm{u}) = \frac{\mathrm{d}_2 .\mathrm{a}(\mathrm{u})}{\mu_{\mathrm{GMM}}} \tag{12.4}$$

The probability density function can be calculated using the kernel density estimate normalization described in Equation 12.5. Let us assume ρ – Peak, KDE- kernel density estimate,

$$a_{KDE}(u) = \frac{d_3.a(u)}{\rho} \tag{12.5}$$

Normalized white matter intensity data normalization of Z-scores can be estimated based on commonly occurring white matter intensity values, as illustrated in Equation 12.6. τ – Tau, e-function, μ_{WS} - intensity segment values, Ω_τ - white stripe set, WS-white stripe.

$$\Omega_\tau = \{a(u) \mid e^{-1}(e(\mu_{WS}) - \tau) < a(u) < e^{-1}(e(\mu_{WS}) + \tau)\} \tag{12.6}$$

Normalize the white stripe data and calculate the corresponding sample standard deviation, as shown in Equation 12.7. Let us assume σ_{WS} - sample standard deviation.

$$a_{WS}(u) = \frac{a(u) - \mu_{WS}}{\sigma_{WS}} \tag{12.7}$$

As shown in Equation 12.8, the white stripe estimates the CSF intensity of technical changes after normalization. Let us assume γ – gamma, t-technical variation, O-unwanted factor, $\gamma O'$ -represents the unknown technical variation, s-matrix of residuals.

$$x_d = \gamma O^t + s \tag{12.8}$$

Calculate the value of the cross sections using singular value decomposition, as shown in Equation 12.9, where x_d -deformable co-registering the data, z-matrix of the right particular vector.

$$x_d = \bigcup \sum z^t \tag{12.9}$$

Based on the equations shown in Equation 12.10, the coefficients were estimated using voxel-wise linear regression on the normalized data, which gave us the following results. Let us assume *raval* - normalized data, γ - coefficient data.

$$a_{raval}(u) = a_{WS}(u) - \gamma_u z_b^t \tag{12.10}$$

As shown in Equation 12.11, the entropy is calculated to obtain the informational advantage of each attribute. Let us assume v-attribute data, R-Pearson's correlation, g-entropy.

$$g(\gamma) = -\sum_{v \in \gamma} R(v)\log_2(R(v)) \tag{12.11}$$

As exposed in Equation 12.12, the merit function calculates the overall properties exhibited by each subset. Let us assume L- number of the attribute, e_Q -subset attribute, $\overline{s_{de}}$ -distribution of the mean attribute, and $\overline{s_{ee}}$ -mean correlation value.

$$\text{Merit}\left(e_Q\right) = \frac{K\overline{s_{de}}}{\sqrt{L + L\left(L-1\right)\overline{s_{ee}}}} \tag{12.12}$$

As shown in Equation 12.13, calculate the weighted values of the attribute standard deviation and the mean separately, where a-measurement size, u-mean of attribute, α - weighted value, and σ – sigmoid function.

$$\overrightarrow{u_a} = \frac{u_a - \mu}{\sigma}\alpha \tag{12.13}$$

As indicated in Equation 12.14, it is imperative to determine whether or not the information in the cell state has been transferred by the value the function returns. B-bio-signal, T- Electromyography point, e_T -output value, u_T - input value, g_{T-1}, u_T -forget late layer, B_e - bio signal function.

$$e_T = \sigma(z_e.[g_{T-1}, u_T] + B_e) \tag{12.14}$$

As the input gate evaluates, the state of the cell will be updated with information between the input gate and the output gate based on the information obtained from the input gate. This is shown in Equation 12.15.

$$a_T = \sigma(z_a.[g_{T-1}, u_T] + B_e) \tag{12.15}$$

The hyperbolic tangent activation function is shown in Equation 12.16 to calculate the added value at the cell location. Let us assume d_T -generate the importance of cell state.

$$\widetilde{d_T} = \text{tang}(z_d.[g_{T-1}, u_T] + B_d) \tag{12.16}$$

Based on Equation 12.17, it is shown that the updated state of a cell can be derived by multiplying the new state by the old state, where d_{T-1} – update the multiplying old state.

$$d_T = e_T * d_{T-1} + a_T * \widetilde{d_T} \tag{12.17}$$

Equations 12.18 and 12.19 show that the element-wise product output values are calculated using the hyperbolic tangent function values. Let us assume the k_T-output value of the sigmoid layer d_T – update cell state value.

$$k_T = \sigma(z_k.[g_{T-1}, u_T] + B_k) \tag{12.18}$$

$$g_T = k_T * \text{tang}(d_T) \tag{12.19}$$

The data transformation parameters are estimated from the masked Z-score normalized data, as shown in Equation 12.20. Let us assume A-size, b-brain mask, A_{ab}^{Main}-output of a brain region, a and b- represent the pixel position, β -Beta, γ - gamma.

$$A_{ab}^{Main} = \left(\gamma.A_{ab}^{ZS} + \beta\right).b_{ab} \tag{12.20}$$

Calculate the discrete values of the dispersion matrix for the reciprocal antenna system as given in Equation 12.21, where a and b- row and column index, f- transmission data, c-unique value.

$$c = \frac{p_f^2 + p_f}{2} \tag{12.21}$$

Equation 12.22 represents the set of components of the normalized data vector. Let us assume

$$\{\log(Q_{ab}(z_L)/(d_{ab}) \mid L = 1,\ldots,q_z, 1 \le a < b \le q_f)\} \tag{12.22}$$

As shown in Equation 12.23, the number of measurement frequencies can be approximated by a normalization constant, where Q- scatting matrix, q_z-number of measured frequencies.

$$d_{ab} = \sqrt{\frac{1}{q_z}\sum_{L=1}^{q_{ab}} | q_{ab}(z_L) |^2} \tag{12.23}$$

In this sense, the normalization constant approximates the scale used to scale the frequency. Preprocessing balances the data and accurate clinical margins to improve accuracy.

12.3.3 Genetic Feature Sequence Algorithm (GFSA)

Furthermore, GFSA can estimate average scores based on natural selection and genetically induced optimization techniques. GFSA can determine the initial population of individuals by randomly generating an initial population of a specific design. Genetic representations can represent a solution to a problem using GFSA. Furthermore, the impact normalized score manipulates the GFSA method to calculate the distance between each gene. A generation terminates when the maximum generation limit is reached or the normalized score value remains constant after several consecutive generations. GFSA is a probabilistic general search method that explores large search spaces. A computational model simulating the GFSA process is implemented in natural selection evolution.

The genetic computational model of Darwin's theory of biological evolution presents a method for finding optimal solutions by simulating natural evolution. Individuals are encoded using real numbers selected from the set of parents by the evolutionary operator's competitive selection, crossover, and transformation. Furthermore, the GFSA can be used to determine the correlation vector of standardized scores of stroke risk factors. An algorithm can be implemented to minimize redundancy in the ordering vector and find the maximum correlation with the output. GFSA performs a genetic movement task after encoding the initial population. To complete the process, specific actions on individual normalized scores can be implemented along with the evolution of ecological normalized assessments and average scores. Crossover operators are crucial in biological manipulation, acting as a mechanism for recombination, a vital process in natural biological evolution. The crossover operator can randomly combine two individuals in the population and create genetically appropriate genes based on their corrected score.

Calculate the class's attribute weights and compatibility scores, as shown in Equation 12.24. Let us assume R-fitness score, $D(w)$ – number of the attribute, $e(w,d,b)$ – fitness value, d-class, w and b-value, and $u(d)$ – value of the attribute.

$$R(w) = d_{ab}\frac{\sum_{d=1}^{D(w)} u(d)z(w_d)e(w,d,b)}{\sum_{d=1}^{D(w)} u(d)z(w_d)} \tag{12.24}$$

Equation 12.25 calculates the weighted multivariate value of the attribute difference. Let us assume H- multivariate value, M-number of attributes, $G_d(u_d,v_d)$ – distance function, w-distance class, z_{wd} - weight of attribute, and z-weight.

$$Z\_H(u,v) = \sqrt{\sum_{d=1}^{M} z_{wd}G_d(u_d,v_d)} \tag{12.25}$$

Equation 12.26 estimates the distance function between a value and an attribute.

$$G_d\ (u_d.v_d) = \begin{cases} 1, & \text{if u or v is unknown,} \\ N\_M_d(u_d,v_d) & \text{if is a qualitive} \\ N_{diff_d}(u_d,v_d) & \text{Otherwise} \end{cases} \tag{12.26}$$

Calculate the normalized distance for the measured attribute using Equation 12.27, where σ_d – standard deviation.

$$N_{diff_d}(u_d, v_d) = \frac{|u_d, v_d|}{4\sigma_d} \tag{12.27}$$

The value of an attribute in the training set is calculated using Equation 12.28. Let us assume the w-number of output classes, q-number of classes, and u, v – values.

$$N\_M_d(u_d, v_d) = \sqrt{\sum_{w=1}^{W} \left| \frac{q_{d,u,w}}{q_{d,u}} - \frac{q_{d,v,w}}{q_{d,v}} \right|^2} \tag{12.28}$$

The weights are normalized by computing the selection probabilities indicated in Equation 12.29. Let us assume a z-weight vector, M-higher parameter.

$$z = \frac{1}{\sum_{a=1}^{W} z_a} \begin{bmatrix} z_1 \\ z_2 \\ \vdots \\ z_W \end{bmatrix} \tag{12.29}$$

As shown in Equation 12.30, calculate the Hamming distance to obtain the similarity between the column vectors corresponding to two genes, where $\sum_L$ – set of all permutations, L-length, and a-element, g- hamming distance.

$$\sigma * = \underset{\sigma \in \Sigma_L}{\operatorname{argmin}} \sum_{a=1}^{L} g(u_a, \sigma_a(v)) \tag{12.30}$$

Calculate the normal set of all permutations as shown in Equation 12.31.

$$\sum_{a=1}^{L} g(u_a, \sigma_a(v)) \tag{12.31}$$

Calculate the total distance between genetic of the optimal recombination normalization, as shown in algorithm 1. Let us assume σ^* – rearrange permutation, L – coverage matrix value, u and b- column indices.

Algorithm 1 for total distance

Start

$$
\begin{array}{l}
Optimal\_n(u,v) \\
\{ \\
\quad for\ a \leftarrow 0\ to\ L \\
\quad\quad for\ b \leftarrow 0\ to\ L \\
\quad\quad\quad G[u_a, v_b] \leftarrow g(u_a, v_b); \\
\quad\quad\quad \sigma* = argmin_{\sigma \in \sum_L} \sum_{a=1}^{L} g(u_a, \sigma_a\ (v)) \\
\quad\quad\quad v \leftarrow \sigma^L(v); \\
\quad\quad End\ for \\
\quad End\ for \\
\}
\end{array}
$$

Stop

Calculate the distance measure for the two genetic features, as shown in Equation 12.32. Let us assume G- distance matrix.

$$
\text{if } u = (u_1, u_2, \ldots, u_L) \,\&\, v = (v_1, v_2, \ldots, v_L), u_T = u_{\wp}, \sigma^* \in \sum\nolimits_{L, \sigma_T^*(v) = v_T} \sum_{a=1}^{L} G(u_a, \sigma_a^*(v)) \leq \sum_{a=1}^{L} G(u_a, \sigma_a(v)) \tag{12.32}
$$

Evaluate the symbolic permutation as shown in Equation 12.33, where s-index.

$$
\sigma_s'(v) = v_{\wp}, \sigma_a''(v) = \begin{cases} \sigma_s'(v) & if\ a = \wp \\ \sigma_T'(v) & if\ a = s \\ \sigma_a'(v) & otherwise \end{cases} \tag{12.33}
$$

Equation 12.34 shows that the Hamming distance can be calculated using a unique measure of the triangle inequality. Let us assume G-distance matrix.

$$\sum_{a=1}^{L} G(u_a,\sigma''_a(v)) = \mathbb{G}(u_T,\sigma''_T(v)) + G(u_s,\sigma''_s(v)) + \sum_{a\neq T, a\neq s} G(u_a,\sigma''_a(v))$$
$$= \mathbb{G}(u_T,\sigma'_s(v)) + \mathbb{G}(u_s,\sigma'_T(v)) + \sum_{a\neq T,\, a\neq s} G(u_a,\sigma'_a(v))$$
$$= \mathbb{G}(u_T,\sigma'_{\wp}(v)) + \mathbb{G}(u_s,\sigma'_T(v)) + \sum_{a\neq T, a\neq s} \mathbb{G}(u_a,\sigma'_a(v))$$
$$= \mathbb{G}(u_s,\sigma'_T(v)) + \sum_{a\neq T, a\neq s} \mathbb{G}(u_a,\sigma'_a(v))$$
$$\leq \mathbb{G}(u_s,u_T) + \mathbb{G}(T,\sigma'_T(v)) + \sum_{a\neq T, a\neq s} \mathbb{G}(u_a,\sigma'_a(v)) \; (\because \textit{triangular inequality})$$
$$= \mathbb{G}(u_s,u_{\wp}) + \mathbb{G}(T,\sigma'_T(v)) + \sum_{a\neq T, a\neq s} \mathbb{G}(u_a,\sigma'_a(v))$$
$$= \mathbb{G}(u_s,\sigma'_s(v)) + \mathbb{G}(u_T,\sigma'_T(v)) + \sum_{a\neq T,\, a\neq s} \mathbb{G}(u_a,\sigma'_a(v))$$
$$= \sum_{a=1}^{L} \mathbb{G}(u_a,\sigma'_a(v)) \leq \sum_{a=1}^{L} \mathbb{G}(u_a,\sigma_a(v)) \; \textit{for all} \; \sigma \in \sum_{L} \tag{12.34}$$

Evaluate the optimal permutation of feasibility as a preserved as Equation 12.35, where ρ – discreate matric.

$$\sigma^{*} \underset{\sigma\in\sum_L}{\text{argmin}} \sum_{a=1}^{L} \rho(u_a,\sigma_a(v)) \tag{12.35}$$

Calculate the discrete matrix for normalization by rearranging it to preserve feasibility, as illustrated in Equations 12.36 and 37.

$$\sigma'_a = \begin{cases} \sigma^{*}_s(v) \text{ if a} = \text{T}, \\ \sigma^{*}_T(v) \text{ if a} = \text{T}, \\ \sigma^{*}_a(v) \text{ othewise} \end{cases} \tag{12.36}$$

$$
\begin{aligned}
&\sum_{a=1}^{L} \rho\left(u_a,\sigma_a'(\mathbb{v})\right)=\rho\left(u_{\wp},\sigma_{\wp}'(\mathbb{v})\right)+\rho\left(u_s,\sigma_s'(\mathbb{v})\right)+\sum_{a\neq \wp,a\neq s}\rho\left(u_a,\sigma_a'(\mathbb{v})\right)\\
&=\rho\left(u_T,\sigma_s^{*}(\mathbb{v})\right)+\rho\left(u_s,\sigma_T^{*}(v)\right)+\sum_{a\neq T,a\neq s}\rho\left(u_a,\sigma_a^{*}(v)\right)\\
&=\rho\left(u_T,\sigma_{\wp}^{*}(v)\right)+\rho\left(u_s,\sigma_T^{*}(\mathbb{v})\right)+\sum_{a\neq \wp,a\neq s}\rho\left(u_a,\sigma_a^{*}(\mathbb{v})\right)\\
&=\rho\left(u_s,\sigma_T^{*}(\mathbb{v})\right)+\sum_{a\neq \wp,a\neq s}\rho\left(u_a,\sigma_a^{*}(\mathbb{v})\right)<1+1+\sum_{a\neq \wp,a\neq s}G\left(u_a,\sigma_a^{*}(v)\right)\\
&=\rho\left(u_T,\sigma_T^{*}(\mathbb{v})\right)+\rho\left(u_s,u_T\right)+\sum_{a\neq \wp,a\neq s}\rho\left(u_a,\sigma_a^{*}(\mathbb{v})\right)\\
&=\rho\left(u_T,\sigma_T^{*}(\mathbb{v})\right)+\rho\left(u_s,u_{\wp}\right)+\sum_{a\neq \wp,a\neq s}\rho\left(u_a,\sigma_a^{*}(\mathbb{v})\right)\\
&=\rho\left(u_T,\sigma_T^{*}(\mathbb{v})\right)+\rho\left(u_s,u_s\right)+\sum_{a\neq \wp,a\neq s}\rho\left(u_a,\sigma_a^{*}(\mathbb{v})\right)\\
&=\sum_{a=1}^{L}\rho\left(u_a,\sigma_a^{*}(\mathbb{v})\right)
\end{aligned}
\tag{12.37}
$$

Estimate the rearranged preserved feasibility for future generations and genetic normalization, as shown in algorithm 2. Let us assume $\mathfrak{F}\wp$_ preserved feasibility, n- normalization, $\mathbb{S}$ - Swap value.

Algorithm 2 for Genetic Normalization

```
Start
    Start
        𝔉℘_n (u, v)
        {
        for a ← 0 to L
            for b ← 0 to L
                if u_a = v_b then
                    𝕊 of v_a & v_b
                End if
            End for
        End for
        }
    Stop
```

In this category, the preserved feasibility is rearranged using the default shortcut. The genetic feature sequence is normalized with the optimal alignment probabilities.

12.3.4 Disease-Prone Factor (DPF)

Furthermore, the BPF method is estimated by fitness evaluation. The fitness function evaluation represents the graph required since the fitness value calculation normally occurs within the BPF. By utilizing the models provided, we can incorporate the optimal fitness function in the center of the image to produce a mean threshold. Additionally, the black dots signify the highest fitness values, while the blue dots indicate the average fitness values. BPF simulates population growth until the strongest survive to determine the ideal constant fitness value. To maximize the mutual information between voxels of the fitness function, the BPF method can estimate a subset of input voxels. All evaluation factors are then sent to BPF to calculate fitness values, which can help select the features of the BPF subset. The BPF method uses a combination of exploitative and exploratory search capabilities to handle large search spaces effectively.

It is based on a heuristic approach inspired by evolution's natural selection process. A brain-prone sequence of binary strings represents a population. Each bit corresponds to a specific trait within the evaluation factor from a hierarchical perspective. The BPF method's quality can be assessed by calculating an overall score for each exercise performance factor. Fitness values are computed for each element using an internal and external ranking of the fitness scale. Channel accuracy and selectivity are considered using a fitness function, balancing these two constraints.

Equation 12.38 calculates the gene's mean squared error (MSE) distance. Let us assume e-fitness function, u-feature signal, L-train data, o-hidden layer, x-bias unit, v-summed with bias, z-weight unit, a and b-fitness neuron values, f-binary sigmoid function, q-neuron, and m-mean square error.

$$(m) = \frac{1}{2}\sum_{a=1}^{2}(T_a - v_a)^2 \tag{12.38}$$

The output value can be calculated using Equation 12.39 as an illustration of how to calculate the binary sigmoid function.

$$e(u) = \frac{1}{1+f^{-u}} \tag{12.39}$$

Calculate the derivative of the binary sigmoid function as indicated in Equation 12.40.

$$e'(u) = e(u)(1-e(u)) \tag{12.40}$$

Calculate the hidden layer weights for each neuron to add biases, as illustrated in Equation 12.41.

$$o\_net_j = x_{b0} + \sum_{q=1}^{q} u_a x_{Lb} \tag{12.41}$$

As shown in Equation 12.42, each output neuron is computed by summing its weights and biases.

$$v\_net_L = z_{L0} + \sum_{j=1}^{T} o_b v_{Lb} \tag{12.42}$$

The target backpropagation of the output neuron calculates the relative position, as shown in Equation 12.43.

$$\Delta z_{L0} = -\delta L \tag{12.43}$$

Compute the biases and weights of the hidden layer output neurons as shown in Equation 12.44.

$$z_{Lb} = z_{Lb}(\text{previous}) + \Delta z_{Lb} \tag{12.44}$$

Equation 12.45 computes the absolute error between the predicted outputs. Let us assume the n-number of the output layer, a-nodes, L-coefficients, e-fitness, c-random number, and d-crossover Operation.

$$e = L\left(\sum_{a=1}^{q} (|\,v_a - k_a\,|)\right) \tag{12.45}$$

Calculate the actual number of crossover processes as per Equation 12.46.

$$d_{Lb} = d_{Lb}(1-c) - d_{ab}c, d_{ab} = d_{ab}(1-c) - d_{La}ca \tag{12.46}$$

Equation 12.47 are illustrated to measure the effect of selected mutations on the gene, where d_{max} – upper bound, d_{min} – lower bound, h-current iteration.

$$d_{ab} = d_{ab} + (d_{ab} - d_{max})e(h), d_{ab} = d_{ab} - (d_{max} - d_{ab})e(h)s \geq 0.5 \tag{12.47}$$

Calculate the upper and lower limits of the random numbers using Equation 12.48.

$$f(g) = Q_2\left(1 - \frac{g}{G_{max}}\right) \tag{12.48}$$

Calculate the curve error as shown in Equation 12.49, where P-power. p_r – Real power, e-error of output power, and P_{out} – output power of test data.

$$f = \frac{|T_{out} - T_s|}{T_s} \tag{12.49}$$

Compute the log-sigmoid function's range output values as shown in Equation 12.50.

$$f(v)=\frac{1}{1+e^{-v}} \tag{12.50}$$

Equation 12.51 evaluates the output neuron for the value obtained from the hidden neuron. Let us assume e-log sigmoid activation function, w_{Bias} -weight origination bias unit hidden layer, w_i *and* w_j – input neuron of the hidden layer, u- input layer.

$$v_q = e\left(z_{Bias_{Input}} + \sum_{a=1}^{q} u_a * Z_{av_a}\right), output = e\left(z_{Bias_{hidden}} + \sum_{b=1}^{q} v_b * z_{ioutput}\right) \tag{12.51}$$

Calculate the amount of error in the output value as shown in Equation 12.52. Let us assume L- number of output nodes, T-number of patterns, a and b – index input and output pattern.

$$m = \frac{1}{oT}\sum_{a=1}^{T}\sum_{b=1}^{L}(output_{ab} - idea_{ab})^2 \tag{12.52}$$

Increasing fitness is evaluated as shown in Equation 12.53.

$$fitness = \frac{1}{1+MSE} \tag{12.53}$$

Estimate the fitness threshold based on the evaluation factor probabilities as shown in Equation 12.54, where s-rank, r-probability population size.

$$T = \frac{r}{1-(1-r)} * (1-r)^{s-1} \tag{12.54}$$

As shown in Equation 12.55, the randomly selected value is computed from the genetic interval.

$$\min\left(u_a^1, u_a^2\right) - \alpha^* \left|u_a^1 - u_a^2\right| \max\left(u_a^1, u_a^2\right) - \alpha * \left|u_a^1 - u_a^2\right| \tag{12.55}$$

Calculate the probability of the mutation program as shown in Equation 12.56.

$$u_{p+1}^1 = \begin{cases} u_p^a + \Delta\left(pu_{upperbound} - u_p^a\right), if\ d = 0 \\ u_p^a + \Delta\left(pu_{lowerbound} - u_p^a\right), if\ d = 1 \end{cases} \tag{12.56}$$

Compute the function of the random variable described in Equation 12.57. Let us assume $\Delta(p,v)$ – random variable, p-current generation, p-maximum generation, u-upper bound minus, c-parameter.

$$\Delta(p,v) = v * (1 - s^{1-\ddagger})^{c} \tag{12.57}$$

Equation 12.58 shows the connection between the output vector and the normalized vector by calculating the conditional entropy function, where s-classification accuracy, p_e – extracted vector, k-length.

$$e = (\alpha \times \gamma) + \beta\left(\frac{|p_e||k_w|}{k_w}\right) \tag{12.58}$$

Equation 12.59 is described to estimate the evolutionary maximum fitness.

$$\sum_{a=1}^{3} zq_a = 1, \sum_{l=1}^{6} zq_l = 1 \tag{12.59}$$

Calculate the weights corresponding to the feature network as shown in Equation 12.60.

$$z_{bl} = z_{qa} \times z_{rl} \tag{12.60}$$

Increasing the accuracy of the measurements is achieved by evaluating the fitness function, as described in Equation 12.61, before the measurement is taken. Let us assume the $\mathcal{Y}$-measured value.

$$(\mathcal{Y}) = \max i_a (\mathbb{A}) \tag{12.61}$$

In this segment, a normalized score can be implemented by optimizing different crossover types. The feature layer's network weights can be optimized based on it.

12.3.5 Decision Tree-Optimized Cuckoo Search (DTOCS)

Furthermore, we use the DTOCS method to select critical features and reduce dimensionality. Decision trees (DTs) are supervised learning techniques used in data classification and regression and for selecting features in a data set. These classifiers are structured like trees, with internal nodes representing data set features, branch decision rules, and leaf nodes describing final decisions. To manage a dataset, it can be divided into subsets to create DTs. While trees are an effective way of organizing data, they cannot be used to describe patients as examples. Each dataset belongs to a distinct category; hence, the approach is supervised rather than unsupervised learning. Employing the same alignment method can help build a tree using the gathered information, leading to further improvement. There are four main data types: continuous, discrete, abstract, and easily predictable. These data types are always easy to read.

A DT has two components: end nodes and leaf nodes. DT is easy to recognize because it simulates humans' stages when creating real-world objects. Prediction accuracy in a dataset evaluates the average of all DTs to determine whether the presence or absence of a particular feature depends on other regions. This helps classify objects into specific categories. Finally, DT has two parts: end nodes, which parse the data, and leaf nodes, which provide the results. Along the tree path, the results of each node test provide enough information to determine its type when classifying a model. The OCS method is driven by the velocity optimization formula, which enables iterative solution space exploration. OCS aims to find optimal answers to optimization problems through duplication, selection, and permutation techniques. The process begins with a set number of cuckoo eggs, which can be generated approximately. The algorithm uses Levy flight techniques to create fresh solutions that avoid getting stuck in local optima. A cuckoo-winged creature lays an egg every so often and then drops it into a nest of its choice.

Egg searches are performed for the least necessary nest locations and excluded from further calculations. The nests were found in a series of less famous nests and were left over from other analyses. The cuckoo egg presents a novel solution. A new account such as cuckoo eggs will provide a technologically advanced solution to replace the old collection and provide a state-of-the-art solution. Likewise, the mass of eggs in the nest is a fitness function of the solution. Cuckoos try to blend in with the eggs already in the nest by copying their appearance, color, spots, and size. If this doesn't work, the host bird will realize that the egg doesn't belong to them. Otherwise, the next-generation cuckoo strategy and process will succeed. The DT classifier's basic structure is shown in Figure 12.4, with its leaf nodes divided into two parts: the former parses the data, and the latter produces the result.

Figure 12.5 shows the components and rules of the DT. The test results at each node provide enough information to categorize a model.

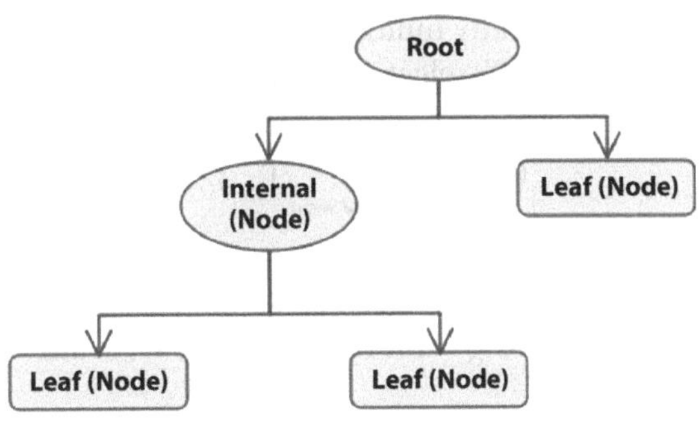

Figure 12.4 Architecture design for decision tree.

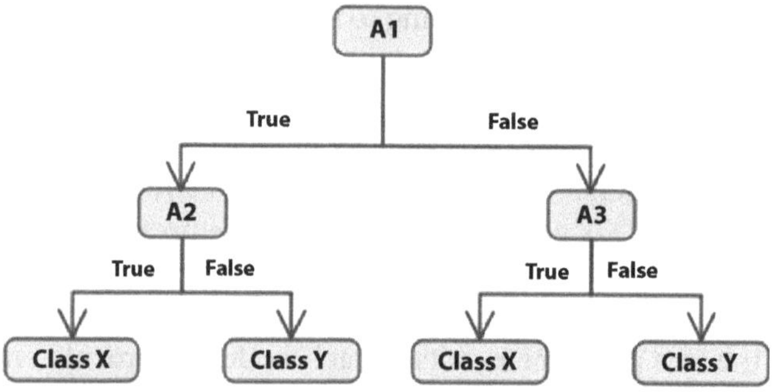

Figure 12.5 Architecture diagram for decision tree element rules.

Calculate an approximation of the likelihood of enhancement using Equation 12.62. Let us assume the L-number of the leaf, the Q-total number of the leaf, and the T-conditional probability.

$$T = \mathcal{Y}\frac{L}{q} \text{ by } \frac{L+1}{q+2} \tag{12.62}$$

Calculate the base rate probability assessment as described in Equation 12.63. Let us assume c-based rate, $\mathbb{M}$ – parameter control.

$$T = \frac{L}{q} \text{byT}' \frac{L+c-\mathbb{M}}{q+\mathbb{M}} \tag{12.63}$$

As shown in Equation 12.64, the information derived from the attribute is calculated. Let us assume T_a – probability of arbitrary Vector, y-Class, $\log_2$-information encode bits, $\mathcal{M}$ - attribute, and i-information.

$$i(y) = -\sum_{a=1}^{\mathcal{M}} T_a \log_2 T \tag{12.64}$$

Calculate the maximum probability multiple entropy as shown in Equation 12.65. Let us assume f- entropy, P-current state, X-selected attribute, and w-attribute.

$$f(p,u) = \sum_{w \in u} T_{(w)} f_{(w)} \tag{12.65}$$

Calculate the differences between the mean entropies based on the characteristic values described in Equation 12.66, where ig - information gain, $\mathbb{A}_L$ – attribute value, and b- subset after the split.

$$ig_{(p,u)} = f_{(p)} - f_{(p,u)}, i_g = f_{(before)} - \sum_{b=1}^{L} f_{(b,after)}, split_{(u,\mathbb{A}_L)} = -\sum_{b=1}^{w} \frac{|u_b|}{|u|} q \frac{|u_b|}{|u|} \tag{12.66}$$

Compute the Gini index using the dataset shown in Equation 12.67. Let us assume $\mathcal{G}$ - Gini value, Q-decision split the database, T_b and $\mathbb{M}$ -probability of the class.

$$\mathcal{G}_Q = 1 - \sum_{b=1}^{\mathbb{M}} T_b \tag{12.67}$$

As shown in Equation 12.68, estimate the split point for the Gini value, where n- is the number of instances.

$$\mathcal{G}_{(split)}(r) = \frac{q_1}{q}\mathcal{G}_{r_1} + \frac{q_1}{q}\mathcal{G}_{r_2} \tag{12.68}$$

Calculate the feature weights for each feature as shown in Equation 12.69.

$$z(\mathbb{A}) = z(\mathbb{A}) - \sum_{b=1}^{L} diff(\mathbb{A}, s, h_b)\mathbb{M}^*L \sum_{w \notin class} \left(\frac{\frac{T_{(w)}}{1 - T_{(class(s))}} \sum_{b=1}^{L} diff(\mathbb{A}, s, \mathbb{M}_b(w))}{\mathbb{M}^*L} \right) \tag{12.69}$$

Compute the information gain ratio for the data set as shown in Equation 12.70.

$$\mathcal{G}_{(s)}(u, \mathbb{A}_L) = \frac{\mathcal{G}_{(u,\mathbb{A}_L)}}{split_{u,\mathbb{A}_L}}, \frac{i_g}{split_i} = \frac{f_{before} - \sum_{b=1}^{L} f_{(u,after)}}{\sum_{b=1}^{L} z_b \log_2 z_b} \tag{12.70}$$

Equation 12.71 calculates the standard number of host nests and the probability of a host bird recognizing a cuckoo egg. Let us assume T_d - cuckoo identification.

$$T_d \in \mathcal{G}_{(s)}[0,1] \tag{12.71}$$

As shown in Equation 12.72, Cuckoo's new solution can be evaluated with collection planes. Let us assume $z_a^{(p+1)}$ - new generation for Cuckoo, α – linked size, $\oplus$ -bitwise multiplication, λ - scalling parameter, and $]\!]$ – levy flight.

$$z_a^{(p+1)} = z_a^p + \alpha \oplus]\!](\lambda) \tag{12.72}$$

Computing the modified cuckoo search algorithm, the size is expressed in Equation 12.73. Let us assume z-distance and $\varkappa_{\mathbb{K}}^{\delta}$ - current iteration.

$$h_g^{\delta+1} = h_g^{\delta} + \varkappa^{\delta+1} \oplus \mathbb{K}(\eta), \varkappa_{\mathbb{K}}^{\delta} = \max * \left(\frac{L^{\delta}}{z} \right) \tag{12.73}$$

As shown in Equation 12.74, compute the current iteration by distance. Let us assume $\mathbb{M}_u$ *and* $\mathbb{M}_v$ – fitness solution, s- value.

$$z = s_{\mathbb{M}_u} - s_{\mathbb{M}_v} \tag{12.74}$$

Compute the cuckoo level optimization described in Equation 12.75, where p- stage and α - length.

$$u(p+1)_a = u(p_a) + \alpha \oplus \mathbb{L}(\lambda) \tag{12.75}$$

Calculate the optimization based on the objective function, as shown in Equation 12.76. Let us assume E-function, $E(u)$ – objective function, k_c – lower bound, and x_c - upper bound.

$$u \in \underset{[k_c, x_c]}{\min} E(u) \tag{12.76}$$

Calculate the value of the objective function for every solution, as outlined in Equation 12.77, where G-decision variable, HNP-host nest population, q-total number of eggs, s-size.

$$HNP = \begin{bmatrix} u_1^1, u_2^1, \ldots, u_G^1 \\ u_1^2, u_2^2, \ldots, u_G^2 \\ u_1^q, u_2^q, \ldots, u_G^q \end{bmatrix} \tag{12.77}$$

Compute the HNP for each row as shown in Equation 12.78.

$$u^a(p+1) = u^a(p) + \alpha \oplus \mathbb{L}(Q, \lambda),\ \mathbb{L}(Q, \lambda) = \frac{\lambda \Gamma(\lambda) \sin(\pi \Gamma / 2)}{Q \geq Q_0 > \pi_0} \frac{1}{Q^{1+\mathbb{A}}} \tag{12.78}$$

Calculate the probability of each solution locally using Equation 12.79. Let us assume g-Heaviside function, ε – generate random number.

$$u^a(p+1) = u^a(p) + \alpha Q \otimes g(T_d - \varepsilon) \otimes \left(y^b(p) - u^l(p)\right) \tag{12.79}$$

Compute the fit of the position vector representing the feature-length as shown in Equation 12.80. Let us assume E-feature, E_k – feature length, y-state vector, $\tilde{y}$ – represents state vector.

$$E = \sqrt{\frac{1}{E_k} \sum_1^{E_k} (y - \tilde{y})^2} \tag{12.80}$$

Estimate the Levy flight using Gaussian distribution is shown in Equation 12.81, where T_a^* – nest population, σ_0, μ – constant value, and w_h - current generation.

$$T_a^* = T_a^{(L+1)} = T_a^{(L)} + \alpha \oplus \sigma_Q, \sigma_Q = \sigma_0 \exp(-\mu \times w_h) \tag{12.81}$$

In this category, a data set's information gain rate is determined by two parameters: the position of the bow of the cuckoo in the environment.

12.3.6 Long Short-Term Memory Gate Multilayer Perceptron Neural Network (LSTM-MLPNN)

Moreover, the LSTM gate MLPNN can be observed using the proposed method to obtain the most suitable values of logical units. This is to process the selected parts of the analytical evaluation. There are several ways in which this unit constant can be used to facilitate the transfer of past data from one step to the next. LSTM-based deep learning overcomes existing neural network limitations, and the gradient fading problem that occurs when errors propagate within the neural network layer is resolved. The multiplier cells within a memory block are known as "gates," which are responsible for controlling the flow of information. The temporary state of the network is stored in memory cells that have self-connections. The original structure has been enhanced with the addition of additional inputs and outputs. It is the input gate that regulates how information and processing enters the cell and how it is arranged, which is based on the sigmoid function and the Tanh function.

The output gate controls the output current. Sigmoid and Tanh functions calculate the remaining network's activation function. Neural network weights are optimized using the backpropagation (BP) algorithm. They achieve this by storing past information in time-step memory cells, which are the building blocks of neural networks. Each gate produces a vector output value by combining the sigmoid and Tanh layers. The MLPNN is a neural network composed of one or more hidden layers and is capable of implementing data input, computation neurons, and storage neurons as well as an output layer.

Figure 12.6 illustrates the MLPNN architecture diagram Also, the proposed MLPNN depends on the number of neurons in each layer and the type of each layer below.

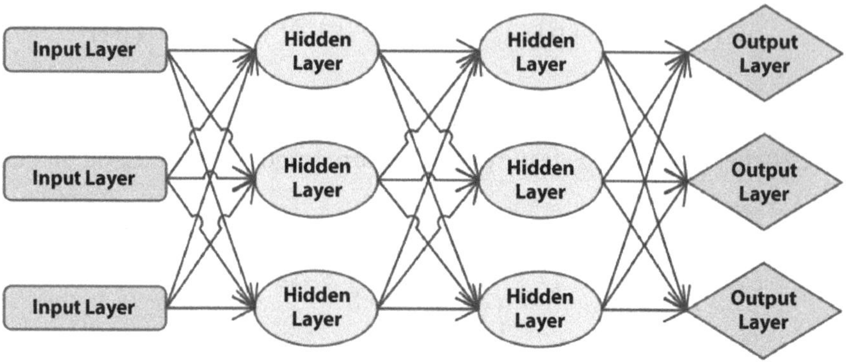

Figure 12.6 Architecture diagram for MLPNN.

Calculate the cell state information value by the activation function described in Equation 12.82. Let us assume e_p - output value, p-point, u_p -input value, and g_{p-1} − forget gate layer.

$$e_p = \sigma_Q\left(z_e\left[g_{p-1}, u_p\right] + c_e\right) \tag{12.82}$$

The input gate is computed using the conditions in Equation 12.83 to determine the updated information.

$$a_p = \sigma\left(z_a\left[g_{p-1}, u_p\right] + c_e\right) \tag{12.83}$$

Compute the hyperbolic tangent of the activation function as shown in Equation 12.84.

$$w_p = \tan_h\left(z_w\left[g_{p-1}, u_p\right] + c_w\right) \tag{12.84}$$

Calculate the new cell position as shown in Equation 12.85. Let us assume, $\tilde{w}$ – cell state, w_p – new cell created.

$$w_p = E_p * w_{p+1} a_p * \tilde{w}_p \tag{12.85}$$

Calculate the output value of the sigmoid cascade as illustrated in Equation 12.86. Let us assume σ – sigmoid function, g_p -generate output value, $tan_h\left(w_p\right)$ - hyperbolic tangent function.

$$\sigma_p\left(g_p\right) = \sigma\left(z_\sigma\left[g_{p-1}, u_p\right] + c_\sigma\right) * \tan_h\left(w_p\right) \tag{12.86}$$

Compute the functional output value of the hidden layer as shown in Equation 12.87. Let us assume p-time, $\vec{g}_p$ and $\overleftarrow{g}_p$ – forward and backwards hidden layer, and v_p – output layer.

$$\vec{g}_p = \sigma\left(Zu^{\vec{g}^{up}} + Z\vec{g}\vec{g}^{\vec{g}_{p-1}} + c_{\vec{g}}\right), \overleftarrow{g_p} = \sigma\left(Zu^{\overleftarrow{g}up} + Z\overleftarrow{g}\overleftarrow{g}^{\overleftarrow{g}_{p-1}} + c_{\overleftarrow{g}}\right), v_p = z_{\vec{g}v}\vec{g}_p + z_{\overleftarrow{g}v}\overleftarrow{g}_p + c_v \tag{12.87}$$

Calculate the weight values for the next connected layer, as shown in Equation 12.88. Let us assume z-weight, z_g, z_c, z_d – diagonal weight values, c-bias function, $^\circ$ -element product, g_a – output activation vector, $\tilde{g}_a$ -cell input and output function, c h, q-input, output and forget gate and tanh-activation function.

$$g_p = \tanh\left(z_{gg}g_{p-1} + z_{ug^{up}}\right), d_a = \sigma\left(z_{d^{ua}} + z_{d^{ga-1+cd}}\right), d_g = \sigma\left(z_{g^{ua}} z_{a^{ga-1+cg}}\right), \tilde{g}_a =$$
$$\tanh\left(z_{g^{ua}} + h_a{}^\circ z_{g^{ga-1}} + c_q\right), g_a = d_a{}^\circ g_{a-1} + \left(1-h_a\right)^\circ \tilde{g}_a \tag{12.88}$$

Estimate the LSTM mathematics behind the entire process is shown in Equation 12.89. Let us assume σ and tanh- is a nonlinear activation function, $k_p, e_p, a_p, \sigma_p, w_p, g_p$ – weight of gate, u -elementwise multiplication, k_p and w_p -represents input vector, g_p and g_{p-1} – current and previous cell output, w_p and w_{p-1} – current and previous memory cell

$$k_p = \tanh\left(w_{p^u}z_k + g_{p-1^u}x_k\right)e_p = \sigma\left(w_{p^u}z_e + g_{p-1^u}x_e\right)a_p = w_{p^u}z_a + g_{p-1^u}x_a,$$
$$\sigma_p = w_{p^u}z_\sigma\ g_{p-1^u}x_\sigma w_p = e_{p^u}w_{p-1} + a_{p^u}, g_p = \sigma_{p^u}\tanh\left(w_p\right) \tag{12.89}$$

Compute the activation function for the sigmoid function as shown in Equation 12.90. Let us assume e-function, $e(o)$ -activation function, and f^{-o} - sigmoidal function.

$$e(o) = g_p \frac{1}{1+f^{-o}} \tag{12.90}$$

Calculate the difference in network outputs as a function of mean squared error (MSE), as shown in Equation 12.90. Let us assume $\mathfrak{M}$ – mean square error, a-data, z-weight vector, u and v – input and output vector, $e(z, u_a)$ - approximating function.

$$\mathfrak{M} = \sum_a \left(v_a e\left(z, u_a\right)\right)^2 \tag{12.91}$$

Calculates the quick propagation weight change, as shown in Equation 12.92, where Q-numeric unstable value.

$$\Delta_z(p) = \frac{Q(p)}{Q_{(p-1)-Q(p)}} \Delta_z(p-1) \tag{12.92}$$

As shown in Equation 12.93, compute the supervised exercise for MLP weight updates, where η – learning rate, δ -local error gradient, α -momentum coefficient, o_a – output, and a-unit.

$$\Delta_{z_{ab}}(p) = \eta\delta_a o_a + \alpha\Delta z_{ab}(p-1) \tag{12.93}$$

Compute the cumulative average error gradient as shown in Equation 12.94. Let us assume $\bar{\delta}(p)$ – derivate of the error surface, ϑ - smoothing constant.

$$\bar{\delta}(p) = (1-\vartheta)\delta(p) + \vartheta\bar{\delta}(p-1) \tag{12.94}$$

Calculating the binary classification loss is shown in Equation 12.95, where t_p – estimated probability, γ – focousibg parameter, α_p - coefficient total weight, and $\mathbb{f}_{\mathbb{k}}$ –function loss, a-class.

$$\mathbb{f}_{\mathbb{k}} = \sum\nolimits_{p=0,1} -\alpha_p(1-t_p)^{\gamma}\log(t_p) \tag{12.95}$$

The estimated value of the optimized multivariate classification loss function is shown in Equation 12.96, where λ - regularization coefficient, $\|z\|^2$ – regularization of weight parameters

$$\mathbb{k} = \sum\nolimits_{a=0}^{3} -\left(1-\hat{t}_a\right)^{\gamma} \log\hat{t}_a + \lambda\|z\|^2 \tag{12.96}$$

Compute the connection weights for the activation function described in Equation 12.97. Let us assume c_a – bias term, z_{ab} – connection weight, a- neuron previous layer.

$$v_a = e\left(\sum\nolimits_{z_{ab}} e_a + c_a\right) \tag{12.97}$$

Estimate the disease risk score for the gene as shown in Equation 12.98. Let us assume h-gene, $Q_G(h)$ – disease gene prediction score, $Q_{qG}(h)$ – non-disease gene prediction score for gene.

$$Q_{(h)} = \frac{Q_G(h)}{\left(Q_G(h) + Q_{qG}(h)\right)} \tag{12.98}$$

In this category, improved loss functions and tuning of weight parameters to enhance model stability and classification accuracy.

12.4 Result and Discussion

A brain stroke dataset is compiled to evaluate the proposed LSTM-gated MLPNN method. Radiologists reviewed the dataset to classify stroke patients accurately. They demonstrated the performance of an LSTM gate trained by MLPNN and the algorithm to predict stroke mortality. Classifiers are evaluated using precision, sensitivity, error rate, F-measure, root mean square recall, and log loss. Compared to alternative systems, the proposed model provides much higher accuracy than alternative systems.

To detect strokes using Python and Jupyter Notebook, Table 12.1 lists the proposed simulation parameters. The brain stroke dataset is collected from Kaggle and includes a bulk dataset that can be used for training and testing to achieve accurate stroke detection.

12.4.1 Performance Matrix

This section evaluates the stroke detection classifier by performance metrics such as precision, sensitivity, precision, F1 score, and recall. Performance metrics are used to measure the performance of the classifier in recognizing and classifying strokes. In order to calculate the metrics, the data used in the training phase as well as the testing stage is used. The performance is then compared to existing state-of-the-art classifiers.

Table 12.1 Simulation parameter.

Simulation limit	**Variable**
Dataset name	Kaggle
No. of datasets	4,982
Training	3,556
Testing	1,426
Language	Python
Tool	Jupyter

To determine the accuracy of sensitivity analysis for detecting brain strokes, the LSTM-gated MLPNN approach proposed by our method, as depicted in Figure 12.7, is utilized. The proposed method has a higher accuracy of 69% than the FES, MWT, and LBE methods obtained from literature analysis. These methods only have an accuracy of 43%.

In Figure 12.8, the proposed method of using LSTM Gates MLPNN for stroke detection shows a significant increase in accuracy. This method involves obtaining stroke data from a dataset and conducting training and testing analyses to determine its accuracy. Compared to other techniques, such as MWT, LBE, and FES, accuracy is less than 52%. However, the accuracy of the proposed LSTM gate MLPNN method improves by up to 73%.

An estimate of F1-score analysis can be obtained using the proposed LSTM Gates MLPNN approach to detect brain stroke accuracy, as shown in Figure 12.9. Based on these

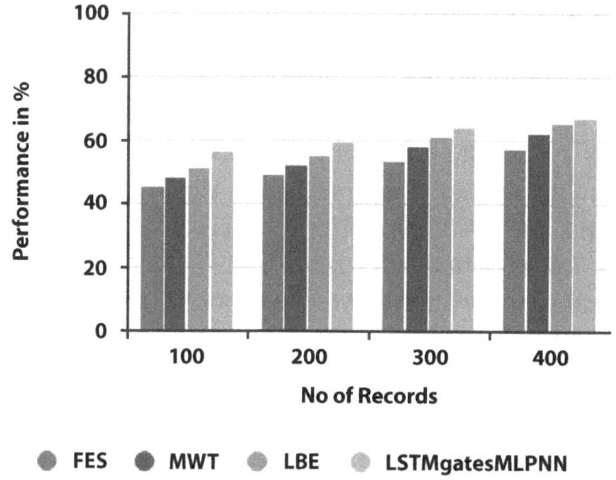

Figure 12.7 Analysis of sensitivity.

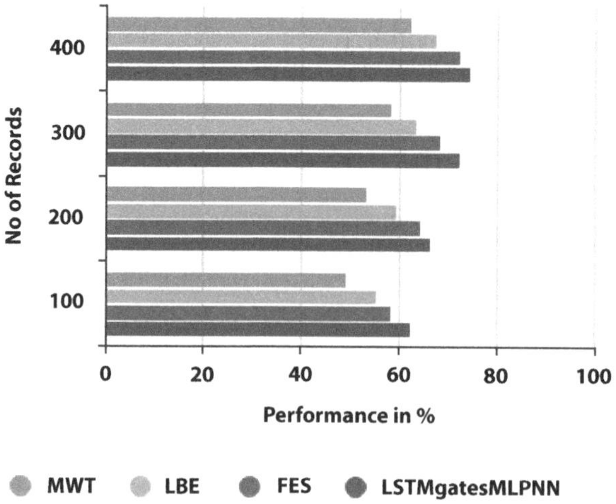

Figure 12.8 Analysis in precision.

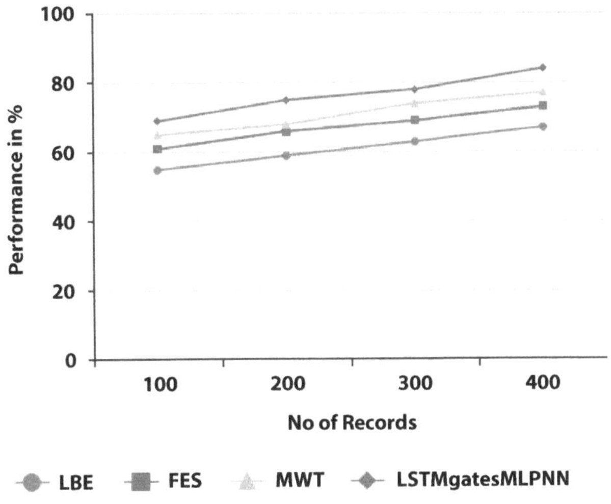

Figure 12.9 Analysis in F1-score.

data, it can be collected from the dataset and processed on a training and testing basis to obtain accuracy for brain lateralization. The F1 score can be analyzed using this. Also, the accuracy of the proposed LSTM Gates MLPNN method increased to 78%. However, the accuracy of the LBE, FES, and MWT methods determined from literature analysis is less than 55% compared to the suggested method.

In Figure 12.10, it was proven that stroke accuracy can be achieved through the use of error rates. The accuracy of the MWT, LBE, and FES techniques increased by 79% due to diminished error rates. When testing the proposed LSTM gate MLPNN method, the error rate accuracy was less than 57%.

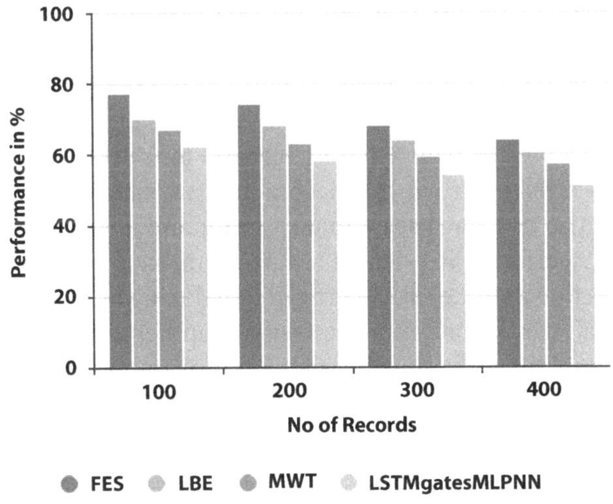

Figure 12.10 Analysis of error rate.

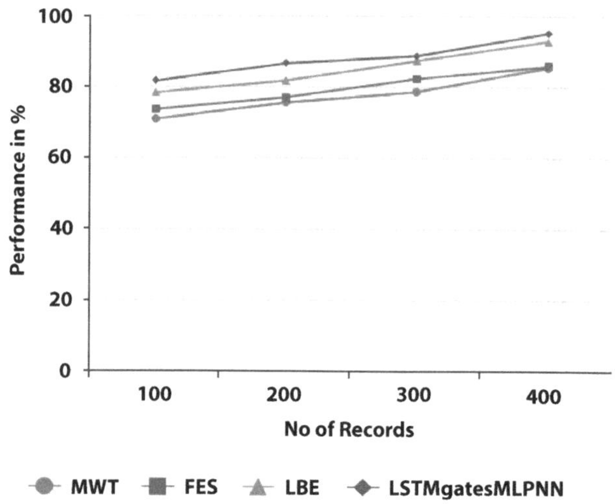

Figure 12.11 Analysis in accuracy.

Figure 12.11 presents a method for collecting stroke data called LSTM-gated MLPNN. This method has been trained and tested for accuracy. When compared to other methods like MWT, LBE, and FES, the proposed method has shown an improvement in accuracy to 95.30%.

12.5 Conclusion

In this segment, we will utilize the LSTM-gated MLPNN method to classify and determine feature metrics variations between medical edges and establish feature thresholds. The first step will be gathering brain stroke data from the Kaggle dataset. We will then develop feature selection and classification systems to examine the disease's characteristics and facilitate its diagnosis. We will also process the data using preprocessing techniques such as transformation and standardization. Next, we will use the GFSA technique to determine normalized brain damage scores. Additionally, we will calculate TPF by conducting physical fitness assessments. We can reduce the aspect ratio for selecting features by employing DTOCS selection. By implementing these techniques, we can utilize uncorrelated feature bias weights to maximize the true disease impact factor and improve recognition accuracy. The proposed model's performance yielded a maximum accuracy of 95.30% compared to another system. Detecting a stroke early is crucial to reducing brain damage and improving recovery chances. Brain strokes are more common than hemorrhagic strokes. Identifying the type of stroke requires evaluating the patient's non-modifiable risk factors and the impact of their specific symptoms.

References

1. Biswas, N., Uddin, K.M.M., Rikta, S.T., Dey, S.K., A comparative analysis of machine learning classifiers for stroke prediction: A predictive analytics approach. *Healthcare Anal.*, 2, 100116, 2022, ISSN 2772-4425.
2. Sundaram, K.D.M., Haritha, G., Abhilash, A., Detection of Brain Stroke Using Machine Learning Algorithm, Quest Journals. *J. Electron. Commun. Eng. Res.*, 8, 4, 25–30, 2022, ISSN (Online): 2321-5941.
3. Tazin, T., *et al.* Stroke disease detection and prediction using robust learning approaches. *J. Healthcare Eng.*, 2021.1, 7633381, 2021.
4. Al-Mekhlafi, Z.G., *et al.* Deep learning and machine learning for early detection of stroke and haemorrhage. *Comput. Mater. Continua*, 72.1, 775–96, 2022.
5. Yalçın, S. and Vural, H., Brain stroke classification and segmentation using encoder-decoder based deep convolutional neural networks. *Comput. Biol. Med.*, 149, 105941, 2022, ISSN 0010-4825.
6. Mridha, K., Ghimire, S., Shin, J., Aran, A., Uddin, M.M., Mridha, M.F., Automated Stroke Prediction Using Machine Learning: An Explainable and Exploratory Study With a Web Application for Early Intervention. *IEEE Access*, 11, 52288–52308, 2023.
7. Mak, J. *et al.*, Detection of Stroke-Induced Visual Neglect and Target Response Prediction Using Augmented Reality and Electroencephalography. *IEEE Trans. Neural Syst. Rehabil. Eng.*, 30, 1840–1850, 2022.
8. Shao, Z., Dou, W., Ma, D., Zhai, X., Xu, Q., Pan, Y., A Novel Neurorehabilitation Prognosis Prediction Modeling on Separated Left-Right Hemiplegia Based on Brain-Computer Interfaces Assisted Rehabilitation. *IEEE Trans. Neural Syst. Rehabil. Eng.*, 31, 3375–3383, 2023.
9. Yu, J., Park, S., Kwon, S.-H., Cho, K.-H., Lee, H., AI-Based Stroke Disease Prediction System Using ECG and PPG Bio-Signals. *IEEE Access*, 10, 43623–43638, 2022.
10. Su, R. *et al.*, autoTICI: Automatic Brain Tissue Reperfusion Scoring on 2D DSA Images of Acute Ischemic Stroke Patients. *IEEE Trans. Med. Imaging*, 40, 9, 2380–2391, Sept. 2021.
11. Choi, W.J., Li, Y., Wang, R.K., Monitoring Acute Stroke Progression: Multi-Parametric OCT Imaging of Cortical Perfusion, Flow, and Tissue Scattering in a Mouse Model of Permanent Focal Ischemia. *IEEE Trans. Med. Imaging*, 38, 6, 1427–1437, June 2019.
12. Zhang, R. *et al.*, An Adaptive Brain-Computer Interface to Enhance Motor Recovery After Stroke. *IEEE Trans. Neural Syst. Rehabil. Eng.*, 31, 2268–2278, 2023.
13. Lutz, Y. *et al.*, Selective Brain Hypothermia for Ischemic MCA-M1 Stroke: Influence of Cerebral Arterial Circulation in a 3D Brain Temperature Model. *IEEE Trans. Biomed. Eng.*, 68, 2, 404–415, Feb. 2021.
14. Yu, H. *et al.*, Effects of Transcranial Direct Current Stimulation on Stroke Based on Brain Functional Networks. *IEEE Trans. Magn.*, 57, 6, 1–4, June 2021, Art no. 5800404.
15. Vaidya, M. *et al.*, Hemicraniectomy in Traumatic Brain Injury: A Noninvasive Platform to Investigate High Gamma Activity for Brain Machine Interfaces. *IEEE Trans. Neural Syst. Rehabil. Eng.*, 27, 7, 1467–1472, July 2019.
16. Guo, L., Khosravi-Farsani, M., Stancombe, A., Bialkowski, K., Abbosh, A., Adaptive Clustering Distorted Born Iterative Method for Microwave Brain Tomography With Stroke Detection and Classification. *IEEE Trans. Biomed. Eng.*, 69, 4, 1512–1523, April 2022.
17. Liu, Y.-C., Yang, Y.-R., Tsai, Y.-A., Wang, R.-Y., Lu, C.-F., Brain Activation and Gait Alteration During Cognitive and Motor Dual Task Walking in Stroke—A Functional Near-Infrared Spectroscopy Study. *IEEE Trans. Neural Syst. Rehabil. Eng.*, 26, 12, 2416–2423, Dec. 2018.

18. Vijayadeep, G. and Rao, N.N.M., A hybrid feature extraction based optimized random forest learning model for brain stroke prediction. *Turk. J. Comput. Math. Educ. (TURCOMAT)*, 11, 3, 1152–1165, 2020.
19. Sundaram, K.D.M., Detection of Brain Stroke Using Machine Learning Algorithm I QUEST JOURNALS-Academia. Edu. (PDF) Detection of Brain Stroke Using Machine Learning Algorithm| QUEST JOURNALS-Academia. Edu (2022).
20. Karthik, R., *et al.* Neuroimaging and deep learning for brain stroke detection-A review of recent advancements and future prospects. *Comput. Methods Programs Biomed.*, 197, 105728, 2020.
21. Mahesh, R., Nandini, N.S., Ravi Kumar, G.K., Automated Prediction of Brain Stroke Disease Classification Using Machine Learning Algorithm Techniques. *Automated Prediction of Brain Stroke Disease Classification Using Machine Learning Algorithm Techniques*, 2022.
22. Sirsat, M.S., FermÃ©, E., CÃ¢mara, J., Machine Learning for Brain Stroke: A Review. *J. Stroke Cerebrovasc. Dis.*, 29, 10, 105162, 2020.
23. Gong, Z., Ding, Y., Chen, Y., Cree, M.J., Wearable Microwave Medical Sensing for Stroke Classification and Localization: A Space-Division-Based Decision-Tree Learning Method. *IEEE Trans. Antennas Propag.*, 71, 8, 6906–6917, Aug. 2023, doi: 10.1109/TAP.2023.3283131.
24. Scapaticci, R., Tobon, J., Bellizzi, G., Vipiana, F., Crocco, L., Design and numerical characterization of a low-complexity microwave device for brain stroke monitoring. *IEEE Trans. Antennas Propag.*, 66, 12, 7328–7338, Dec. 2018.
25. Merunka, I., Massa, A., Vrba, D., Fiser, O., Salucci, M., Vrba, J., Microwave tomography system for methodical testing of human brain stroke detection approaches. *Int. J. Antennas Propag.*, 2019, 1–9, Mar. 2019.
26. Salucci, M., Gelmini, A., Vrba, J., Merunka, I., Oliveri, G., Rocca, P., Instantaneous brain stroke classification and localization from real scattering data. *Microw. Opt. Technol. Lett.*, 61, 3, 805–808, Mar. 2019.
27. Karadima, O. *et al.*, Experimental validation of microwave tomography with the DBIM–TwIST algorithm for brain stroke detection and classification. *Sensors*, 20, 3, 840, Feb. 2020.
28. Choi, Y.A., Park, S., Jun, J.A., Ho, C.M.B., Pyo, C.S., Lee, H. *et al.*, Machine-learning-based elderly stroke monitoring system using electroencephalography vital signals. *Appl. Sci.*, 11, 4, 1–18, Feb. 2021.
29. Yu, J., Park, S., Kwon, S.H., Ho, C.M.B., Pyo, C.S., Lee, H., AI-based stroke disease prediction system using real-time electromyography signals. *Appl. Sci.*, 10, 19, 1–19, Sep. 2020.
30. Subudhi, A., Dash, M., Sabut, S., Automated segmentation and classification of brain stroke using expectation-maximization and random forest classifier. *Biocybern. Biomed. Eng.*, 40, 277–289, 2020.
31. Jayachitra, S. and Prasanth, A., Multi-Feature Analysis for Automated Brain Stroke Classification Using Weighted Gaussian Naïve Bayes Classifier. *J. Circuits Syst. Comput.*, 30, 2150178, 2021.
32. Vijayadeep, G. and Rao, N.N.M., A hybrid feature extraction based optimised random forest learning model for brain stroke prediction. *Turk. J. Comput. Math. Educ. (TURCOMAT)*, 11, 3, 1152–1165, 2020.
33. Meinel, A., Sosulski, J., Schraivogel, S., Reis, J., Tangermann, M., Manipulating Single-Trial Motor Performance in Chronic Stroke Patients by Closed-Loop Brain State Interaction. *IEEE Trans. Neural Syst. Rehabil. Eng.*, 29, 1806–1816, 2021.
34. Karadima, O., Lu, P., Sotiriou, I., Kosmas, P., Experimental Validation of the DBIM-TwIST Algorithm for Brain Stroke Detection and Differentiation Using a Multi-Layered Anatomically Complex Head Phantom. *IEEE Open J. Antennas Propag.*, 3, 274–286, 2022.
35. Li, S. and Manogaran, G., Design and Implementation of Networked Collaborative Service System for Brain Stroke Prevention and First Aid. *IEEE Access*, 7, 14825–14836, 2019.

36. Alqadami, A.S.M., Bialkowski, K.S., Mobashsher, A.T., Abbosh, A.M., Wearable electromagnetic head imaging system using flexible wideband antenna array based on polymer technology for brain stroke diagnosis. *IEEE Trans. Biomed. Circuits Syst.*, 13, 1, 124–134, Feb. 2019.
37. Salucci, M., Polo, A., Vrba, J., Multi-step learning-by-examples strategy for real-time brain stroke microwave scattering data inversion. *Electronics*, 10, 1, 95, 2021.
38. Coli, V. *et al.*, Detection of simulated brain stroke using microwave tomography. *IEEE J. Electromagn. RF Microw. Med. Biol.*, 3, 4, 254–260, Dec. 2019.
39. Sailasya, G. and Kumari, G.L.A., Analyzing the performance of stroke prediction using ML classification algorithms. *Int. J. Adv. Comput. Sci. Appl.*, 12, 6, 539–545, 2021.

13

Hybrid Approaches: Combining Deep Reinforcement Learning with Other Techniques

M. T. Vasumathi[1]*, Manju Sadasivan[2] and Aurangjeb Khan[2]

[1]*New Horizon College of Engineering, Bengaluru, India*
[2]*CMR University, Bengaluru, India*

Abstract

The fusion of digital twin technology with reinforcement learning (RL) constitutes a groundbreaking paradigm at the forefront of technological innovation. This chapter encapsulates the essence of a comprehensive exploration into integrating RL algorithms with digital twin frameworks, forging a dynamic synergy poised to revolutionize industries and applications across the spectrum. Digital twins, intricate digital replicas of physical entities or processes, have recently witnessed remarkable adoption, transforming industries such as manufacturing, healthcare, urban planning, and more. Their capacity to simulate, monitor, and predict real-world systems has proven invaluable, yet their true potential awaits further enhancement. Reinforcement learning, a dynamic approach to machine learning, is renowned for its ability to enable intelligent agents to make optimal decisions in complex, ever-changing environments. The synergy between digital twins and RL augments their capabilities, empowering these digital counterparts with the ability to adapt, learn, and optimize within their virtual domains. From predictive maintenance in manufacturing to personalized patient care in healthcare, digital twins serve tools for decision support, process optimization, and innovation. This research explores the fundamental principles of reinforcement learning (RL) and its transformative impact on evolving digital twins into dynamic learning systems. Practical cases exemplify how RL-driven digital twins optimize manufacturing, predict patient responses in therapies, and revolutionize urban planning. This study introduces an Internet of Things (IoT) driven digital twin (DT) framework for automotive manufacturing. It guides visionaries and industry leaders in optimizing efficiency and adaptability using reinforcement learning (RL) for digital twins.

Keywords: Reinforcement learning, digital twins, predictive maintenance, simulation, Internet of things

**Corresponding author*: vasumathi_mt@yahoo.com
M.T. Vasumathi: ORCID: https://orcid.org/0000-0003-4658-3267
Manju Sadasivan: ORCID: https://orcid.org/0000-0003-1991-3996
Aurangjeb Khan: ORCID: https://orcid.org/0000-0003-3988-2613

Shubham Mahajan, Pethuru Raj and Amit Kant Pandit (eds.) Deep Reinforcement Learning and Its Industrial Use Cases: AI for Real-World Applications, (301–324) © 2024 Scrivener Publishing LLC

13.1 Introduction

Digital twin (DT) technology, often referred to as digital avatars, digital masters, or digital shadows, is a cutting-edge technology that enables the seamless translation of the physical world into the digital realm, facilitating real-time interaction between the two. This innovation effectively transcends the limitations imposed by real-world environmental factors and empowers the digital realm to extend its influence into the physical world while responding to real-world changes [1]. Currently, DTs are characterized by three primary functions:

a) Aggregating and fusing data from various aspects of physical objects, enabling high-fidelity, real-time mapping of these objects.
b) Facilitating the coexistence and coevolution of digital representations with their physical counterparts throughout the entire lifecycle of these objects.
c) Providing the means to describe, optimize, and control physical objects.

The idea of digital twins initially surfaced in the United States military aerospace sector but has since extended its influence across a wide range of industries, spanning transportation, industrial manufacturing, smart education, and more. Digital twins serve as crucial components in simulation, monitoring, assessment, forecasting, enhancement, control, and numerous other applications. As illustrated in Figure 13.1, DT technology exhibits close associations with several other technological innovations. Consequently, DTs are acknowledged as a fundamental technology that drives the digital evolution of businesses, firmly establishing themselves as a prominent and highly coveted asset in both industrial and academic circles [2].

13.1.1 Digital Twin—Introduction

The digital twin concept was introduced in 2002 by Dr. Michael Grieves, and NASA had been practicing similar twinning ideas since the 1960s for space programming. Despite early efforts, the concept gained recognition in 2002 with the integration of real and virtual space elements. The integration of digital and physical parts has remained consistent, evolving into the term "digital twin" in 2010 by NASA's John Vickers. A digital twin comprises the physical part, the digital part, and the connection facilitating data flow between the two [3].

A digital twin model is a live, data-powered depiction of a physical entity, enabling real-time monitoring, simulation, and analysis. It serves as a virtual representation of a physical object, system, or process, meticulously crafted using digital data and technology.

Developing a digital twin involves blending data from various sources—physical, manufacturing, operational, and analytics. Sensors gather data, optimizing performance with IoT, AI, ML, and spatial network graphs. The digital twin learns from continuous data exchange, incorporating insights from sensors, experts, and historical usage. Analytics are applied for anomaly detection, performance improvement, and long-term planning [4].

13.1.2 Model of a Digital Twin

Figure 13.1 illustrates a straightforward model that visualizes the continuous exchange of information within the DT. Data originating from the real-world environment is seamlessly

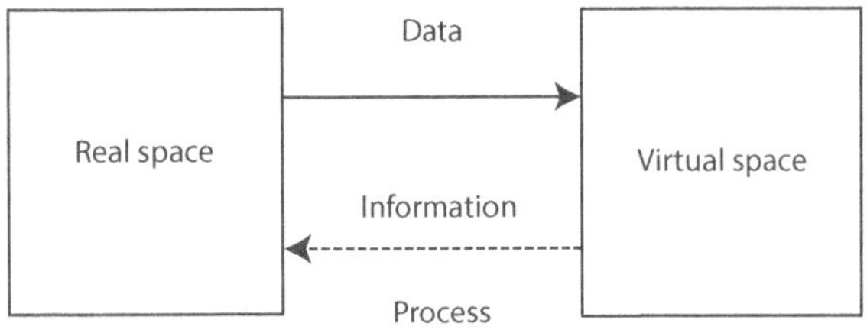

Figure 13.1 General model of a digital twin [5].

integrated into the virtual realm, where the DT processes and refines it. Subsequently, the enhanced virtual data is fed back into the real world, contributing to the updating of physical entities. This two-way information flow fosters a comprehensive connection between the physical realm and the digital domain, ensuring that both spaces remain in sync and well-informed [5].

13.1.2.1 Steps Involved in Building a Digital Twin Prototype

Building a digital twin prototype involves a comprehensive approach, combining technological expertise, data integration, and iterative development to achieve a functional and effective representation of the physical system [6]. The basic steps involved in building a digital twin prototype are as follows:

i) Define purpose and scope: Clearly articulate the objectives of the digital twin prototype, specifying its intended applications. Begin with a focused scope and gradually expand based on evolving needs.
ii) Data integration and model development: Gather and integrate relevant data from diverse sources. Develop a 3D model that accurately represents the physical system. Prioritize key details and choose appropriate technologies for data integration.
iii) Real-time connectivity and analytics: Establish real-time connections between physical assets and their digital counterparts using sensors. Implement analytics tools to interpret data, focusing on anomaly detection, performance analysis, and predictive modeling.
iv) Testing, visualization, and deployment: Conduct thorough testing to ensure functionality and reliability. Develop a user-friendly interface for visualizing data. Plan a deployment strategy considering scalability and integration. Provide training and documentation for seamless implementation and ongoing use.

13.1.3 Application Areas of Digital Twins

Digital technology offers not only the potential for accelerating innovation in industrial production but also a profound transformation of the conventional industrial system. It injects fresh vigor into traditional production methods. Today, with the orchestration of the industrial Internet, industrial intelligence is gradually taking form, and digital twins stand as the linchpin in steering industrial intelligence. Currently, this technology primarily finds applications in "high-value" sectors, such as rocket manufacturing and oil and gas extraction systems. With the widespread adoption of 5G and industrial IoT platforms, DT technology is poised for extensive utilization on a global scale. The industry is presently ushering in the era of Industry 4.0, where DT will play a pivotal role. The main application areas of DT are explained in detail below.

13.1.3.1 Digital Twin in Medical Field

DTs have a multifaceted role in healthcare, encompassing patient-specific modeling for personalized treatment and drug optimization, as well as aiding in surgical planning and training. They support disease modeling and drug development through *in silico* clinical trials and biomarker identification. DTs enable remote patient monitoring, improving continuous care for chronic conditions and enhancing telemedicine. Additionally, they are pivotal in creating customized rehabilitation plans, analyzing patient movement, and promoting mental health by monitoring emotional states and offering therapeutic simulations. These virtual models assist in diagnostics, treatment recommendations, healthcare system optimization, and even the customization and real-time adjustment of prosthetics and assistive devices. Yet, the imperative of safeguarding patient data security and privacy remains paramount to maintain trust and adhere to healthcare regulations [7].

13.1.3.2 Digital Twin in Smart City

Digital twins for smart cities are advanced digital representations of urban environments, integrating data from various sources to enhance planning, management, and decision-making for sustainable, efficient, and livable cities. Here are a few essential elements and uses of digital twins within the realm of smart cities: They aid in designing and optimizing infrastructure projects, such as transportation networks, utilities, and green spaces, to improve resource allocation and sustainability. They enable the simulation of public transport routes and schedules for efficiency and improved commuter experiences. They track energy consumption in buildings and across the city, helping to identify areas for energy efficiency improvements. They monitor air quality and pollution levels, providing data for real-time environmental management and health protection measures. They help optimize waste collection routes, reducing costs and environmental impact. They enhance urban security through the integration of surveillance systems and real-time monitoring. They aid in simulating and preparing for various disaster scenarios, allowing for better disaster response planning. They support the integration of smart grid technologies to manage and distribute energy more efficiently. DTs for smart cities are crucial for the efficient management of urban resources and the improvement of the well-being of citizens, as smart city initiatives continue to grow, digital twins will play an increasingly significant role in shaping the urban landscapes of the future [8].

13.1.3.3 Digital Twin in Sports

Digital twins are rapidly transforming the sports arena by harnessing cutting-edge technology to elevate athlete performance, streamline training protocols, and enhance fan participation. Their applications are multifaceted, including athlete performance analysis through detailed biomechanical modeling and real-time feedback, as well as injury prevention and rehabilitation featuring risk assessment and personalized recovery plans. They play a significant role in optimizing training regimens by monitoring workload and predicting performance outcomes. In the realm of sports equipment, digital twins simulate gear performance and expedite virtual prototyping. Moreover, they enrich the fan experience by providing immersive virtual stadium tours and real-time in-game analytics. DTs aid in game strategy and coaching by simulating opponent tactics and evaluating athletes' decision-making abilities. Additionally, they monitor athletes' vital signs, sleep patterns, and overall health while contributing to talent scouting and sports science research. In the context of virtual sports and esports, DTs create lifelike virtual athletes and optimize their performance, underscoring the dynamic evolution of this technology in enhancing sports engagement and performance [9].

13.1.3.4 Digital Twin in Smart Manufacturing

A digital twin in smart manufacturing represents a virtual counterpart of a real manufacturing system, process, or product. It has become increasingly important in the era of Industry 4.0 and the (IoT). Digital twins serve multiple crucial purposes, including real-time tracking and control of physical manufacturing processes through data collected from sensors, enabling simulation and understanding for optimization and issue prediction, supporting predictive maintenance by identifying maintenance needs in advance, ensuring quality control by comparing digital representations to specifications, managing the entire product lifecycle from design to end-of-life considerations, optimizing supply chain processes, and resource management. These digital twins facilitate collaborative decision-making by offering a common platform for various stakeholders to work with real-time and historical data, ultimately enhancing the efficiency, flexibility, and competitiveness of smart manufacturing systems. Creating digital twins involves a combination of IoT sensors, data analytics, cloud computing, and simulation software to build dynamic, accurate models that closely mirror their physical counterparts [10].

The applications of digital twin technology are poised to revolutionize various industries and sectors. From smart manufacturing and healthcare to urban planning and beyond, digital twins offer a powerful tool for real-time monitoring, predictive analysis, and decision-making. They enable businesses and organizations to optimize processes, reduce costs, improve product quality, and enhance sustainability. As this technology continues to advance and mature, its potential for innovation and transformation across a wide range of fields is only set to grow. Embracing digital twins is not just a strategic choice but a necessity for staying competitive and agile in the increasingly data-driven and interconnected world of the future [11].

13.2 Digital Twin Technologies

Digital twin technologies employ a range of tools and technologies to create and manage virtual replicas of physical objects, systems, or processes. Figure 13.2 shows the various technologies that may be applied to build the digital technology.

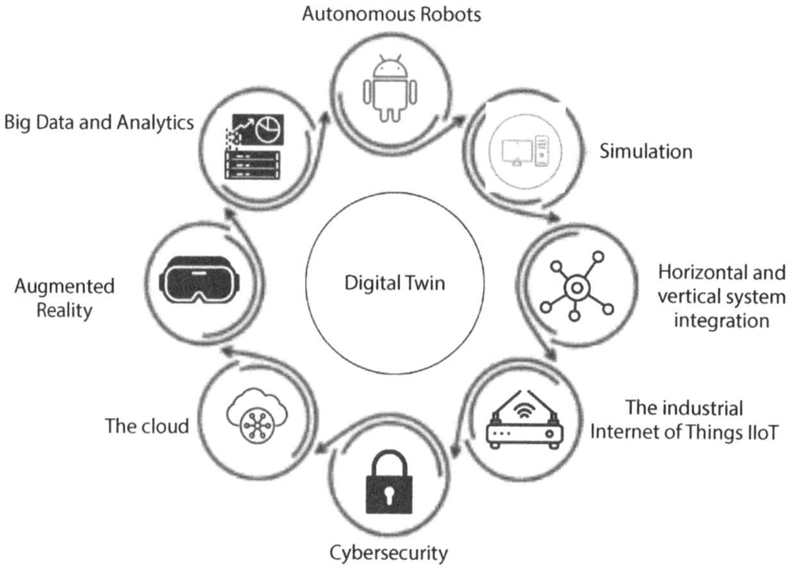

Figure 13.2 Technologies used in digital twin.

13.2.1 Data Acquisition and Sensors

Data acquisition sensors are pivotal in the establishment and maintenance of digital twins as they gather real-world data essential for precise digital renditions. Within the domain of digital twins, a range of common sensor types are employed, such as IoT sensors, cameras, RFID and NFC sensors, biometric sensors, and electromagnetic sensors, each serving distinct purposes. IoT sensors, including temperature, humidity, pressure, and motion sensors, capture environmental data and relay it to the digital twin. Cameras and imaging sensors play a fundamental role in visual representation creation, applied in tasks like remote monitoring and quality control. RFID and NFC sensors excel in tracking and identifying objects through unique tags or codes. Biometric sensors, like fingerprint scanners and facial recognition cameras, aid in identification and security in human-centric applications. Environmental sensors, monitoring factors such as air quality, prove indispensable in smart buildings, cities, and environmental monitoring within digital twin environments. Furthermore, sensors like vibration, acoustic, GPS, location, pressure, force, and motion sensors find substantial utility in manufacturing and industrial scenarios, detecting movement, measuring environmental conditions, and ensuring structural health. The selection of data acquisition sensors hinges on the specific use case and the nature of data required for the digital twin, often involving the combination of these sensor types to create a comprehensive and faithful digital representation of physical objects and systems [12].

13.2.2 Data Analytics and Machine Learning

Data analytics and machine learning are pivotal in enhancing digital twins. They are instrumental in collecting, cleaning, and integrating data from diverse sources, such as sensors and IoT devices, establishing a comprehensive dataset. Machine learning monitors real-time data from physical assets, identifying anomalies and enabling predictive maintenance. It creates advanced simulations, optimizing processes and resource allocation. Anomaly detection and pattern recognition further refine the digital twin's performance. Machine learning also personalizes the user experience and optimizes energy usage. In healthcare, it enables real-time patient monitoring and early diagnosis. Additionally, these technologies enhance security. Together, they empower digital twins to boost efficiency, productivity, and decision-making across domains [13].

13.2.3 Cloud Computing

Cloud computing is indispensable for the development and operation of digital twins, offering a range of vital advantages. These include scalability, enabling digital twins to expand or contract as required, making it especially relevant for complex systems. Cloud platforms provide secure, scalable data storage for historical and real-time data. They offer powerful computing resources, crucial for running advanced digital twin models and analytics. Accessibility allows digital twins to be accessed from anywhere, enabling remote management and monitoring. Collaboration is simplified, and cost-efficiency is enhanced through a pay-as-you-go model. Robust security, integration, and seamless updates and maintenance contribute to digital twin reliability. Cloud computing also offers advanced analytics and artificial intelligence services, facilitating predictive and prescriptive analytics. With a global reach and backup and disaster recovery options, cloud computing ensures digital twins are agile, secure, and ever-ready for real-world applications [14].

13.2.4 Other Technologies

Several other technologies and tools are integral to the development and utilization of digital twins. 3D modeling and simulation software create detailed digital replicas and analyze their behavior. Augmented reality (AR) and virtual reality (VR) enhance visualization and interaction, enabling immersive training, maintenance, and remote monitoring. Edge computing minimizes latency by processing data closer to IoT sensors. Communication protocols like MQTT and CoAP facilitate efficient data transfer between IoT devices and the digital twin. Robust cyber security measures, including encryption and access control, safeguard digital twins and their data. Blockchain technology secures data integrity, offering an immutable ledger for tracking changes. Databases store historical and real-time data, with options like relational, NoSQL, or time-series databases. Human–machine interfaces (HMIs) provide user interaction and real-time data visualization through dashboards, mobile apps, or web interfaces [15].

13.3 Integration of RL and Digital Twin

RL, a subset of machine learning, focuses on optimizing actions to maximize rewards within a given context. Unlike supervised learning, it operates without predefined answers, relying on experiential learning through trial and error. This method accumulates data for decision-making, making it suitable for systems requiring numerous autonomous decisions. A widely used model in reinforcement learning is the Markov decision process

(MDP), encompassing key components: states, actions, transition probabilities, rewards, and policy. The agent's goal is to determine the optimal policy (π^*) that maximizes cumulative rewards, achieved through algorithms like Q-learning and policy gradient methods that balance exploration and exploitation.

The convergence of RL and DT technology is a pivotal advancement in smart systems and autonomous decision-making. This synergy leverages digital twins and reinforcement learning to enable virtual models to learn and adapt, with transformative potential across industries, enhancing system autonomy and efficiency.

The integration of RL with DT technology can take several forms.

a) Training RL Agents With Digital Twins

RL agents can undergo training within the controlled environment of a digital twin. This approach ensures safe and cost-effective learning before RL agents are deployed in real-world scenarios. For instance, in the realm of robotics, a DT can replicate the physical environment and the robot's interactions, enabling RL agents to refine their control strategies without exposing the physical robot to potential risks.

b) Optimizing Real-World Processes

DTs can create digital replicas of real-world systems, such as manufacturing plants or supply chains. RL algorithms can then be utilized to fine-tune and optimize processes within this virtual environment. Within this setup, RL agents can explore different strategies and actions offered by the DT, ultimately identifying the most efficient and cost-effective means of achieving specific objectives.

c) Real-Time Control and Adaptation

RL agents can seamlessly integrate with DTs, enabling real-time control and adaptability of physical systems. The continuous data flow from the physical system *via* the DT empowers RL agents to make informed decisions and real-time adjustments. This capability is particularly beneficial in autonomous systems, where RL agents can dynamically adapt to changing conditions, such as autonomous vehicles adjusting their behavior based on real-time traffic data.

d) Predictive Maintenance

DTs can actively monitor the condition of physical assets, such as machinery and equipment. RL algorithms can predict maintenance needs and schedule them for optimal downtime reduction. The RL agent learns from historical data and real-time inputs from the DT to proactively make maintenance decisions.

e) Energy Efficiency

RL algorithms can be harnessed to optimize energy consumption across diverse systems, including building management and industrial processes. The DT can simulate various control strategies, enabling RL agents to learn and identify the most energy-efficient approaches.

f) Risk Mitigation
In critical sectors like healthcare and aerospace, RL agents can collaborate with DTs to detect and mitigate risks. Through simulation of different scenarios and learning from these simulations, RL agents bolster safety and support decision-making in complex, high-stakes environments.

The fusion of RL and digital twin technologies empowers industries to enhance efficiency, reduce risks, and make data-informed decisions across a wide range of applications.

13.3.1 Motivation for Combining Digital Twin and RL

Reinforcement learning offers sophisticated decision support and automation for operations, drawing insights from experience to excel in specific tasks and deliver impressive outcomes. The merging of RL and DT technology opens up an intriguing frontier with vast potential across different industries. DTs serve as virtual duplicates of physical systems or processes, while RL involves agents learning to make decisions through interactions with their environment. The DT acts as the virtual setting, enabling the RL agent to explore and acquire optimal strategies without impacting the real physical system.

Employing an RL agent enables task automation and provides decision support for determining the next course of action. This not only reduces costs but also optimizes quality consistently throughout the day with undivided attention [16]. Figure 13.3 shows a few methods through which these technologies can be combined.

i) Optimizing processes: RL can be applied to enhance control algorithms in a digital twin, leading to increased efficiency in processes. The RL agent learns through interactions with the virtual representation, and the refined policies can subsequently be implemented in the actual real-world system [17].

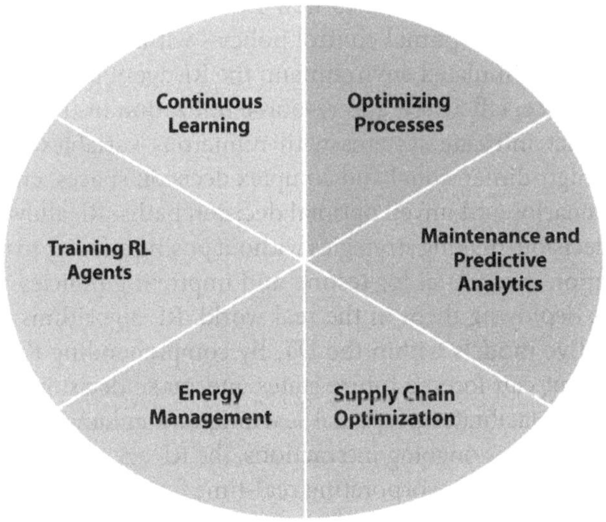

Figure 13.3 Various areas of RL and DT integration.

ii) Maintenance and predictive analytics: RL algorithms have the capability to acquire optimal maintenance schedules through interactions with a DT simulating the aging and degradation of physical components. This can result in improved predictive maintenance strategies, minimizing downtime, and prolonging the lifespan of equipment.
iii) Supply chain optimization: DTs have the capacity to simulate the complete supply chain, and RL can be utilized to discover optimal decision-making policies for tasks such as inventory management, order fulfillment, and logistics. This has the potential to boost efficiency and cut down on costs.
iv) Energy management: RL can optimize energy usage in buildings or industrial processes using data from a DT. This is especially beneficial in the context of smart cities and sustainable development.
v) Training RL agents: DTs offer a secure and regulated setting for training RL agents. This is particularly crucial in situations where real-world training might be costly, hazardous, or time-consuming.
vi) Continuous learning: RL algorithms can iteratively learn and adjust through feedback from the real-world system. The DT acts as a testing ground for evaluating and perfecting policies before implementing them in the real environment.

13.3.2 How RL Enhances Decision-Making Within Digital Twins

RL improves decision-making in DTs by learning optimal strategies through interactions with a simulated environment. RL agents trained within a DT can apply their knowledge when employed in comparable real-world systems [18]. This notion of transfer learning facilitates quicker adaptation and enhances decision-making in novel environments. In what ways RL enhances decision-making within the realm of DTs are shown in Figure 13.4.

RL empowers DTs to respond to evolving conditions. The RL agent assimilates knowledge from experiences in the virtual environment, enabling it to dynamically modify decision-making processes in response to new information or alterations in the system. RL can be applied to identify optimal control policies within a digital twin. Through trial and error learning in the simulated environment, the RL agent unveils strategies that result in enhanced performance, efficiency, and resource utilization in the actual system. Digital twins frequently depict intricate systems with numerous variables and interactions. RL excels in managing high-dimensional and complex decision spaces, enabling it to navigate through intricate scenarios and unveil optimal decision paths. RL allows the DT to experiment with various decision-making strategies without posing any risk to the physical system. This secure exploration is essential for testing and improving policies within a controlled environment before deploying them in the real world. RL algorithms can learn decision-making from predictive models within the DT. By comprehending the system's evolution over time, the RL agent can foresee future states and make decisions that consider long-term consequences. RL facilitates perpetual learning and enhancement. With the digital twin serving as a platform for ongoing interactions, the RL agent can consistently fine-tune its decision-making policies by incorporating real-time feedback and adapting to changing system dynamics. DTs enable the generation of diverse scenarios. RL agents can undergo training to handle a range of situations, including uncommon or extreme events, enhancing

Figure 13.4 Reinforcement learning's role in digital twins.

the resilience of decision-making processes in response to unexpected challenges. RL can enhance the distribution of resources in a system depicted by a DT. Whether it involves managing energy usage, allocating inventory, or scheduling maintenance tasks, RL learns to make decisions that optimize efficiency and reduce costs. RL agents, once trained within a DT, can apply their acquired knowledge when deployed in analogous real-world systems. This idea of transfer learning facilitates quicker adaptation and enhances decision-making in novel environments.

RL improves decision-making in DTs by offering a learning framework that adjusts to intricate, dynamic systems. It enables the identification of optimal strategies, risk-free exploration, and ongoing enhancement, ultimately resulting in more intelligent and adaptable decision-making in the actual world.

13.4 Challenges of Using RL in Digital Twins

The fusion of RL and DTs holds the promise of transforming the way we conceptualize, manage, and enhance intricate systems. This connection between the virtual and physical realms creates opportunities for more resilient and intelligent decision-making across diverse domains. Although the fusion of RL and DTs presents promising prospects, there are various challenges linked to this combination.

a) Computational complexity: RL algorithms can require a lot of processing power, particularly when working with intricate digital twin representations of complicated systems. Complex models may take a long time and a lot of processing power to train.
b) Data requirements: Large volumes of data are frequently needed for RL to train effectively. It can be difficult to provide realistic and varied training data for digital twins, particularly when there is a dearth of real-world data.
c) Accuracy: The success of RL hinges on the precision of the models embedded in the digital twin. If the virtual portrayal deviates significantly from the actual system, the acquired strategies might struggle to adapt effectively to the real-world setting.
d) Simulation fidelity: The fidelity of the digital twin's simulation can impact how well RL policies transition to reality. Simulations may miss certain intricacies and uncertainties inherent in the real system, creating a disparity between simulation and reality.
e) Safety concerns: When learning inside digital twins, RL agents may experiment and discover things that could prompt them to do risky or bad activities outside of the virtual world. It is quite difficult to ensure the safety of RL-driven judgments, especially in high-risk applications.
f) Real-time constraints: Certain applications necessitate making decisions in real time, like control systems. It can be difficult to educate RL agents to make judgments rapidly and effectively given the limitations of real-time processing.
g) Ethical considerations: Biased behaviors or unforeseen effects may result from RL judgments. It is essential to ensure ethical decision-making in digital twins, and it is a continuous struggle to overcome any biases in models and training data.

13.5 Digital Twin Modeling with RL

Creating an appropriate digital twin representation for RL involves several essential steps.

Initially, it is imperative to construct a virtual world that faithfully emulates the real-world scenario. This entails not only crafting the environment itself but also preparing the actor or learner that will interact within it. The actor is subsequently educated to address the specific problem at hand. Once the training phase reaches a satisfactory level of proficiency, the transition to the mirrored system is initiated. During this phase, the environment shifts from merely simulating the consequences of the agent's decisions to a mode where it mirrors the real-world problem as a cyber-replica. In the following stage, the actor's responsibility is to suggest actions using the representations provided by the digital twin. The ultimate decision-maker retains the authority to either adhere to these recommendations or deviate from them.

Several critical elements must be incorporated into the environment to enable this process. At each time step (denoted as "t"), the domains are in charge of collecting the state, computing a reward associated with that state, and updating the state in response to actions taken. The state retrieval mechanism, in particular, is tasked with gathering essential data from the relevant assets, with the nature of the data being contingent on the specific problem being addressed. The reward system is typically customized to suit the specific objectives of the RL task, with the goal to optimize the objective function. Rewards are sourced directly from the environment and are computed based on the state.

The interaction between the agent and the environment unfolds over multiple time steps. To begin, the agent collects information about the state (state(t)) and formulates an action (action(t)) based on its evolving policy. The environment records the action at time (t) and evaluates its outcomes. Afterward, the environment progresses by one time step, furnishing the reward at time (t + 1) and the revised state at time (t + 1) to the agent. This cyclical procedure persists until a predetermined termination condition is satisfied.

To transform the environment and agent into a digital twin, an essential step involves the introduction of a model adapter. This adapter's primary role is to convert raw data from real-world sources into a format that the agent can effectively utilize. Subsequently, the actions generated within the digital twin environment are transmitted to an adapter, which transforms them into instructions that can be executed in the actual physical world. In this context, the environment effectively assumes the role of a digital twin, and the agent takes on the role of the decision-maker, evaluating and acting upon the various possibilities.

[19]. Figure 13.5 shows the simple representation of digital twin modeling with RL.

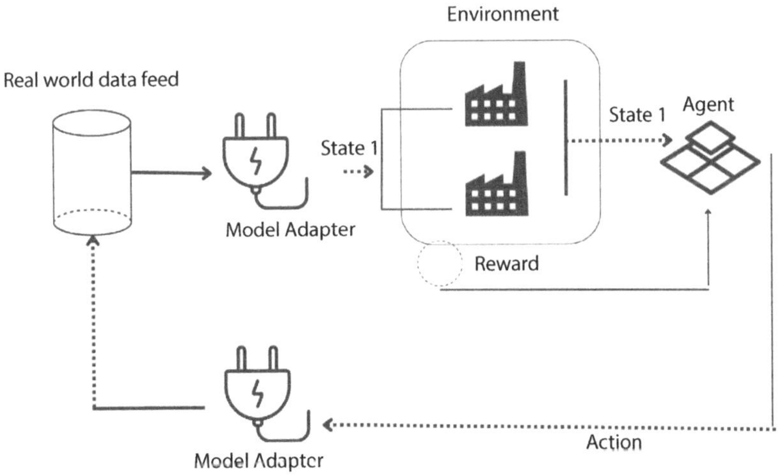

Figure 13.5 DT and RL integrated model [19].

13.6 Technology Underlying RL-Based Digital Twins

The utilization of RL in digital twins represents a novel and thrilling application within the realm of digital twin technology. Digital twins, serving as virtual representations of physical objects or systems, have found application across diverse industries such as manufacturing, healthcare, and smart cities, facilitating the simulation and optimization of real-world processes. When RL is integrated into digital twins, it adds a dynamic and adaptive component, allowing the digital twins to make autonomous decisions and learn from its interactions with the physical system it represents. RL is a machine learning methodology in which an agent acquires the ability to make a series of decisions through interaction with an environment. The agent is rewarded or penalized for its actions, and its objective is to acquire a policy that maximizes the cumulative reward over time [20, 21].

13.6.1 Integration of RL with Digital Twins in Four Stages

When RL is integrated with digital twins, it allows the digital twin to actively control and optimize the physical system it represents. It works in 4 stages. Stage 1 as the agent which is a software component responsible for making decisions and taking actions within the virtual model. Stage 2 as the environment represents the physical system or process. The agent interacts with this environment by sending actions and receiving observations and rewards. Stage 3 as a reward function which is defined to provide feedback to the agent. It specifies the goals or objectives of the system, and the agent receives rewards based on how well it achieves those objectives. The RL algorithm seeks to identify a policy that maximizes the anticipated cumulative reward. In stage 4, the state space encompasses all potential states of the system, while the action space encompasses all feasible actions the agent can undertake.

13.6.2 Tools and Libraries for Developing RL-Based Digital Twins

Developing RL-based digital twins requires a combination of tools and libraries to create the digital twin models and to implement reinforcement learning algorithms.

13.6.2.1 Simulation and Digital Twin Platforms

Various tools and platforms are available for simulation and implementation. Platform 1 as Unity3D, which is a widely used game engine that can be repurposed for creating realistic simulations and digital twin environments. The Unity ML-Agents toolkit is an extension for implementing reinforcement learning in Unity environments. Platform 2 as Gazebo which is an open-source 3D simulation environment commonly used in robotics. It is suitable for creating digital twin models of robotic systems and environments. Platform 3 as OpenAI Gym which provides a simple and consistent interface for creating RL environments. While it may not handle 3D simulations directly, we can integrate it with other tools to create RL-based digital twins. Platform 4 as Simulink which is from MathWorks and is a popular platform for modeling and simulating dynamic systems. It is suitable for digital twin development in fields like engineering and control systems. Platform 5 as Roboschool which is an open-source simulation environment that integrates with OpenAI Gym. It focuses on robotic control tasks.

13.6.2.2 *Reinforcement Learning Libraries*

Some of the common libraries for implementing RL-based digital twins are [22]:

1) OpenAI Baselines is a collection of high-quality RL algorithms implemented in Python, including DQN, PPO, A3C, and more. It can be used as a starting point for RL-based digital twin development.
2) Stable baselines which is a set of high-level RL algorithms built on top of OpenAI Baselines, offering a more user-friendly API and extended features.
3) TensorFlow and TensorFlow Agents (TF-Agents), TensorFlow is a popular deep learning framework, and TF-Agents provides a comprehensive set of tools and algorithms for RL. We can build custom RL models using TensorFlow.
4) PyTorch, a deep learning framework, is becoming increasingly popular for RL. It offers flexibility and ease of use.

Other popular libraries for data collection and processing are pandas and numpy. Matplotlib and TensorBoard are mainly used for data visualization and analysis. And at the end we can use cloud services like Amazon SageMaker which offers a cloud-based platform for building, training, and deploying RL models, making it easier to scale our RL-based digital twin.

13.6.3 Integration with Existing Systems and IoT Devices for RL Deployment

Integrating RL-based digital twins with existing systems and IoT devices involves bridging the virtual world of the digital twin with the physical world of our devices and systems. This integration enables the digital twin to interact, control, and learn from real-world systems through IoT sensors and actuators.

RL deployment for digital twins poses unique challenges due to the dynamic nature of RL algorithms and the need to simulate and optimize physical systems in real-time. We need to create a robust architecture of a digital twin which should support data synchronization, state representation, and action execution. The key factors for the integration and the deployment after finalizing the digital twin architecture are:

13.6.3.1 *Data Collection and Sensor Integration*

Gathering data involves utilizing IoT sensors to capture pertinent information from the physical system and transform the physical quantities into a digital format. These sensors could measure temperature, pressure, humidity, machine status, or any other data critical to our digital twin's operation. After gathering data from various sensors, we carry out data preprocessing and cleaning to guarantee its quality and pertinence to the digital twin. This process might involve tasks such as filtering, normalization, and addressing missing data [23].

13.6.3.2 Communication and Data Ingestion

IoT Gateway or edge devices like raspberry pi are used to collect data from sensors, preprocess it, and transmit it to the cloud or local server for further processing. These gateways act as intermediaries between the sensors and our digital twin system. At the application layer, Implementation of message queues or publish-subscribe systems (e.g., MQTT, Apache Kafka) are used to efficiently transmit data between IoT devices and our digital twin application.

13.6.3.3 Digital Twin Integration

We can host our digital twin model on cloud servers or local infrastructure. These servers will receive and process data from IoT edge devices. Generally the edge devices like Raspberry Pi or other IoT gateways are used to filter and send the sensor data to the cloud for further processing and analysis. We need to ensure that our digital twin can ingest real-time data from IoT devices, update its internal state, and provide predictions or take actions in response to this data.

13.6.3.4 RL Integration

We need to Implement RL algorithms within our digital twins to make autonomous decisions and optimize the physical system. After that we need to map the state space and action space of our RL agent to the real-world state and actions of the physical system. This mapping ensures that the digital twin's decisions are relevant and can be executed.

13.6.3.5 Control and Actuation

Implementing actuators that can carry out actions recommended by the digital twin. These actuators could be motors, fan, light, valves, or any devices that affect the system's behavior. We also need to ensure the measurement of the implementation safety mechanisms and constraints to prevent risky or harmful actions in case the digital twin makes suboptimal decisions.

13.6.3.6 Implementation of Feedback and Learning Process

We define a reward mechanism for the RL agent based on the system's objectives. The reward provides feedback to the agent and guides its learning process. Simultaneously, incorporating mechanisms for the digital twin to consistently acquire knowledge and adjust to alterations in the physical system enhances its decision-making capabilities over time.

13.6.3.7 Dashboard for Alert and Visualization

Designing a dashboard that provides insights into the performance of our digital twin and the physical system is required to be created using the IoT cloud which is the urgent requirement for monitoring the results of data analysis as well and at the same time for alerting systems to notify operators of critical issues of the system. Data visualization tools like matlab, power Bi or tableau are used to display real-time data in graphical format from IoT devices and the decisions made by the digital twin, making it easier to understand system behavior.

13.6.3.8 Ensuring the Security and Authentication

Putting in place security measures to safeguard the data exchanged among IoT devices and our digital twins, as well as the digital twin itself, is fundamental for ensuring the safety and security of the system. Robust authentication mechanisms are necessary to guarantee that only authorized entities can engage with our digital twin and IoT devices.

13.7 Industry-Specific Applications: A Case Study of DT in a Car Manufacturing Unit

Utilizing a DT for a car manufacturing unit offers an efficient means of supervising and enhancing its operations. This entails the creation of a virtual depiction of the manufacturing facility and its products on a screen, all driven by real-time data collected from a range of sensors. Through this seamless integration, the activities within the unit can be adeptly managed and optimized in real time.

The responsibilities within a car manufacturing unit encompass a spectrum of critical tasks, including overseeing the production of car parts, assembling these components into finished products, ensuring top-tier quality, analyzing production timelines, and establishing an autonomous production environment. A DT proves to be an invaluable tool in providing a comprehensive visualization of the entire production process. This encompasses everything from the manufacturing of car parts to their assembly, the precise determination of the required quantity and color for achieving the ideal final product in the shortest possible time, the assessment of the necessary service engineers required for monitoring and testing the end product and more.

The inclusion of the Internet of Things (IoT) within the digital twin framework introduces solutions to common managerial inquiries that often arise in manufacturing units. These inquiries revolve around efficiently managing fluctuations in demand, optimizing waiting periods, accurately calculating the required quantities of color and parts to meet specific demands within predefined timeframes, and ascertaining the essential service personnel needed to attain desired service levels. The integration of IoT-based digital twins adeptly addresses these concerns, streamlining the entire manufacturing process.

Even with the presence of robotic systems designed to automate production and assembly tasks, there has been a noticeable void in terms of real-time predictive analysis and management systems for monitoring and controlling manufacturing unit activities. The remedy to this gap lies in harnessing the capabilities of sensors, IoT gateways, and cloud computing technologies to establish a comprehensive and efficient control and monitoring system for car manufacturing units [23].

13.7.1 IoT Components Required for Creating Digital Twin for the Manufacturing Unit

The car manufacturing unit's physical environment will be visualized on the screen using digital twins. These digital replicas facilitate the management of essential variables, including component quantities, manufacturing and assembly durations, product testing, resource utilization, and inventory levels. Achieving this transformation relies on digitalization and data analysis, drawing from information gathered by sensors integrated into the unit. This data empowers us to display pertinent information on the screen through digital twin representations, optimizing the unit's workspace and enabling real-time operations. The elements necessary for implementing the proposed architecture of digital twins for the car manufacturing unit include

Sensors: Cameras, LDR, temperature, ultrasonic and IR sensors are the basic requirements for the systems to capture the real time data from the physical environment of the manufacturing unit. Data from these sensors gets collected at the edge devices for the primary analysis and cleaning of the data at the edge layer as well as for the implementation of the automation [24]. Table 13.1 shows the various sensors that may be employed in any manufacturing unit [25].
IoT gateways: IoT gateways are generally the edge devices in IoT which collects and analyzes the real time data coming from the sensors and send them after cleaning and filtering to the cloud for further analysis, very commonly and best suited IoT gateway for this type of implementation is Raspberry Pi which efficiently allows us to integrate camera and other sensors and perform data processing at the edge layer. Other than Raspberry Pi, NodeMCU, ESP32, and other routers can be used as gateways to facilitate the working of the system.

IEEE-based wireless networks for communicating between the sensors, Edge devices and the cloud. There are various wireless technologies designed for this purpose by IEEE specially based on IEEE 802.15.4 which can help us integrate the sensors wirelessly with the edge nodes, integration of the sensors wirelessly with nodes using various RF technologies are explained in the paper [26].

13.7.2 Architecture of the Proposed Digital Twin for Car Manufacturing Unit

A proposed architectural framework consisting of four layers, as depicted in Figure 13.6, aims to replicate the functions of a car manufacturing unit. The initial layer, known as the sensor tier, encompasses an array of components, including Node MCUs, IR sensors, light sensors, temperature sensors, smoke sensors, and cameras. These devices establish connections with a Raspberry Pi through both Wi-Fi and USB interfaces. This layer's primary goal is to gather essential data from the actual surroundings of the manufacturing unit.

The second layer, identified as the IoT gateways tier, comprises edge devices such as the Raspberry Pi and Wi-Fi routers. The Raspberry Pi serves as a compact computer equipped with Python programs housing algorithms responsible for tasks like counting the quantity of parts at specific locations, identifying the types of parts needed at any given moment, and specifying color requirements. It processes the gathered data at the local edge layer and

Table 13.1 Sensors used in the manufacturing unit.

Description	**Sensors**
Webcams will be used to capture the real-time live videos from the industry to simulate the working of the whole system. Multiple webcams will be used to get the streams from all the corners of the industry. It will be integrated with the Raspberry Pi boards to process the videos in real time.	
LDR sensors will be used in the manufacturing unit to check the light and the light intensity. It will be integrated with a NodeMCU board to check the light and ensure the proper light in the manufacturing unit for better visibility.	
DHT 11/22 sensors will be used to check the temperature of the manufacturing unit and manage the overall temperature. It will be integrated with a NodeMCU board to check the temperature threshold and manage the cooling system of the unit.	
IR sensors will be used to sense the motion in the manufacturing unit to implement the automation of the robotic hands.	
Ultrasonic sensors will be used to measure the distance between the various components to implement the automation of the robotic hands and other components for the proper working. It will be integrated with raspberry pi to complement robotic hands and other actuators.	

subsequently transmits the processed information to the cloud, which can be hosted on platforms like AWS or Google.

The third layer encompasses the cloud environment, where an IoT cloud channel is designed to accept data points from sensors *via* the edge device. This tier is the site for the execution of analytics services on the received data, generating recommendations guided by predefined business logic.

The fourth and final layer, recognized as the application tier, features a tailor-made web application serving two primary functions. Firstly, it provides a platform for displaying digital twins in a unified interface. Secondly, it presents the outcomes of data analytics conducted within the cloud in a distinct panel. Data analytics can be executed on any of the available IoT cloud platforms, such as AWS IoT Analytics, Thingspeak, or Google IoT Cloud. Dashboards, featuring the necessary data fields, can be established on these cloud platforms, and the information seamlessly integrated into the customized web application, enabling viewing through multiple panels. The flow of operations of the proposed work is shown in Figure 13.6.

13.7.3 Challenges and Opportunities in the Implementation of DTs for Car Manufacturing

DT implementation for a car manufacturing unit is a challenging project which has lots of chances and challenges due to the dynamic RL algorithms, complexity and accuracy required in the automotive industry. Some of the challenges are highlighted below:

a) Data integration and interoperability: The unit involves numerous machines, sensors, and software systems. Integrating these data sources into a coherent digital twin can be challenging. It is crucial to guarantee interoperability throughout various systems and devices.

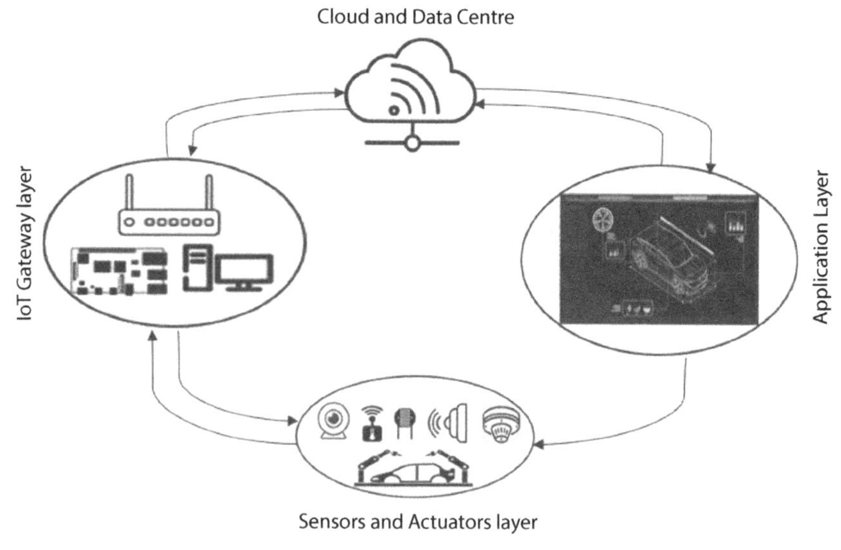

Figure 13.6 Architecture of IoT-based DT for car manufacturing unit.

b) Data volume and velocity: The volume and velocity of data generated in a car manufacturing process can be overwhelming. Managing, processing, and analyzing this data in real-time is a significant challenge.
c) High quality and tolerance: Cars require precise manufacturing with high tolerances. Ensuring that the digital twin accurately simulates and monitors these tolerances is crucial.
d) Complexity of multistage processes: The unit involves numerous stages, from design and prototyping to assembly and quality control. Managing a digital twin that spans all these stages is always complex.
e) Simulation loyalty: Achieving high-loyalty simulation in a digital twin is challenging but critical for accurately modeling the manufacturing process. Real-world conditions, materials, and physics must be accurate.
f) Security: Protecting sensitive data is paramount in car manufacturing. It is difficult yet essential to ensure the security of the digital twin and the data it consumes.

Along with challenges, digital twin implementation offers lots of opportunities like:

a) Manufacturing optimization: A well-implemented digital twin can optimize manufacturing processes, identify inefficiencies, and reduce waste, leading to cost savings and increasing productivity.
b) Quality: Digital twins enable the real-time monitoring of the manufacturing process, helping to detect defects and inconsistencies early in the process, and ultimately improving product quality.
c) Predictive maintenance of machineries: Using IoT sensors and analytics, digital twins can predict when machineries need maintenance or replacement, it reduces downtime and increases the overall machineries efficiency.
d) Virtual prototyping and testing: Digital twin allows virtual prototyping and testing, which can accelerate productivity and reduce the requirement for physical prototypes.
e) Customization: Digital twins can support mass customization in the automotive industry, allowing customers to tailor their vehicles to their preferences while maintaining efficiency in the manufacturing process.
f) Supply chain optimization: Digital twins can extend beyond the factory floor to optimize the entire supply chain, ensuring the right parts and materials are available at the right time.
g) Environmental sustainability: Digital twins will help reduce the environmental impact of car manufacturing by optimizing processes, reducing waste, and ensuring efficient resource utilization of the unit.

13.8 Conclusion

The fusion of digital twin technology with RL represents a transformative synergy that promises to revolutionize various industries and domains. This integration leverages the power of simulation, data-driven decision-making, and real-time control to unlock new levels of efficiency, safety, and innovation. By training RL agents within the controlled environments of digital twins, organizations can ensure safe and cost-effective learning, reducing the risks associated with real-world experimentation. This approach is particularly vital in domains like robotics and autonomous systems. The optimization of real-world processes through DTs and RL enables industries to fine-tune their operations, identifying the most efficient and cost-effective strategies for achieving specific objectives. This lowers operating expenses while simultaneously increasing performance. Real-time control and adaptation, facilitated by RL agents in conjunction with digital twins, empowers systems to dynamically respond to changing conditions. Whether in autonomous vehicles or manufacturing processes, this capability enhances adaptability and responsiveness. Predictive maintenance, energy efficiency, and risk mitigation benefit immensely from this integration. RL agents, informed by digital twins, can predict maintenance needs, optimize energy consumption, and bolster safety measures through proactive risk assessment. In essence, the combination of digital twin and RL technologies offers a comprehensive solution for decision-making, optimization, and control in complex systems. It empowers industries to harness the advantages of virtualization, data-driven insights, and real-time adaptation, ultimately driving progress and efficiency across a wide spectrum of applications. As these technologies continue to evolve, the potential for innovation and improvement in our increasingly complex world is boundless.

References

1. Fuller, A., Fan, Z., Day, C. and Barlow, C., "Digital Twin: Enabling Technologies, Challenges and Open Research," in *IEEE Access*, vol. 8, pp. 108952–108971, 2020, doi: 10.1109/ACCESS.2020.2998358.
2. Kaishu Xia, Christopher Sacco, Max Kirkpatrick, Clint Saidy, Lam Nguyen, Anil Kircaliali, Ramy Harik, A digital twin to train deep reinforcement learning agent for smart manufacturing plants: Environment, interfaces and intelligence, *Journal of Manufacturing Systems*, Volume 58, Part B, 2021, Pages 210–230, ISSN 0278-6125, https://doi.org/10.1016/j.jmsy.2020.06.012.
3. Grieves, M. and Vickers, J., Origins of the digital twin concept. *Florida Institute of Technology*, 8, 3–20, 2016.
4. Singh, M., Fuenmayor, E., Hinchy, E.P., Qiao, Y., Murray, N., Devine, D., Digital Twin: Origin to Future. *Appl. Syst. Innov.*, 4, 2, 36, 2021, https://doi.org/10.3390/asi4020036.
5. Z. Wang, 'Digital Twin Technology', Industry 4.0 - Impact on Intelligent Logistics and Manufacturing. *IntechOpen*, Mar. 18, 2020. doi: 10.5772/intechopen.80974.
6. Segovia, M. and Garcia-Alfaro, J., Design, Modeling and Implementation of Digital Twins. *Sensors*, 22, 14, 5396, 2022, https://doi.org/10.3390/s22145396.
7. Sun, T., He, X., Li, Z., Digital twin in healthcare: Recent updates and challenges. *Digit. Health*, 9, 20552076221149651, 2023 Jan 3, doi: 10.1177/20552076221149651, PMID: 36636729; PMCID: PMC9830576.
8. Guo, J. and Lv, Z., Application of Digital Twins in multiple fields. *Multimed. Tools Appl.*, 81, 26941–26967, 2022, https://doi.org/10.1007/s11042-022-12536-5.
9. Lukač, L., Fister, I., Jr, Fister, I., Digital Twin in Sport: From an Idea to Realization. *Appl. Sci.*, 12, 24, 12741, 2022, https://doi.org/10.3390/app122412741.
10. Qi, Q., Tao, F., Zuo, Y., Zhao, D., Digital Twin Service towards Smart Manufacturing. *Proc. CIRP*, 72, 237–242, 2018, ISSN 2212-8271, https://doi.org/10.1016/j.procir.2018.03.103.

11. Yao, J.F., Yang, Y., Wang, X.C. *et al.*, Systematic review of digital twin technology and applications. *Vis. Comput. Ind. Biomed. Art*, 6, 10, 2023, https://doi.org/10.1186/s42492-023-00137-4.
12. Harshavardhan, B., Reddy, D., Joseph, C., Sensor types and its applications. *Int. J. Pharm. Technol.*, 8, 20172–20180, 2016.
13. Li, X., Liu, H., Wang, W., Zheng, Y., Lv, H., Lv, Z., Big data analysis of the Internet of Things in the digital twins of smart city based on deep learning. *Future Gener. Comput. Syst.*, 128, 167–177, 2022, ISSN 0167-739X, https://doi.org/10.1016/j.future.2021.10.006.
14. Wang, E., Tayebi, P., Song, Y.-T., Cloud-based Digital Twins Storage in Emergency Healthcare. *2023 IEEE/ACIS 21st International Conference on Software Engineering Research, Management and Applications (SERA)*, Orlando, FL, USA, pp. 331–335, 2023, doi: 10.1109/SERA57763.2023.10197705.
15. Tao, F., Xiao, B., Qi, Q., Cheng, J., Ji, P., Digital twin modeling. *J. Manuf. Syst.*, 64, 372–389, 2022, ISSN 0278-6125, https://doi.org/10.1016/j.jmsy.2022.06.015.
16. Farazi, N.P., Zou, B., Ahamed, T., Barua, L., Deep reinforcement learning in transportation research: A review. *Transp. Res. Interdiscip. Perspect.*, 11, 100425, 2021, ISSN 2590-1982, https://doi.org/10.1016/j.trip.2021.100425.
17. Kor, M., Yitmen, I., Alizadehsalehi, S., An investigation for integration of deep learning and digital twins towards Construction 4.0. *Smart Sustain. Built Environ.*, 12, 3, 461–487, 2023, https://doi.org/10.1108/SASBE-08-2021-0148.
18. Cronrath, C., Aderiani, A.R., Lennartson, B., Enhancing Digital Twins through Reinforcement Learning. *2019 IEEE 15th International Conference on Automation Science and Engineering (CASE)*, Vancouver, BC, Canada, 293-298, 2019, doi: 10.1109/COASE.2019.8842888.
19. Martin, G. and Oger, R., A Reinforcement Learning Powered Digital Twin to Support Supply Chain Decisions. *HICSS2022 - Hawaii International Conference on System Sciences*, Hawaii, United States, pp. 2291–2299, Jan 2022, ff10.24251/HICSS.2022.287ff. Ffhal-04074719f.
20. Mahadevan, S., Average Reward Reinforcement Learning: Foundations, Algorithms, and Empirical Results. *Mach. Learn.*, 22, 159–195, 1996, https://doi.org/10.1023/A:1018064306595.
21. Park, K.T., Son, Y.H., Ko, S.W., Noh, S.D., Digital Twin and Reinforcement Learning-Based Resilient Production Control for Micro Smart Factory. *Appl. Sci., 11*, 2977, 2021, https://doi.org/10.3390/app11072977.
22. Sanghi, N., *Deep Reinforcement Learning with Python*, Edition Number 1, Apress, Berkeley, CA, 2021, https://doi.org/10.1007/978-1-4842-6809-4.
23. Vasumathi, M.T., Khan, A., Sadasivan, M., Ramamoorthy, U., Digital Twins—A Futuristic Trend in Data Science, Its Scope, Importance, and Applications, in: *Expert Clouds and Applications. ICOECA 2022*. Lecture Notes in Networks and Systems, vol. 673, I.J. Jacob, S.K. Shanmugam, I. Izonin (Eds.), Springer, Singapore, 2023, https://doi.org/10.1007/978-981-99-1745-7_58.
24. Karakra, Abdallah & Fontanili, Franck & Lamine, Elyes & Lamothe, J. & Taweel, Adel., Pervasive Computing Integrated Discrete Event Simulation for a Hospital Digital Twin. 1–6, 2018. 10.1109/AICCSA.2018.8612796.
25. Sehrawat, D. and Gill, N.S., Smart Sensors: Analysis of Different Types of IoT Sensors. *2019 3rd International Conference on Trends in Electronics and Informatics (ICOEI)*, Tirunelveli, India, pp. 523–528, 2019, doi: 10.1109/ICOEI.2019.8862778.
26. Khan, A. and K, J., RF Technologies for Connecting Various Constrained Devices in IoT Systems. *Int. Res. J. Adv. Sci. Hub*, 5, 05S, 254–259, 2023, doi: 10.47392/irjash.2023.S034.

14

Predictive Modeling of Rheumatoid Arthritis Symptoms: A High-Performance Approach Using HSFO-SVM and UNET-CNN

Anusuya V.[1], Baseera A.[2], Dhiyanesh B.[3]*, Parveen Begam Abdul Kareem[4] and Shanmugaraja P.[5]

[1]*IT, Ramco Institute of Technology, Rajapalayam, Tamil Nadu, India*
[2]*School of CSE, VIT Bhopal University, Bhopal, India*
[3]*CSE, SRM Institute of Science and Technology, Vadapalani Campus, Chennai, Tamil Nadu, India*
[4]*CSE, Taibah University, Yanbu, Saudi Arabia*
[5]*IT, Sona College of Technology, Salem, Tamil Nadu, India*

Abstract

Increasing health defects and diseases make human life very dangerous and serious. In this way, rheumatoid arthritis (RA) is a chronic disease affecting the immune system and damaging organs like cardiovascular and heart tissues. The main problem is that disease properties do not show disease symptoms and contain various feature dimensions. To resolve this problem, we propose applying Honey Scout Forager Optimized SVM Features Selection (HSFO-SVM) based on UNET-gated convolution neural network (UNET-CNN) to predict symptoms earlier. C-score and cross-fold validation are used to normalize the RA dataset. The Disease Impact Risk Score (DIRS) is evaluated based on decision margins. With DIRS support, Honey Scout Forager creates the feature web structure to find relational feature margins. It selects them through the support vector membership rule function. Using a convolutional neural network, the set features are logically processed into a UNET-gated unit. To balance bias weight training under the tested samples, the classifier intakes feature clusters by class. The classification unit finalizes the risk margin feature limits to categorize the risk of the patients by class. The proposed system demonstrates a prediction accuracy of up to 96%, a recall rate of 95.2%, and an F-measure of 94.6%, producing high performance with a low false rate of 3.5% and time. The proposed classifier performs better predictions than the previous methods to achieve high RA prediction accuracy. The results section shows the proposed system's accuracy with different comparison methods to achieve high performance.

***Keywords*:** Optimization, features selection, neural network, artificial intelligence, rheumatoid arthritis, hypertension

**Corresponding author*: dhiyanu87@gmail.com

Shubham Mahajan, Pethuru Raj and Amit Kant Pandit (eds.) Deep Reinforcement Learning and Its Industrial Use Cases: AI for Real-World Applications, (325–348) © 2024 Scrivener Publishing LLC

14.1 Introduction

Health facts and diseases are becoming commonplace in the modern fast-food world. Attention to various diseases and facts affects human life through disease discoveries. Identifying unique disease properties is difficult because of the feature conditions. Rheumatoid arthritis (RA)-oriented disease is the deadliest disease, causing sudden death and affecting the heart. Heart valve disease can cause cardiovascular valve damage, and hypertension can cause left ventricular hypertrophy. Identifying the nature of the disease is difficult because artificial intelligence performs the data analysis. After all, artificial intelligence fulfills the optimization requirements to identify risks.

There are several common structures that can be affected by rheumatoid arthritis (RA), a chronic, debilitating, autoimmune, and inflammatory disease of several organs. It happens when the immune system cannot differentiate between cells and tissues. The ANFIS model was developed to forecast RA in human test subjects. A comprehensive approach enabled the technique to accurately identify rheumatoid arthritis in people with 93.5% accuracy. RA is diagnosed in humans by looking for 12 symptoms, including age, stiffness, joint dysfunction, ESR, CRP, leukocytes, and uric acid, which can all be reflected in other bodily fluids. ANFIS is compared to other classification methods such as the naive Bayes, Bagging, and KNN methods in this article.

ANN is a powerful tool to understand how the brain processes and performs information, which is why it is inspired by the biological neural system. Applications of AI and ML in research and healthcare include the diagnosis of various diseases, the management of chronic (chronic) ailments, the provision of healthcare services, and the diagnosis of drugs and medicines, respectively. More people experience rheumatic diseases than any other type of sickness, often known as musculoskeletal pain. These diseases interfere with daily activities. Identifying people sensitive to rheumatic diseases is crucial for quality of life. It is designed for people of all ages; however, women prefer it more often. Many of the symptoms of this illness resemble those of other diseases. Therefore, it is challenging to find. Additionally, diagnostic tools are expensive and sophisticated. To study rheumatoid arthritis, machine learning algorithms predict RA with four parameters. Shortly, artificial intelligence (AI) will enhance rheumatic disorder prognostication.

The most common cause of death in the world today is heart disease. Deaths due to heart conditions have become a concern in today's modern world. This is when one person per minute is incarcerated and dies as a result of a heart condition. It is now extremely challenging to anticipate a disease in its early stages. Machine learning can provide precise and early disease diagnosis in the healthcare industry. Leveraging traditional machine learning approaches, we aimed to find correlations between different variables in a data set in this study. We wanted to leverage those connections to accurately forecast the likelihood of heart disease. The results indicate that, compared to other machine learning algorithms, random forest offers higher accuracy in a shorter time for prediction.

Python manipulates data sets using random forest algorithm or ML algorithm. This technique forecasts the future and averts death based on patient records from earlier encounters. It is demonstrated in this study that the random forest method, an established machine learning technique that has provided good results in the forecasting of cardiac disease, can be used to develop a predictive system. Data from CSV files containing patients' records can be read. Data access is followed by surgery, which creates a functional myocardial stage. Higher efficiency and precision, better adaptability, and higher success rates are all benefits of the suggested approach.

It is urgent that we develop a technique for detecting myocardial infarction symptoms early and preventing the RA since it is impractical for the average person to routinely undergo complicated testing like an ECG, a straightforward and accurate technique for determining the possibility of heart disease is required. Throughout the world, cardiovascular diseases have been identified as a leading cause of death, accounting for almost a quarter of all deaths. Clinical data analysis is challenging in terms of predicting cardiovascular diseases based on clinical data. There is considerable potential in leveraging the massive data generated by the healthcare industry to make better decisions and forecasts by using ML and NN in order to harness the massive amount of data produced. As we have demonstrated, contemporary machine learning models use a number of features. This study focuses on enhancing the cardiovascular disease prediction characteristics of the prediction system by using machine learning techniques.

Create another prescient device for identifying mitigating epitopes that suppress the immune response and calming epitopes that reduce inflammation. The first peptide grouping was encoded because of three-layered representations (amino-corrosive linkage, positional scoring grid, physicochemical properties, and improved techniques, data extraction, and investigation). We propose a component choice model that uses numerous element extraction models to build various base classifiers from different element portrayals [1]. A series of defuzzification strategies can be selected from the library of defuzzification strategies to assess the accuracy and reliability of the framework. After consulting with a specialist, the information collected is put together through the review process, and the results are presented as bogus positives and misleading negatives. The most extreme outcomes show that the proposed framework matches the Ayurveda master's consequences. Application/enhancement: The proposed framework will guide specialists in arranging doses of medications and medicines for patients. This dubiously based framework will be shown as a learning device for Ayurvedic and Vedic experts [2]. The main problem is that disease properties do not show disease symptoms, which contain various feature dimensions. Research contributes to identifying RA diseases effectively at a high precision rate with the support of feature selection and classification based on applying Honey Scout Forager Optimized SVM Features Selection (HSFO-SVM) with UNET-gated convolution neural network (UNET-CNN) to predict symptoms earlier.

14.1.1 Novelty of the Research

- This paper aims to improve RA disease prediction accuracy using the UNET CNN algorithm. C-score and cross-fold validation are used to normalize the RA dataset.
- The disease impact risk score (DIRS) is evaluated based on decision margins. With DIRS support, the Honey Scout Forager creates the feature web structure to find relational feature margins. It selects them through the support vector membership rule function.
- The set features are logically processed into a UNET-CNN-gated unit, and then a convolutional neural network is used to classify them. In order to

balance the bias weight training under the tested samples, the classifier consumes the feature clusters by class.
- To categorize risks and set risk margins, the classification unit comes up with feature limits that address risk categories.
- The features that are identified by this system are those located between invariant medical margins in order to determine the cause of the disease.
- The RH-A dataset is collected from the standard medical library to evaluate the results based on the optimization technique. A confusion matrix was used to conduct testing and training in order to validate the results.

14.2 Related Work

The precision and authenticity of the framework are examined by comparing the results of various dispersion techniques. Information is collected using the study strategy, and after talking with specialists, the outcomes are highlighted as evident and bogus. The great results created by the proposed technique correspond with the effects of the Ayurvedic master. Application/upgrade: The proposed framework will direct doctors to customize treatment for their patients. This flexible framework is a learning instrument for Ayurvedic and Vedic specialists [3].

Rheumatoid arthritis is one of those infections whose cause is obscure. Investigating clinical information extraction helps in early illness analysis and treatment. Another positioning calculation called CS-Lift is proposed, which utilizes the Cuckoo search calculation to present the AdaBoost calculation. The trial results show that the CS-Lift calculation works on Adaboost's exactness in foreseeing rheumatoid joint pain [4]. A way to deal with follicular lymphoma arrangements is introduced. This order grades given properties taken as information data, for example, stage, stage bunch, growth size, and execution status. The arrangement shows how rapidly lymphoma spreads in the patient's body, which is important for specialists to treat the illness. This proposed framework is a multi-objective multifacet perceptron (MLP), which utilizes a help vector machine for order with numerous bits, like straight, quadratic, and outspread premise capability (RBF), which groups the information into particular follicular levels [5].

This paper contrasts various portrayals of clinical archive highlights, such as client-indicated terms, remarkable idea identifiers, packs of words, and bigram highlights. Execution examinations are conducted at the highest quality level for all job sets and under various job determination conditions [6]. The proposed calculation is integral to the stochastic mystery of TGA deserting. What is more, DGA tackled the issue because of the intricacy of wellness capability and the model articulation limit in the hereditary calculation. Trial results utilizing MNIST information brought about an exact calculation of 75.3%. If, by some stroke of luck, the dropout calculation is used, the accuracy is 41.4%. DGA has been demonstrated to be better than surrender alone [7].

The bone limits are characterized by a functioning shape model. This provides a measurable model of bone shape and nearby designs. Joint positions are recognized through neighborhood direct planning in light of surface highlights. We investigated radiological pictures of five RA hands. The collocation gauge is 92% accurate. In addition, programmed explorations lessen the need for skilled laborers. Also, it is feasible to perform far-off examinations and prescriptions [8].

Explicit elements are disengaged from pre-handled information through straight-out hypothesis-part examination. The element extraction limit depends on the arrangement. Classifier capabilities are deducted from the base. Input information is organized and classified. The arrangement is performed by neurosomal mental guides with qualities. Calculation—this arrangement gives high accuracy. Infection can be detected from a grouped information collection with a minimum to no effort. Data on various types of joint inflammation is provided. Consequently, it helps anticipate and identify joint pain [9].

A model that predicts the number of readmissions within 30 days of discharge from the hospital has been developed. In this study, we analyzed 1 lakh diabetic patients with 50 elements, using AI calculations, such as strategic relapse, choice trees, arbitrary woodlands, adaboost, and xgboost. We achieve the highest accuracy of 94% by utilizing irregular woods over any remaining calculations. The results of this study are empowering and may help caring suppliers develop their operations [10].

It has been found that the clinical laboratory tests utilized in the proposed examination procedure can accurately predict arthritis with 94.03%, 96.00%, and 93.51% accuracy rate, respectively, while the review accuracy with 30 clinical examinations utilizing crude information and sequencing is 93.51%. The direct element choice method (SFFS) [11] is used to investigate virtual individual collaborators, email spam, online misrepresentations, traffic determining, interpersonal organization personalization, and others. Concerning the calculations, we will discuss about genuine Bayes calculations, brain organizations, KNN calculations, direct relapse models, strategic relapse models, etc. Considering the ongoing need, we investigate semantic division calculations to distinguish Coronavirus 19 [12].

There was a significant difference between the predictions made using the LSTM model, which achieved 82.38% accuracy, and the predictions made using the SVM model, which achieved 79.48% accuracy. For instance, in quiet 2, the highest LSTM accuracy was 98.89%, while the highest level of accuracy achieved with straight SVM was 80% [13]. There is a normal characterization accuracy of 95.08% and 93.47% for all parcel conventions in the mechanized framework. The manual framework precisions were 94.06% and 92.02%, respectively. Mechanized ML scoring yielded 98.05% and 97.53% for all-over walls, compared to manual gamble appraisal strategy [14].

Some well-known arrangement and relapse models are evaluated with the proposed fluffy BLS, compared with a few cutting-edge smooth and silky brain models. The results demonstrate that the fluffy BLS beats the other models that were involved. In addition, fluffy BLS has the advantage of having fewer fuzzy guidelines and requiring less preparation time as compared with neuro-fluffy models, which is beneficial for the standard blast problem in some ways [15]. Taguchi's exploratory plan strategy focused effectively on the ideal blend of its client-specific boundaries. A multi-facet perceptron and the proposed dynamic network model (DNM) are used in the simulations to conduct tests on 14 points, including characterization, estimation, and expectation. Considering the results of the pro posed learning calculation, we believe that the proposed approach is likely to work well in preparing DNMs [16].

The precision level of the instrument reached 90%, and the results were consistent with clinical specialists' assessments. Considering the actual results in practice, we accept that the proposed choice help apparatus can assist GPs in the exact and reasonable diagnosis of

RA patients. In this way, the risk of misdiagnosis or delay is diminished, and the patient's illness can be kept from progressing past cutting-edge stages [17]. Multiobjective streamlining has become inextricably popular over the past 5 years as a result of decay-based transformative multiobjective improvement calculations (emotional). According to recent research, its adequacy is primarily determined by Pareto-front (PF) designs, which have proven to be effective. There are several reasons for this effect. One of these is that the benchmarks and sub-issue plan frameworks do not adjust well to the various properties of the issue within the deterioration strategy [18].

NB with CFW has shown to outperform NB in a number of tests, and any remaining cutting-edge, highlighted channels of weighting are compared for correlation in order to find out which one is most accurate. In contrast to specific weighting containers for NB streamlining, CFW's fundamental benefits are its low computational intricacy (no inquiry required) and the straightforward nature of its model. Besides these, we applied CFW to message order and made significant improvements [19]. Revolutionizing medical care through artificial intelligence and Web of Things applications gives a superior comprehension of what artificial intelligence means for medical care and clinical productivity. This will enable us to develop our results further. It centers on a comprehensive and complete AI prologue. Covering points like patient consideration, digital actual frameworks, and telemedicine, this elite asset is unique for clinic heads, clinical experts, government authorities, advanced education understudies, instructors, curators, specialists, and teachers [20].

A classifier outfit system is a gathering of classifiers whose results are separately joined using a weighted calculation. This is normally done to keep patients from becoming more extreme. In addition, for quality characterization, the DNCM-ICSA calculation is utilized. The consequences of the recently developed classifier are dissected, agreeing with measurements including accuracy, review, F-measure, and accuracy [21]. This paper additionally surveys customary AI strategies distributed around 2015 and their application to diagnosing these illnesses. The form distinguishes between open issues and examination holes. We accept that profound learning will help generate specialists and experts to foresee the course of the illness, foster therapy designs, and evaluate their potential advantages [22]. Machine advancement as a field of manufactured consciousness is progressively being used in medicine to help patients and doctors. Developing informational indexes gives a phenomenal premise for applying AI strategies that take advantage of the experience of the past [23].

Manual appraisal of different SNPs and joint associations is extremely time-consuming for clinicians. Using AI-based factual information to identify the impact of joint inflammation on epistasis in the elderly, this study investigates the additional factual AI and advance strategies that have been developed so far. Computational models and their exploratory consequences were also examined [24]. Artificial knowledge in medical services is a general term used to depict the utilization of AI programming and calculations, or manufactured brainpower (simulated intelligence), to imitate an individual's discernment in the examination and comprehension of clinical and medical services information. building or surpassing human abilities by giving better approaches to analyze, treat, or anticipate illnesses. In particular, simulated intelligence is the capacity of PC calculations to reach estimated determinations from issued information alone [25]. Understanding accurate medication inspirations, traps, and constraints is fundamental to understanding and applying these techniques. The significance of acquiring individuals' and patients' trust can be recognized, yet developments in manufacturing brainpower, which must be accomplished through human mediation or through a system in the vicinity called expanded artificial intelligence, require the establishment of a setting [26].

Given that profound learning strategies might assume a significant role coming soon for rheumatology, rheumatologists should grasp the techniques and presumptions fundamental at present generally utilized in deep learning calculations, their constraints, and the scene of profound learning research—improvement of calculations and the eventual fate of clinical choice help apparatuses. The most effective utilization of profound learning in rheumatology should be founded on rheumatologists' clinical experience so that calculations can be fostered that address the most significant rheumatological problems [27].

The proposed model, working in a major information climate called AI-Based Group Scientific Methodology (MLEAA), comprises two stages: learning and expectation. In the learning stage, information is handled through a diagram-decrease system in Hadoop. The properties of the highlights are directed towards the expectation stage. The expectation period of the proposed MLEAA approach comprises three distinct calculations: AbaBoost, SVM, and ANN, and the last forecast esteem is determined by the democratic technique. Besides improving results through this review, it is extremely helpful for the early diagnosis of rheumatoid joint pain [28].

The ML and DL major information environment consists of two stages—namely, the learning stage and the forecast stage—which is based on AI-based Outfit Logical Methodology (MLEAA). The learning stage is handled by a diagram decrease system in Hadoop, and the element credits are in the expectation stage. The forecast period of the proposed MLEAA approach comprises three unique calculations: AbaBoost, SVM, and ANN, and the last expectation value is determined in light of casting a ballot strategy. It is extremely helpful for detecting early signs of rheumatoid joint pain and improving test results [28]. In order to estimate the fundamental parameters of the classifier using these ensemble classifiers, k-NN and weights with arbitrary values are used. As a part of the arrangement, 10-fold cross-approval of the information is performed with execution measures like accuracy, exactness, and the reciprocal order of correlation being used to assess your work. The upsides of these measurements were contrasted with the underlying means and a few gathering classifiers. This enhancement looks at the comparison of the base classifiers with the outfit classifiers, yielding critical improvements [29].

14.2.1 Challenges and Problem Identification Factor

- A major problem is that RA disease exhibits different properties than feature findings in medical observations, so it is difficult to identify the disease's nature and its symptoms.
- Increased feature dimension: The RA dataset contains more outlier forms of feature dependence, which leads to low precision accuracy. Most existing systems failed to identify the feature–relation margin factor of RA to identify disease risks, leading to poor accuracy.
- The majority of bias adjustments due to error rate cause a low learning level and an F1 score, which increases the false rate.

- It is more complex to find the feature support weight due to the dimensionality factor because considering more features leads to classifications with low accuracy and low precision and a higher recall F1 score with an increased time complexity.

14.3 HSFO-SVM Based on LSTM-Gated Convolution Neural Network (LSTMG-CNN)

Rheumatoid arthritis (RA) is one of the most common heart diseases. This disease mostly affects the respiratory system and the heart tissues. However, the most common disease of RA is heart failure, a critical disease of heart function. A famously dangerous disease is heart failure. This disease affects everyone, and our heart functions change when we are under abnormal conditions, so that person must be admitted to the hospital. Many studies need to be done to determine this disease's seriousness. Below is the diagram, Figure 14.1, of the proposed system of heart disease prediction, particularly for RA-based heart diseases.

For that, it should collect relevant data and employ the classification method. For that, preprocessing, feature selection, and classification should be done. By preprocessing tabs, they are normalized before they are displayed. C-score and cross-fold validation methods are used for preprocessing. It is used to select features for the Honey Scout Forager model using the Honey Scout Forager Optimized SVM Feature method. The LSTMG-CNN method is also employed in the classification of classified applications as well as in other classified applications.

14.3.1 C-Score and Cross-Fold Validation

In order to predict rheumatoid arthritis (RA) heart disease from C-scores, the probability of contracting RA heart disease must be determined. The probability of finding the C-score for prediction and classification depends on this score. This is used to normalize the data. The C-Score works pairwise, which means rows and columns of function data, and finds the highest probability of heart disease prediction. Here fetch the content using U1 and U2

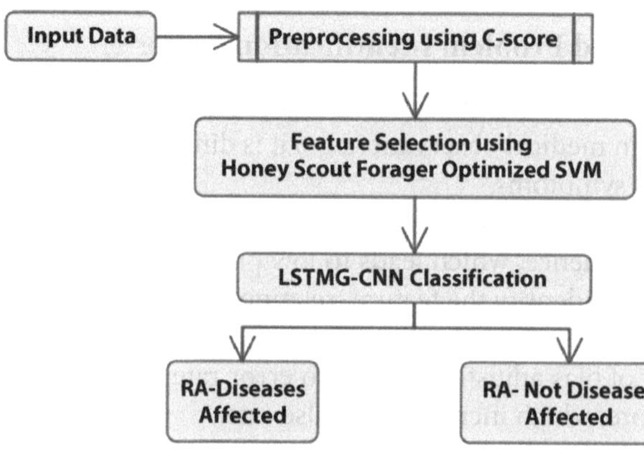

Figure 14.1 Architecture diagram.

inputs. The performers prepare punctuation and duplicate words and numbers. The dataset was created using Kaggle datasets of heart.csv files. We need to find the feature similarity score (CSS) based on CSV features in the files.

The CSV term-wise total

$$T(j)=\begin{cases}1+\sum_{i=1}^{2} CSV_i,\ if\ CSV_1\ or\ CSV_2=0\\ 2+\sum_{i=1}^{2} CSV_i,\ if\ CSV_2\ or\ CSV_2=0\end{cases} \tag{14.1}$$

Computation of probability term wise CSV_1 *and* CSV_2

$$p=\begin{cases}\dfrac{1+\sum_{i=1}^{2} CSV_i(j)}{\sum_{j=1}^{n} T(j)},\text{if } CSV_1 \text{ is } 0\\[2ex] \dfrac{\sum_{i=1}^{2} CSV_i(j)}{\sum_{j=1}^{n} T(j)},\text{otherwise}\end{cases} \tag{14.2}$$

Estimated values

$$< CSV_1 \geq = PXT(j) \tag{14.3}$$

Error-values

$$\text{error}(j)=\frac{\langle CSV_1(j)\rangle - CSV_1(j)}{\sqrt{T(j)XPX(1-P)}}. \tag{14.4}$$

Weight

$$Wj=\sqrt{T(j)} \tag{14.5}$$

The content similarity score is

$$CSS_{u1u2}=\frac{\sum_{j=1}^{n}(\text{error}(j))^2 X(w(j)}{\sum_{j=1}^{n}\sqrt{T(j)}} \tag{14.6}$$

To determine the cross-validation of the rheumatoid arthritis (RA) heart disease prediction, apply Equation 14.6 as shown above. One of the most commonly used data regression techniques for estimating real prediction error and modifying model parameters is cross-validation. There are a number of types of cross-validation and methods of data reconstruction which are commonly used in this article.

There are a number of techniques that are used for assessing and contrasting learning algorithms, and one of these is cross-validation, which consists of splitting the data into two sets: one set for learning or training the algorithm, and an additional set of tests for evaluating the algorithm's performance. It is not only through multiple cross-validation rounds that each data point has an opportunity to be validated but also through the combination of training and validation sets that each sample has a chance to be assessed. One method to calculate candidates' predictive performance is cross-validation. Here we employ two forms of validation for our training and testing. Data-splitting techniques are used for validation research.

Here CV is the cross-validation of the splitting schemes, and K is the common-only fold; k is a distinct pair of training and testing.

The score estimated K-fold cross-validation (CV)

$$S_k = \frac{1}{n}\sum_{i=1}^{n} L(y_{i,} y_i^{-k[i]}) + CSS_{ulu2} \tag{14.7}$$

Here i=1, n indexes the data points, k is index of the fold, and script – [i] indicates the training fitted data point.

There is often an upward bias in cross-validated score estimates obtained through cross-validation. The training set is smaller than the entire data set, so the expected loss amount is overestimated. However, this bias can easily be adjusted to make the model more accurate. Point-wise bias cross-correction ki and different between sample loss $L\left(y_i, y_i\right)$ and average training fold K. The cross-validation estimation error is E_{cv}.

$$E_{cv} = \sum_{i=1}^{n} L\left(y_{i,} \int y_i^{-k[i]}\right) \tag{14.8}$$

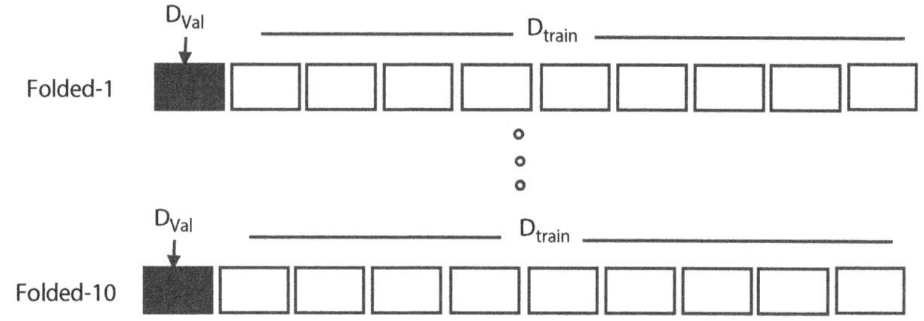

Figure 14.2 Cross-fold validation.

Two different classifiers are compared in the *T*-test with cross-validation and variance correction in Figure 14.2 Cross-fold validation. As the training sets overlap in different cross-validation folds, the standard t-100 test fails to meet the assumption of independence, which leads to an underestimation of variance in the training sets when doing cross-validation tests. Validation of T after variance correction is as follows:

$$T = \frac{\frac{1}{kr}\sum_{j=1}^{k}\sum_{j=1}^{r}\left(a_{ij} - b_{ij}\right)}{\sqrt{\left(\frac{1}{kr} + \frac{n2}{n1}\right)S^2}} \sim t_{kr} - 1 \tag{14.9}$$

Here T is the distribution of validation, and k.r-1 is the degree of freedom a_{ij} b_{ij} is the number of the classifier; it is the cross-validation of the functions, and S^2 is the variance of cases in the validation set.

14.3.2 Honey Scout Forager Optimization

Data optimization is a very significant problem for data collection, and before feature selection, data optimization must do that work. The problem is solved using the Honey Scout Forager Optimized algorithm to optimize the datasets. The following Honey Scout Forager algorithms have been optimized for use.

The Honey Scout Forager Optimization algorithm is initialized randomly.

$$x_{i,j=x_{min,j}} + \text{rand}(0,1)(x_{max.j} - x_{max,j}) + T \tag{14.10}$$

Here i=1,2,3,...SN is the population size, and j=1,2,D is the population dimension, and rand (0,1) is the random number for the range [0,1], and here Xmin.j and Ymax.j are the parameters to optimize according to their lower and upper bounds.

To solve problems associated with maximum optimization and minimum optimization, the optimization function fitness needs to be used, which corresponds to the following:

$$fit = \begin{cases} \frac{1}{1+obj}, obj > 0 \\ 1 + |obj|, obj < 0 \end{cases} \tag{14.11}$$

Then, calculate probability next, which is defined as

$$P_i = \frac{fit}{\sum_{i=1}^{SN} fit_i} \tag{14.12}$$

There is a population limit for the solution i, but the random-wise problem solves the population limits by first finding the solution of the solution i that has the best fitness. The Honey Scout Forager Optimization algorithm solved the optimization of the dataset problem solved.

14.3.3 Feature Selection Using SVM

One of the methods used in ML is the support vector machine which we need for feature selection and support with the Honey Scout Forager Optimization algorithm. The dataset could be filtered by selecting feature values. Filter process: unwanted data has been removed or filtered.

Let $(x_i, y_i) |<|< N < T$ is training of N each sample $x_i \in R^d$ each label class $y_i \in \{-1,1\}$

Here d is the dimension of feature selection. As a result of this, we have to decide what w and b are so that we can

$$y_i \left(w.x_i + b + T\right) > 0, i = l \ldots N + P_i \tag{14.13}$$

In this rescale of w and b so that

$$Tmin_{;<j<N}^{y_i(w.x_i+b)>0,i=l\ldots N} + P_i \tag{14.14}$$

So the close-set point Equation 14.1 hyperplanes of distance

$$\sum T.y_i(w.x_i + b) > l + P_i \tag{14.15}$$

Here find the optimal separating hyperplane and closed set of distance w

$$\frac{1}{||w||} = \sum T \frac{1}{2} ||w||^2 + P_i \tag{14.16}$$

Where minimizing amount of SVM is under constraints and $||w||^2$ is under linear constraints, Equation 14.2 is achieved with the multipliers. The multiplier associated with the selection of the features of the model is referred to as $\alpha = (\alpha_1 \ldots \alpha_N)$ N non-negative multiplier

$$W(\alpha) = \sum_{i=0}^{N} \alpha_i - \frac{1}{2} T \sum_{i=0}^{N} \alpha_i, \alpha_j, y_i, y_j x_i . x_j + P_i \tag{14.17}$$

Here $\sum_{i=0}^{N} y_i \alpha_i$ is achieved the standard quadratic programming, Method. We denote the $\alpha^0 = (\alpha^0{}_1 \ldots \alpha^0{}_N)$ is solution for maximum problem found. Here SVM (w^0, b^0) the following expression is

$$w_0 = P_i + \sum_{i=0}^{N} T.\alpha^0 y_i x_i \tag{14.18}$$

Honey Scout Forager Optimization with SVM are points for which $\alpha_i > 0$ is satisfy Equation 14.18 with equality.

From Equation 14.6, hyper decision plane is to be written as

$$f(x) = \text{sign}.T\left[\sum_{i=0}^{N} \alpha^0 y_i x_i + b^0\right] + P_i \tag{14.19}$$

The input is mapped. The non-linear high-dimensional feature has been selected here Replace x is feature selection $\Phi(x)$, taken Equation 14.5 combined here.

$$W(\alpha) = \sum_{i=0}^{N} \alpha_i - \frac{1}{2} + P_i + \sum_{i=0}^{N} \alpha_i, \alpha_j, y_i, y_j \Phi x_i . \Phi x_j \tag{14.20}$$

Here $k = \Phi x_i . \Phi x_j$ *is training algorithm of mapping* Φ. Symmetric Position kernel K(x,y) it mapped with existing Mercer's theorem As shown by the mapping Φ

$$K(x, y) = T.\Phi x.\Phi y + P_i \tag{14.21}$$

It has been decided to showcase kernel K as the featured algorithm. This is a training algorithm that reduces the likelihood of error, as in Mercer's complaint.

$$W(\alpha) = \sum_{i=0}^{N} \alpha_i - \frac{1}{2} T.\sum_{i=0}^{N} \alpha_i, \alpha_j, y_i, y_j K(x_i, y_i) + P_i \tag{14.22}$$

Here the decision function becomes

$$f(x) = \text{sgn}\left[\sum_{i=0}^{N} \alpha^0 y_i x_i + x_i x + b\right] + P_i \tag{14.23}$$

Construction of Honey Scout Forager Optimization support with vector machines for feature selection: A good multi-class approach is required when working with multiple classes, such as in object identification and image categorization. Honey Scout Forager Optimization used with point x is the argument of $f(x)$. The highest support vector machine feature selection steps

$$f1 = \min_{w^{ij}, b_{ij}} \frac{1}{2}(w^{ij}.w^{ij}) \tag{14.24}$$

$$f2 = +C\sum_{t}(\xi_{ij}^{ij}T) \tag{14.25}$$

$$f3 = f(x) + W(\alpha) \tag{14.26}$$

$$f_{ea} = f1 + f2 + f3 \tag{14.27}$$

Equation 14.27 outlines the presumption that every image has just one label, meaning that it belongs to just one class. However, since an image's contents are not distinct, it can be feature-selected into several classes for heart disease prediction. Multi-class learning can be strengthened and expanded to accommodate the selection of multiple labels.

In Figure 14.3, we explain how the maximum support can be achieved from feature selection based on the importance of the features for the prediction of RA disease. The feature importance was determined based on the similarity between the RA disease label and each feature. The higher the similarity between the feature and the RA disease label, the higher the feature importance and the greater the support.

14.3.4 UNET-CNN Classification

One of the most well-liked deep learning algorithms is UNET-CNN. It uses both apparent and concealed layers, each serving a different purpose. Neural networks are categorized using two layers. Features in the visible (v1, v2, and v3) are selected and hidden layers h1, h2, and the random initialization weight value W of the data in Figure 14.4.

Given the values v and h, the energy function of UNET-CNN is given by the following equation:

$$E(v,h) = -\sum_{i=0}^{n} a_i v_i - \sum_{j=0}^{m} b_i v_i - \sum_{i=0}^{n}\sum_{j=0}^{m} vh_j + f(w) + 1 \tag{14.28}$$

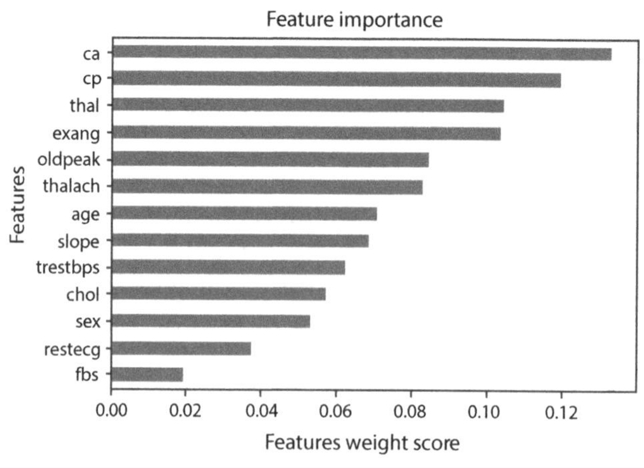

Figure 14.3 Feature selection process.

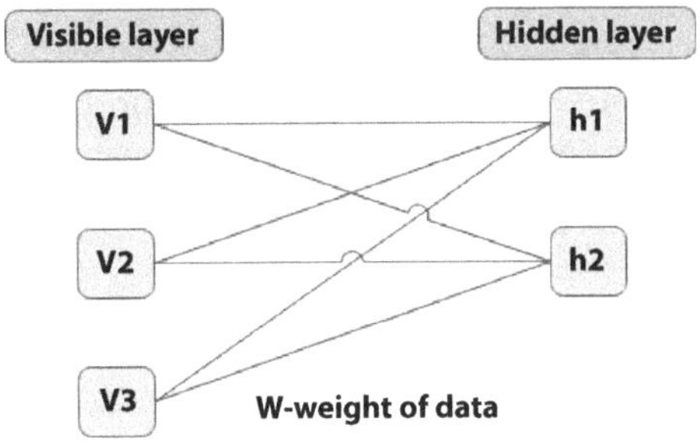

Figure 14.4 CNN diagram.

Here $w_{i,j}$ is the weight of the matrix, and jth is the hidden elements.
Probability distribution p (v)

$$P(V) = \frac{1}{Z}\sum_{h} e^{-E(v,h)} + f(w) + f_{ea} \tag{14.29}$$

Probability distribution P= (h)

$$P(V) = \frac{1}{Z}\sum_{v} e^{-E(v,h)} + f(w) + f_{ea} \tag{14.30}$$

HSFO-SVM are the establishing joint probability of feature labels

$$P(v,h_1,h_2 \ldots h_l) = P(v|h_1)P(h_2 \mid h_1)\ldots P(v|h_1)P(h_l \mid h_{l-1})P(h_{l-1}|h_l) + f_{ea} \tag{14.31}$$

$$W = \sum [(P(v,h_1,h_2 \ldots h_l))k + f(w) + f_{ea}] \tag{14.32}$$

Each window is convolutionally filtered so that the scalar values ri are obtained for each window, corresponding to each window in the convolution:

$$l_i = \sum W * g(x_1, x_2 \ldots x_n + u) \epsilon R + f_{ea} \tag{14.33}$$

Practically, one frequently utilizes more channels, u1….. un, so this can be addressed as a vector increased by a framework U and added with the b term:

$$l_i = \sum g(x_i.U + b) + f_{ea} \tag{14.34}$$

With $r_i \in R^l, x_i \in R^{(k*d)}, U \in R^{(k,d*l)}$ and $b \in R^l$

The way the data was presented here was sequentially for data flows and inside a particular window. Convolution is applied to data and yields m vectors w; it is also applied to post tags and yields m equal vectors for shape, making a total of m vectors.

$$M_i = \int f_{eal:m} + v_{l:m} + l_i \tag{14.35}$$

Or by concatenation

$$M_j = \int [f_{eal:m} + POS_{l:m} + v_{l:m}] + l_i \tag{14.36}$$

This function supports the result of vectors from various complexities of dimension v. Here M_i is the past heart problem, and M_j is the present heart problem. Result: The calculation also considers other organizational layers and treats them as a single HSFO-SVM during preparation, and Figure 14.5 shows the final classification results of heart failure disease predictions.

Figure 14.5 defines the following parameters: high smoking, blood pressure, cholesterol, hypertension, the presence and absence of high sugar, human weight, height, age-old peak, etc.; heart disease predictions based on the SVM classification method; and results in analysis based on art disease prediction. The feature extraction data allows for easy information matching and performance improvement at past and current levels.

	cp	thalach	exang	oldpeak	slope	thal	target
1224	3	60	1	1.4	2	2	1
1439	1	72	0	0.0	1	2	1
1790	2	77	0	0.0	2	2	0
4771	0	85	0	0.0	2	2	1
117	0	120	1	2.5	1	3	0
4417	0	75	0	0.6	2	1	0
1023	0	159	0	0.0	2	2	1
4186	0	66	0	1.0	1	3	1
595	0	161	0	0.0	2	3	0
786	0	163	0	0.2	1	3	0

Figure 14.5 Classification of rheumatoid arthritis (RA) heart disease prediction.

14.4 Result and Discussion

Results and discussion about the heart disease prediction Our proposed method compares various algorithms, including ensemble learning, decision trees, and logical regression. The task performs with precision and recall, F1-score, and proposed accuracy. The data is from the Kaggle dataset of the RA heart.csv file from Kaggle sites. Table 14.1 shows the implementation parameters.

This image shows the types of parameters used in the heart disease prediction dataset below. All of these parameters are indicators of heart disease and can be used to make a prediction about the risk of developing heart disease.

Below is a confusion matrix, which shows whether the disease is available or not. The heart disease prediction dataset, and below is a confusion matrix that shows whether it is available. Figure 14.6 illustrates the confusion matrix conditions for calculating accuracy, precision, recall, and F1-score using the confusion matrix conditions

According to the UNET-CNN algorithm, Figure 14.7 describes the true positive indicator of the prediction of positive detection. This algorithm has high rates of heart disease prediction.

This indicates that the prediction model is very accurate in predicting heart disease. The recall performance measures how well the model can detect true positives, and the precision measures how well the model can identify true negatives, both of which are important indicators for the model's accuracy. Figure 14.8 explains precision and recall for heart disease prediction. The average prediction is 96.2%, and the average recall performance is 95.2%.

$$\text{precision} = \frac{True\ positive}{True\ positive + False\ positive} \tag{14.37}$$

Precision measurement value is $\text{precision} = \text{to } 96.2\%$,

Table 14.1 Simulation parameters.

Parameter	Values
Language	Python
Tools	Anaconda
Dataset	RA heart disease
Records size	30 kb
Total records	1,000

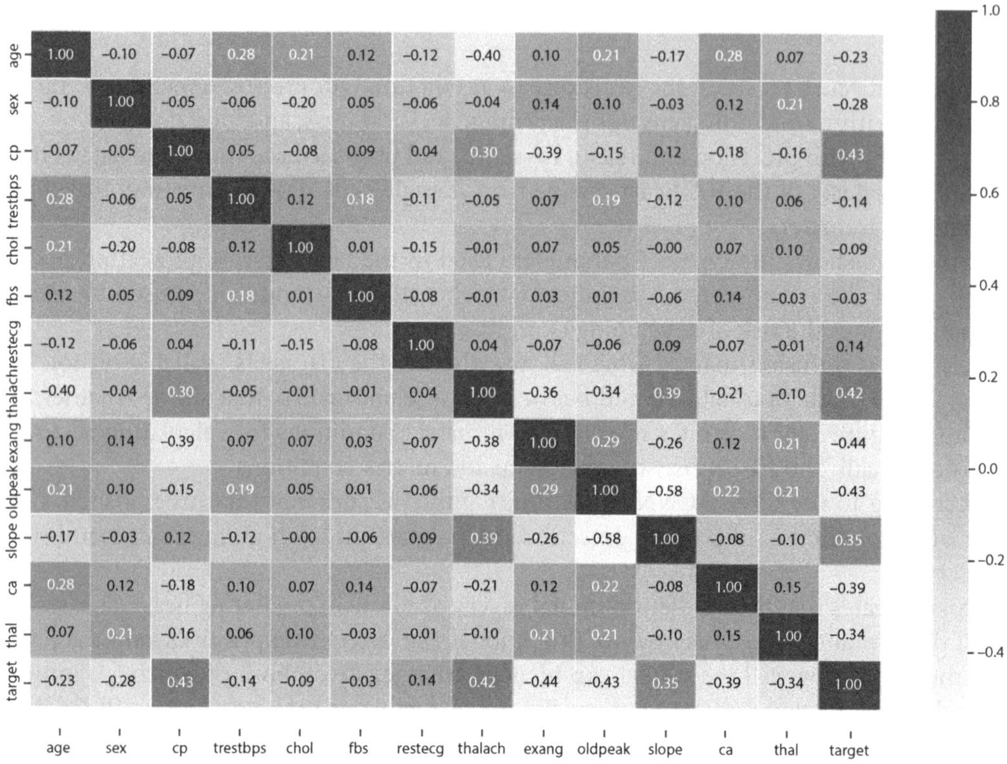

Figure 14.6 Confusion matrix.

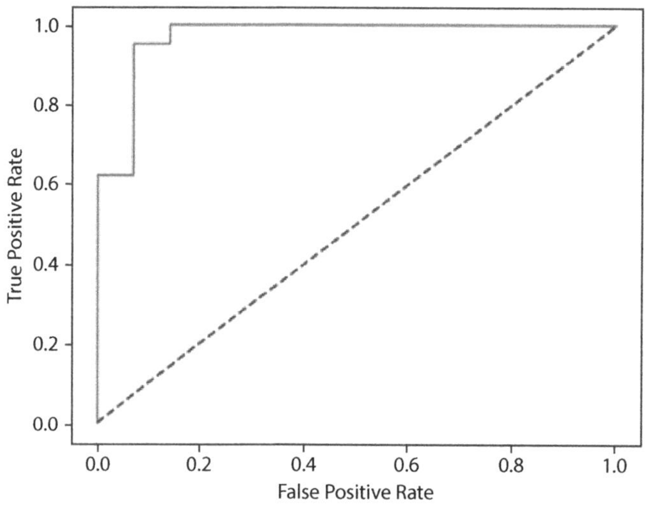

Figure 14.7 True positive rate.

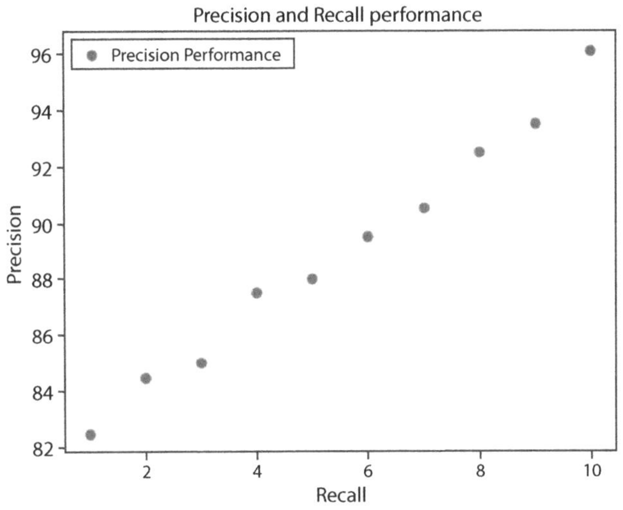

Figure 14.8 Precision and recall performance.

$$\text{Recall} = \frac{Trueostive}{true\ postive + False\ negative} \tag{14.38}$$

Recall measurement values is recall = 95.2%.

Figure 14.9 illustrates the performance of the F1 measure in predicting RA heart disease. The precision and recall values were multiplied, summed, and then divided by the upper and lower values to create the F1 measure for heart disease prediction using the DP learning approach. The 94.6% F1-measure value was determined lastly.

$$F1 = \frac{\text{precision} \times \text{recall}}{\text{recall} + \text{precision}} \tag{14.39}$$

F1 measure values= 94.6%

This algorithm is a proposed system; it compares various methods. The following algorithms, ensemble, decision tree, and regression forest, are compared to the proposed UNET-CNN system.

As illustrated in Figure 14.10, the overall precision and recall of heart disease prediction can be seen in Figure 14.10 as well as the F1-measure performance calculation accuracy. The comparison of the results is based on recent approaches based on testing and training validation strategies. The results section shows the proposed method's accuracy compared with other methods.

Figure 14.11 and Table 14.2 discuss algorithm accuracy, decision tree accuracy (94.5%), ensemble accuracy (82.5%), decision tree accuracy (94.5%), and the proposed UNet-CNN system (96%). Finally, accuracy is given by the UNet-CNN algorithm of the machine learning method.

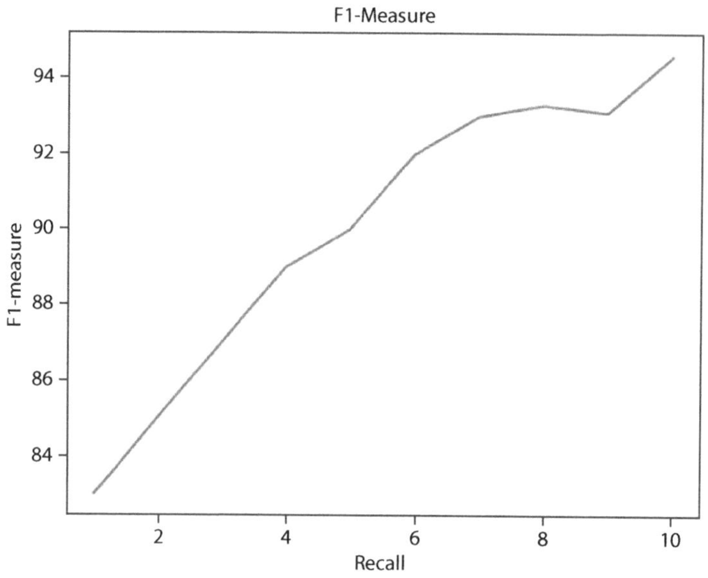

Figure 14.9 F1-measure performance.

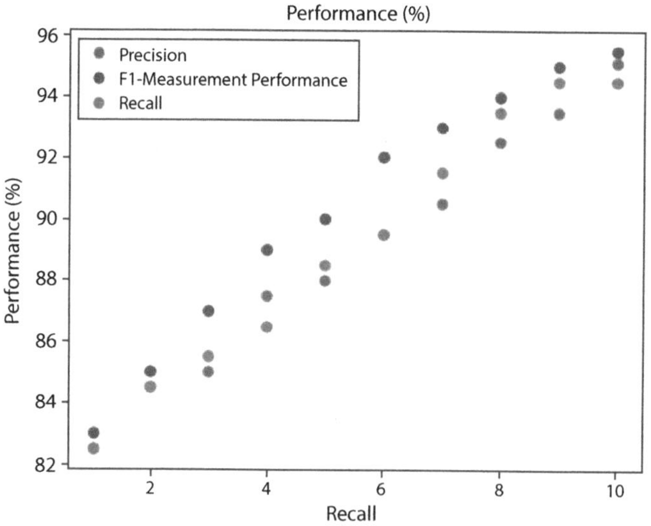

Figure 14.10 Overal precision and recall, F1-measure performance.

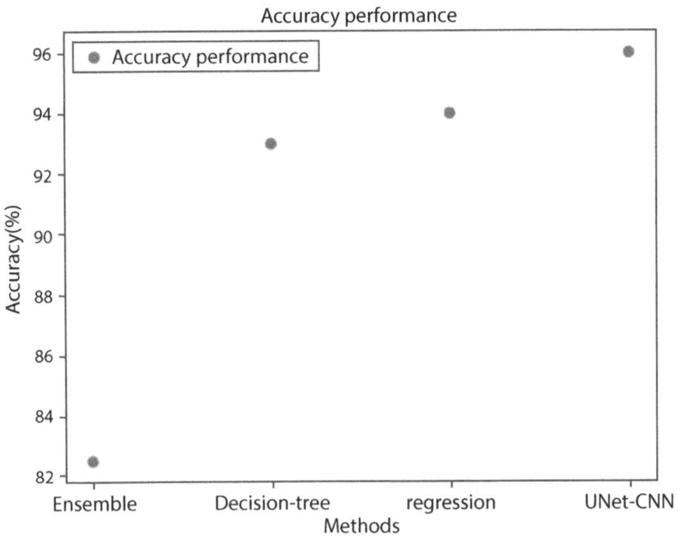

Figure 14.11 Accuracy performance.

Table 14.2 Proposed system accuracy.

Method	**Accuracy (%)**
Ensemble	82.5
Decision tree	94.5
Regression	95.1
UNet-CNN	96

14.5 Conclusion

Rheumatoid arthritis (RA) prediction disease prediction and classification to find the disease's accuracy, preprocessing using the HSFO method to filter the dataset, SVM-based dataset optimization and selection processes, and finally, classification using the UNET-CNN-based algorithm. The proposed system demonstrates a F-measure of 94.6%, a precision rate of 96.2%, a recall rate of 95.2%, and a prediction accuracy of up to 96%. Strong performance with a temporal complexity of 3.5% and a low false rate. To attain high prediction accuracy, the suggested classifiers should perform better than earlier techniques. The accuracy of the suggested system is displayed in the results section, using various comparison techniques to highlight the performance attained. Finally, the proposed algorithm has the highest accuracy produced by the UNET-CNN algorithm.

References

1. Zhang, J., Zhang, Z., Pu, L., Tang, J., Guo, F., AIEpred: An Ensemble Predictive Model of Classifier Chain to Identify Anti-Inflammatory Peptides. *IEEE/ACM Trans. Comput. Biol. Bioinf.*, 18, 5, 1831–1840, 1 Sept.-Oct. 2021.
2. Kaur, R., Madaan, V., Agrawal, P., Fuzzy Expert System to Calculate the Strength/Immunity of a Human Body. *Indian J. Sci. Technol.*, 9, 44, 1–8, 2016.
3. Kaur, R., Madaan, V., Agrawal, P., Singh, S.K., Kaur, A., Fuzzy Expert System for Identifying the Physical Constituents of a Human Body. *Indian J. Sci. Technol.*, 9, 28, 1–8, July 2016.
4. Shiezadeh, Z., Sajedi, H., Aflakie, E., Diagnosis of rheumatoid arthritis using an ensemble learning approach. *Comput. Sci. Inf. Technol. (CS & IT)*, 5, 15, 139–148, 2015.
5. Kaur, K., Singh, S.K., Agrawal, P., Goyal, A., Classification of Follicular Lymphoma Grades using Support Vector Machine. *Int. J. Adv. Eng. Res. (IJAER)*, RIP Publications, 10, 8, 20441–20449, 2015.
6. Lin, C., Miller, T., Dligach, D., Savova, G., Feature Engineering and Selection for Rheumatoid Arthritis Disease Activity Classification Using Electronic Medical Records. *ICML Workshop on Machine Learning for Clinical Data Analysis*, 2012.
7. Park, J.-G. *et al.*, Dropout Genetic Algorithm Analysis for Deep Learning Generalization Error Minimization. *Int. J. Adv. Culture Technol.*, 5, 2, 74–81, 2017.
8. Bhisikar, S.A. and Kale, S.N., Automatic analysis of rheumatoid Arthritis based on statistical features. *2016 International Conference on Automatic Control and Dynamic Optimization Techniques (ICACDOT)*, pp. 242–245, 2016.
9. Shanmugam, S. and Preethi, J., Study of Early Prediction and Classification of Arthritis Disease using Soft Computing Techniques. *Int. J. Res. Eng. Appl. Manage. (IJREAM)*, 3, 5, 1–13, 2017.
10. Sharma, A., Agrawal, P., Madaan, V., Goyal, S., Prediction on Diabetes Patient's Hospital Readmission Rates. *International Conference on Advanced Informatics for Computing Research (ICAICR-19) ACM-ICPS*, 2019.
11. Olatunji, S.O. and Alansari, A., A Novel Ensemble-Based Technique for the Preemptive Diagnosis of Rheumatoid Arthritis Disease in the Eastern Province of Saudi Arabia Using Clinical Data. *Comput. Math. Methods Med.*, 2022, 1, 2022.
12. Singh, S., Anand, A., Mukherjee, S., Choudhury, T., Machine Learning Applications in Decision Intelligence Analytics, in: *Decision Intelligence Analytics and the Implementation of Strategic Business Management*, p. 163, 2022.
13. Ashour, S., El-Attar, A., Dey, N., El-Kader, H.A., El-Naby, M.M.A., Long short term memory based patient-dependent model for fog detection in parkinson's disease. *Pattern Recognit. Lett.*, 131, 23–29, 2020.
14. Araki, T., Jain, P.K., Suri, H.S., Londhe, N.D., Ikeda, N., El-Baz, A., Shrivastava, V.K., Saba, L., Nicolaides, A., Sha Que, S. *et al.*, Stroke risk stratification and its validation using ultrasonic echolucent carotid wall plaque morphology: a machine learning paradigm. *Comput. Biol. Med.*, 80, 77–96, 2017.
15. S. Feng and C. L. P. Chen, Fuzzy Broad Learning System: A Novel Neuro-Fuzzy Model for Regression and Classification. *IEEE Trans. Cybern.*, 50, 2, 414–424, 2018.
16. Gao, S., Zhou, M., Wang, Y., Cheng, J., Yachi, H., Wang, J., Dendritic neuron model with effective learning algorithms for classification approximation and prediction. *IEEE Trans. Neural Networks Learn. Syst.*, 30, 2, 601–614, 2018.
17. Salmeron, J.L., Rahimi, S.A., Navali, A.M., Sadeghpour, A., Medical diagnosis of rheumatoid arthritis using data driven pso–fcm with scarce datasets. *Neurocomputing*, 232, 104–112, 2017.
18. Wu, M., Li, K., Kwong, S., Zhang, Q., Zhang, J., Learning to decompose: a paradigm for decomposition-based multiobjective optimization. *IEEE Trans. Evol. Comput.*, 23, 3, 376–390, 2019.

19. Jiang, L., Zhang, L., Li, C., Wu, J., A correlation-based feature weighting filter for naive Bayes. *IEEE Trans. Knowl. Data Eng.*, 31, 2, 201–213, 2018.
20. Sharma, B., Sille, R., Bansal, A., Chaturvedi, S.K., A Comprehensive Review on AI Techniques for Healthcare, in: *Revolutionizing Healthcare Through Artificial Intelligence and Internet of Things Applications*, p. 187, 2023.
21. Chithra, B. and Nedunchezhian, R., Dynamic neutrosophic cognitive map with improved cuckoo search algorithm (DNCM-ICSA) and ensemble classifier for rheumatoid arthritis (RA) disease. *J. King Saud Univ. - Comput. Inf. Sci.*, 34, 6, 3236-3246, 2020.
22. Ur Rehman, Z., A Review of Arthritis Diagnosis Techniques in Artificial Intelligence Era: Current Trends and Research Challenges - ScienceDirect. *Neurosci. Inf.*, 2, 4, 100079, May 2022.
23. Hügle, M., Omoumi, P., van Laar, J.M., Boedecker, J., Hügle, T., Applied machine learning and artificial intelligence in rheumatology. *Rheumatol. Adv. Pract.*, 4, 1, rkaa005, 2020.
24. Manavalan, R. and Priya, S., Rheumatoid arthritis identification using epistasis analysis through computational models. *Biomed. Biotechnol. Res. J.*, 4, 8–15, 2020.
25. Yu, K.-H., Beam, A.L., Kohane, I.S., Artificial intelligence in healthcare. *Nat. Biomed. Eng.*, 2, 719–31, 2018.
26. Stafford, I.S., Kellermann, M., Mossotto, E., Beattie, R.M., MacArthur, B.D., Ennis, S., A systematic review of the applications of artificial intelligence and machine learning in autoimmune diseases. *NPJ Digit. Med.*, 3, 30, Mar 2020.
27. McMaster, C., Bird, A., Liew, D.F.L., Buchanan, R.R., Owen, C.E., Chapman, W.W. *et al.*, Artificial intelligence and deep learning for rheumatologists: A primer and review of the literature. *Arthritis Rheumatol.*, 74, 12, 1893-1905, 2022.
28. Shanmugam, S. and Preethi, J., Design of Rheumatoid Arthritis Predictor Model Using Machine Learning Algorithms, in: *Cognitive Science and Artificial Intelligence*, SpringerBriefs in Applied Sciences and Technology, 2018.
29. Sundaramurthy, S., S.C., Kshirsagar, P., Prediction and Classification of Rheumatoid Arthritis using Ensemble Machine Learning Approaches. *2020 International Conference on Decision Aid Sciences and Application (DASA)*, Sakheer, Bahrain, pp. 17–21, 2020.

15

Using Reinforcement Learning in Unity Environments for Training AI Agent

Geetika Munjal[1] and Monika Lamba[2*]

[1]Amity University, UP, Noida, Uttar Pradesh, India
[2]The Northcap University, Gurugram, Haryana, India

Abstract
An intelligent AI agent capable of performing tasks in a virtual environment is developed using reinforcement learning techniques. This chapter serves to demonstrate how machine learning and artificial intelligence methods can be employed to deploy an AI agent across settings, effectively addressing a wide range of challenges. By utilizing an AI agent, the need for developing agents for each unique problem encountered in diverse environments is eliminated. This approach transforms the AI agent into an entity that can be trained and adapted to scenarios, enabling it to effectively solve specific problems presented in each situation. The utilization of AI agents enhances resilience and adaptability in dynamic environments, leading to optimized resource allocation (including time, money, and energy) and increased human innovation. To accomplish this, the tools utilized are Unity 3D Engine, Python programming language, PyTorch framework, and ML agents.

***Keywords*:** Reinforcement learning, AI agent, machine learning, artificial intelligence, unity, environment

15.1 Introduction

Machine learning has a subfield called reinforcement learning. The technique of teaching machine learning models to execute actions or choose paths sequentially is known as reinforcement learning. The secret is to act sensibly to gain the most from a particular situation. Many computer programmers and robots use it to determine the best sequence of actions to take in a specific situation. Unlike supervised learning (SL), which uses the answer key as part of the training data, the agent, which makes decisions on what to do to finish the task at hand, is the main component of reinforcement learning (RL). The model is trained in supervised learning using the previously recognized correct response. If a training dataset is missing, it will unavoidably learn new skills through experience [1, 2].

**Corresponding author*: missmonikalamba@gmail.com

Shubham Mahajan, Pethuru Raj and Amit Kant Pandit (eds.) Deep Reinforcement Learning and Its Industrial Use Cases: AI for Real-World Applications, (349–374) © 2024 Scrivener Publishing LLC

The research "Applying Reinforcement Learning in Unity Situations for Training an AI-Agent" focuses on using machine learning, particularly the idea of reinforcement learning, for training an intelligent AI agent that can accomplish many objectives in various virtually simulative environments. The environment can be thought of as a streamlined digital workspace that simulates real-world scenarios and the challenges that our AI agent will try to overcome. The AI agent can be characterized as an agent (robot or model) that does not naturally belong to the environment and engages in external interactions with the environment to address the problem. By "externally," we mean that the AI agent is truly ignorant of the regulations governing how the environment functions. This differs from conventional agents that interact with simulated environments, such as game AIs, which are programmed to be a part of the environment itself. These agents have an awareness of the environment's internal norms, which helps in completing the tasks.

The intelligent agent develops the skills necessary to complete a task in an uncertain and potentially complex environment. An AI frequently finds itself in a position reminiscent of video games in real life. The computer then opts to employ the trial-and-error approach to generate a likely solution to the issue at hand. The AI gets either positive or negative feedback for the choices it makes and the deeds it does to train it to do what the programmer desires. Maximizing the overall favorable feedback is its main goal. RL is a form of learning that is thus founded on engagement with the environment. It is expanding quickly and creating a wide range of learning algorithms that may be applied in many different situations.

By deploying a single, versatile AI agent, you open the door to a host of benefits, with cost reduction and heightened innovation taking center stage. Bringing together various functions into one unified agent not only streamlines how resources are used but also cuts down on the need for multiple specialized agents, leading to reduced development costs, lower computational demands, and easier ongoing maintenance. Beyond these practical advantages, a versatile AI agent promotes innovation by adapting to different tasks and scenarios. Its ability to swiftly switch between functions supports quick prototyping and experimentation, giving a boost to the innovation process. This approach not only offers flexibility, scalability, and simplified maintenance but also provides a comprehensive solution that optimizes resource allocation. It encourages collaboration across different fields and ensures consistency in decision-making. Ultimately, a single, versatile AI agent emerges as a potent and efficient tool, aligning seamlessly with the dynamic landscape of artificial intelligence applications.

The amalgamation of the mentioned tools and resources creates a potent toolkit that excels in various domains. Unity 3D Engine, celebrated for its versatility in game development, provides an intuitive interface and a rich collection of assets accessible through its extensive store. Python, a widely used and easily readable programming language, distinguishes itself with its versatility and benefits from a large, supportive community. PyTorch, known for its dynamic capabilities in machine learning, enhances the toolkit's prowess in handling complex data-driven tasks. ML agents, seamlessly integrated into Unity environments, bring the ability to train intelligent agents, further expanding the toolkit's utility. This cohesive amalgamation of tools forms a versatile foundation with broad applicability, extending its impact across a diverse range of domains. In the dynamic realm of game development, this comprehensive combination, including Unity 3D for its versatility in crafting immersive experiences, Python for its adaptable scripting capabilities, PyTorch for dynamic machine learning, and AI agents like ML agents for intelligent character development, lays the groundwork for creating captivating and interactive virtual worlds. Simultaneously, the utility of this powerful toolkit transcends the gaming landscape, delving into the intricate terrains of advanced machine learning projects. In the context of machine learning, the dynamic capabilities of PyTorch coupled with the adaptability of Python offer a robust environment for researchers and practitioners to explore and innovate. This comprehensive foundation not only caters to the creative intricacies of game development but also provides a solid footing for addressing the complexities inherent in cutting-edge machine learning endeavors.

Unity 3D is geared towards those in game development and content creation, providing a platform for crafting diverse experiences in 2D, 3D, AR, and VR. Python, known for its versatility, attracts a broad audience, including programmers, developers, and data scientists, finding applications in web development, data science, and artificial intelligence. PyTorch, tailored for machine learning researchers and practitioners, stands out for its dynamic framework, particularly in deep learning experiments. AI agents like ML agents, seamlessly embedded in Unity, meet the needs of game developers seeking to craft intelligent non-player characters and researchers employing machine learning in various tasks, highlighting adaptability and learning capabilities. Each tool has its unique audience, collectively spanning game development, general programming, data science, and advanced machine learning applications, reflecting the diverse landscape of creative and technical pursuits.

The chapter describes the creation of design and writing rules for various goals and surrounding using Unity 3D, developing an AI agent and utilizing reinforcement learning.

15.2 Literature Review

In order to exploit a numerical performance measure which represents a long-term aim, a system is taught how to govern itself using the reinforcement learning paradigm. RL and SL are fundamentally dissimilar because the learner only receives partial feedback on the predictions. Additionally, the projections might change how the regulated system will operate in the future, which might have a long-term impact. It is possible to understand the benefits and drawbacks of learning algorithms by using reinforcement learning. RL is gaining a lot of favorable consideration as an outcome of the broad range of real-world issues which it may address, including those in artificial intelligence, operations research, and control engineering. This chapter focuses on RL algorithms that are built on the robust notion of dynamic programming. Presenting a fairly comprehensive list of learning-related difficulties, define key terms, provide a long list of cutting-edge algorithms, and investigating the theoretical merits and limitations of each.

The concept of generalization is a key issue in deep reinforcement learning [21]. To that purpose, new developments in deep RL will surely advance the current trend of differentiating explicit algorithms so that they can be incorporated into a specific kind of neural network and trained end-to-end. As a result, algorithms may develop richer, more intelligent structures that are better suited for reasoning at a higher level of abstraction, enabling them to handle an even wider range of applications than they now can. Although much work in the area of temporal abstraction is still needed, smart structures may be used for hierarchical learning.

The introduction of deep learning has knowingly improved the state-of-the-art in several arenas of ML, including tasks like object detection, speech recognition, and language translation [3, 6, 10].

15.3 Machine Learning

ML was primary brought into presence by Arthur Samuel in the 1959. Arthur Samuel stated that artificial intelligence "gives computers the capability to learn short of being explicitly programmed". Thus, ML may now be referred to as the field of study that enables computers to have the capacity for learning without being exposed to explicit programming. By giving our computers the ability to learn, as is clear from the field's name, machine learning (ML) helps them to become more human-like. Contrary to expectations, ML is actively being used in a lot more contexts than one may think.

15.3.1 Categorization of Machine Learning

Three main categories can be used to broadly classify ML implementations (Figure 15.1). The classification will be applied to the various learning methods based on the nature of the signal or response presented to the relevant learning system. Thus, the classification is discussed below.

15.3.1.1 Supervised Learning

An algorithm learns from previously acquired, verified data and the target responses associated with it, which can take the form of string labels or numeric values, in order to predict the proper reaction when faced with similar new conditions.

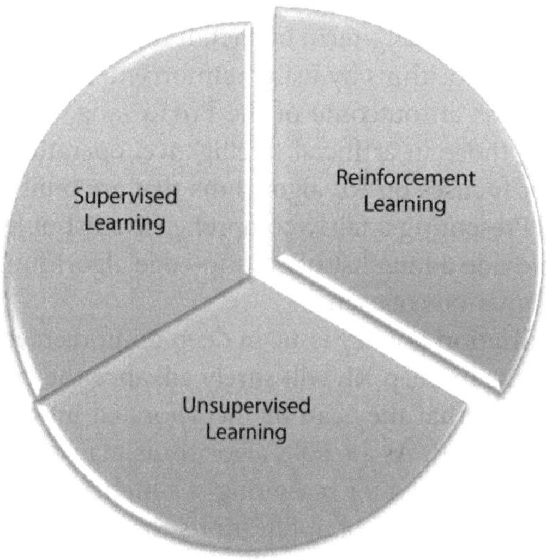

Figure 15.1 Different types of machine learning techniques.

15.3.1.2 Unsupervised Learning

An algorithm learns from raw data without any accompanying responses, allowing the algorithm to discover the data patterns on its own. These algorithms' main goal is to restructure data in order to make it more valuable.

15.3.1.3 Reinforcement Learning

When providing an algorithm with data without specific labels and feedback that is either positive or negative depending on the required solution. This process is mostly related to applications where the algorithm itself is in charge of making choices and accountable for the effects of those choices.

The difference between various categorizes of machine learning are given in Table 15.1. There exists another form of learning, although not commonly defined or taken into practice, but who knows maybe in the future it may become a sprawling area of study. This method of learning is calling semi-supervised learning. This is the case wherein we feed incomplete training data to the algorithm. The training set is incomplete because a few of the target outputs are missing.

15.3.2 Classifying on the Basis of Envisioned Output

When considering the desired output of the system, add another criterion for classifying machine learning tasks.

Table 15.1 Comparison of supervised, unsupervised, and reinforcement ML.

	Reinforcement learning [11]	**Supervised machine learning [9]**	**Unsupervised machine learning [10]**
Motive	Learn a series of actions	Determine results	Discover underlying
Rationale	Focuses on interacting with the environment	Focuses on learning by labeled data	Focuses on learning by unlabeled data without any guidance
Data kind	No predefined data	Labeled	Unlabeled data
Supervision	No	Extra	No
Example	Q-Learning, SARSA etc.	SVM, KNN, logistic regression, etc.	K-means, Apriori, etc.
Application	Self-driving cards, gaming, etc.	Forecast sales, risk evaluation, etc.	Anomaly detection, recommendation system, etc.

15.3.2.1 Classification

This is the case where, when inputs are divided into multiple labels, and the system must create a model that links one or more of these labels to unknown inputs.

15.3.2.2 Regression

This is the case where, when the outputs are continuous rather than discrete.

15.3.2.3 Clustering

This is the case where, when a set of inputs is to be divided into a set of labels. Unlike classification however, these labels are not known prior to processing.

15.3.3 Artificial Intelligence

Artificial intelligence is the process of giving something "intelligent" characteristics. The reasons why it is necessary for humans to create and refine AI are

- Creating expert systems that can teach, model, explain, and provide advice to users is the aim.
- Assisting machines in solving complex problems the same way humans do by applying them as computer-friendly algorithms.

According to the traditional definition of intelligence, a being must be capable of learning, reasoning, solving problems, perceiving, and having linguistic intelligence. However, in order to solve our most challenging problems, artificial intelligence has created a wide range of tools that simulate these abilities. A few of these tools are languages, probabilistic reasoning methods, neural networks, search and optimization, logic, classifiers, and probabilistic learning approaches. Autonomous vehicles, medical diagnosis, art creation, mathematical theorem proof, search engines, gaming, spam filtering, virtual assistants, image recognition in photographs, judicial decision forecast, and targeted online advertising are all examples of how AI is being used in the real world.

15.4 Unity

Unity Technologies created Unity, a multi-platform game development engine. It was released in June 2005 as a Mac OS X-exclusive game engine at Apple Incorporated's WDC (Worldwide Developers Conference) [3]. However, as time went on native support for other platforms was added. Currently (since 2018) the engine has been expanded to work on and with more than 25 different platforms including Windows, Linux, Android etc. While working on a project based on the Unity Engine, one needs to be aware of the following components that come together to create an amazingly creative and immensely powerful workspace:

15.4.1 Unity Hub

It is a standalone program that streamlines the way of finding, downloading and handling Unity Projects and applications.

15.4.2 Unity Editor

It is made up of multiple sub-windows, which all serve different purposes. The most commonly used are: Project Browser, Inspector, Game View and Hierarchy.

15.4.3 Inspector

It reveals the specifics of every Game Object, and modified. This is where the creator can change values to get the best feel for their game. The Inspector displays all the components attached to an object (e.g., scripts, Colliders, Sound, and Physics).

15.4.4 Game View

This provides complete feedback which helps the developer to test their improvements without having to wait to build which execute their project on the target platform.

15.4.5 Scene View

It the screen where the game is designed. The planner is able to drag and drop assets from a view of the building. Familiar 3D handle controls and grid snapping allow the user to bring their items down to the pixel in the ideal location.

15.4.6 Hierarchy

The window in Hierarchy includes a list of all the items in the current scene. This list is changed periodically anytime an item is inserted in the scene. In this window the developer

Figure 15.2 Overview of an environment.

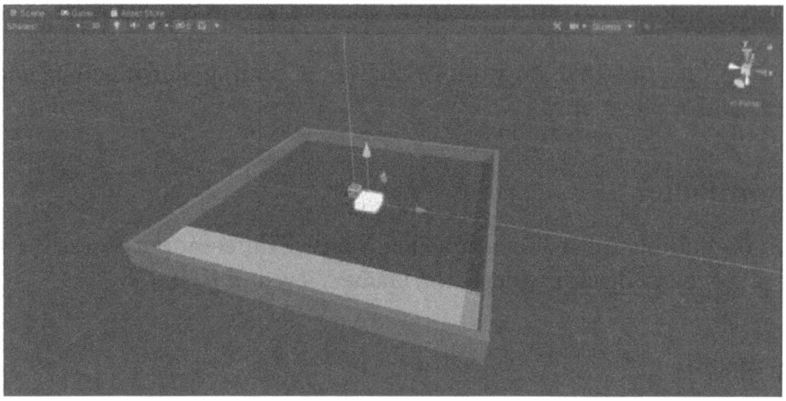

Figure 15.3 Selecting the object in unity.

will delegate parents or children by moving an object on top of another object. Enables nesting and structuring of Game Objects within the scene.

15.4.7 Project Window

The Project window shows all your project-related files and is the primary way to search and locate materials and other project data within your program.

Figures 15.2 and 15.3 show the overview of environment and selecting object in Unity respectively.

15.5 Reinforcement Learning and Supervised Learning

Few parameters along which differentiate between reinforcement learning and the more commonly used supervised learning, so as to be able to gauge a clearer understanding of the definition of reinforcement learning in the given context [4]. These parameters are discussed below:

- Decision style: For decision making in Reinforcement Learning, the process itself helps us in taking our decisions sequentially.
- Works on: Reinforcement Learning, generally, has its working paradigm based on the various ways in which interaction with the environment happens. On the other hand, when talking about supervised learning, its working paradigm is based the pre-existing examples or the given sample data.
- Dependency on decision: In the case of Reinforcement Learning, the decisions are dependent. Therefore, we should ideally give proper labels to all the decisions that are dependent. On the other hand, in SL, the choices are

autonomous of each other, so ideally labels are to be given for each and every decision.

- Best suited: Reinforcement Learning is best suited in the case where it works better in AI where human interface is prevalent. On the other hand, for supervised learning, it is mostly suited in cases where it is functioned with an cooperating software applications.

A relevant example of an application of Reinforcement Learning would be the making of an AI player for a Chess game. A relevant example of the use case of supervised learning would be to train an AI agent for the purpose of object recognition.

15.5.1 Positive Reinforcement

When an event occurs as a result of a desirable behavior, the strength and frequency of that behavior are maximized. Positive reinforcement has the advantage of increasing performance and sustaining changes for a longer period of time. The main disadvantage of positive reinforcement is that it appears that too much reinforcement might lead to situations where there is a "overload of states," which leads to a decrease in results [5, 6].

15.5.2 Negative Reinforcement

It might be explained as the reinforcement of a behavior as a result of preventing or avoiding a harmful circumstance. Positive reinforcement's benefits include maximizing behavior and enabling resistance of the required minimum level of performance. The fundamental drawback of negative reinforcement is that it only seems to produce evidence that is sufficient to meet the minimum behavior standard [5, 6].

15.5.3 Model-Free and Model-Based RL

"Model" here refers to a representation for a computer simulation of the various environmental dynamics. In other words, the model seeks to determine the likelihood of a shift $T(s1|(s0, A))$ from the pair of current state s0 and action A to the next successive state s1. The agent will know precisely how probable it would be to enter a certain state, taking into account the present state and activity, if the probability of transition is effectively learned [7, 8].

On the other hand, model-free algorithms primarily rely on trial-and-error learning to update the knowledge they have. Because of this, these algorithms don't require any additional storage to accommodate all possible combinations of actions and states.

Theoretical comparison of Reinforcement Learning is described in Table 15.2. Table 15.3 describes the difference between real and unreal Engine.

Table 15.2 Theoretical comparison of RL algorithms.

Technique	Characterization	State space	Action space	Policy
A3C	Asynchronous advantage actor-critic algorithm	Continuous	Continuous	On
DQN [16]	Deep Q Network	Continuous	Continuous	Off
Monte Carlo	Every visit to Monte Carlo	Discrete	Discrete	On/off
NAF [13]	Q-learning with normalized advantage functions	Continuous	Continuous	Off
Q-Learning [12]	State-action-reward-state	Continuous	Continuous	Off
Q-Learning—Lambda [14]	State-action-reward-state with eligibility traces	Discrete	Discrete	Off
SARSA [13]	State-action-reward-state-action	Discrete	Discrete	On
SARSA - Lambda [15]	State-action-reward-state-action with eligibility traces	Discrete	Discrete	On
PPO [17]	Proximal policy optimization	Continuous	Continuous	On
SARSA—Lambda [15]	State-action-reward-state-action with eligibility traces	Discrete	Discrete	On
TRPO [17]	Trust region policy optimization	Continuous	Continuous	On

Table 15.3 Contrast between unreal engine and unity engine.

Parameters	Unreal engine	Unity engine
Description	Source available game engine [18]	Cross-platform game engine [19]
Creator	Epic games	Unity technologies [20]
Developmental programming languages	C#	C++ or JavaScript
Features	A robust multiplayer framework, VFX, and particle simulation	2D improvements, animation, creating snapshots
Source code	Open-source	Not open-source
Learning curve	It is difficult to learn, but once mastered, it is extremely powerful	Easy and fun to learn as it has an intuitive interface
Graphics	Graphics are of AAA quality	The graphics are good, but unreal is superior

15.6 Proposed Model

Reinforcement learning is one of the many concepts that come under the expansive umbrella of Machine Learning. RL could be best described as being about taking the most suitable action so as to capitalize on the rewards obtainable in a particular environment. RL is presently being employed by various machines & software to work towards coming up with the best possible decisions, behavior or path that ought to be undertaken in a specific environment. In the SL concept, the agent training data has access to the solution to the problem, so the model is trained with access to the right answer, whereas in the RL concept, there is no solution key and the reinforcement agent chooses what to do on its own to successfully complete the task. Amazingly, the agent is obligated to learn from its experiences because there isn't a training dataset available.

The Proposed Model for creating a virtual environment on Unity is shown in Figure 15.4. The detail description of the proposed model is described as follows:

15.6.1 Setting Up a Virtual Environment

Setting up a virtual environment is important to be able to fully utilize all the functionality of python for the purposes of our project, and to make sure any changes we make are reflected properly in our project code. This process is to be done in the cmd.exe and involves navigation to the installation folder and then running the required command.

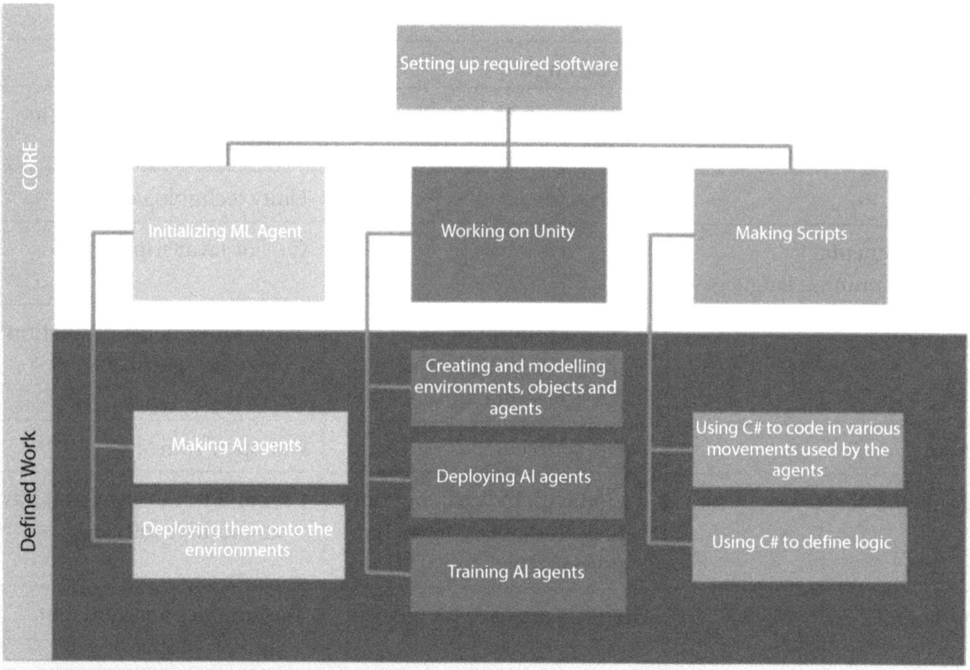

Figure 15.4 Flowchart of the proposed model.

15.6.2 Setting Up of the Environment

Setting up of the environment involves processes like modeling and shaping various three-dimensional shapes to mold the vision of a virtual dimension that we have in our mind. All the three-dimensional models need to be properly placed by hand and their physical properties have to be calibrated while keeping their purpose in mind. Figure 15.5 describes creating scene in environment.

Figure 15.5 Creating a new scene for the environment (PushBlock).

15.6.2.1 *Creating and Allocating Scripts for the Environment*

Unity's integrated development environment (IDE) supports three programming languages: Boo, C#, and JavaScript. Developers can implement the same material using any one of these scripting languages, regardless of their preferences. In this chapter, using C# for scripting functions into the application. We have made extensive use of C# scripting and Microsoft Visual Studio for our application. Scripts may be described as functions that we are programming game objects to perform for the user while playing the game. Let us glance over a few of the scripts used to make the game work as designed.

Scripts are one of the most important parts of creating a functional environment that behaves exactly as we want it to. This is because scripts may be described as the guidelines or set of rules that the object onto which the script is allocated, is expected to follow.

15.6.2.2 *Creating a Goal for the Agent*

The process of goal creation is an important one as it outlines the actual purpose of the environment that we have created. A goal can be described as the outcome or objective that an Agent has to undertake in order to fulfill its purpose. The entire process of machine learning to which the Agent is subjected to is all done so that it may fulfill its goal. Creating a goal is a two-pronged process: one is the creation of the model in the environment, which may be an area or an object or even a point a space; the second is the programming and allocation of scripts to inform the environment that the specified object is the goal.z. Example of creating goal area in shown in Figure 15.6.

15.6.2.3 *Reward-Driven Behavior*

Learning by interaction is at the heart of RL. An RL agent engages in interactions with its surroundings and learns to modify its behavior in response to rewards after witnessing the

Figure 15.6 Creating a goal area (in green).

effects of its choices. One of the basic tenets of RL is based on this paradigm of trial-and-error learning, which has its roots in behaviorist psychology [10]. The general term used are defined as:

1. Action (A): It is all the probable decisions/steps that an agent can take
2. State (S): It is the present state prevalent in the environment.
3. Reward I: It is an instant return sent back by the environment so as to analyze the consequences of the previous action performed.
4. Policy (π): It is the policy that an agent pays to decide its next action grounded on its present state.
5. Value (V): It is the predicted long-term return with total reward (discount), as compared to the short-term reward R. $V\pi(s)$ may be defined as the predicted long-term return of the current states that are underneath policy π.
6. Q-value (Q): Q-value, also called action-value is similar to Value, but different in the fact that it takes into consideration an additional parameter, the current action a. $Q\pi(s,a)$ refers to the long-term return of the current state s, taking action underneath policy π.

15.7 Markov Decision Process

Reinforcement Learning may be formalized as a Markov decision process (MDP), that consists of:

- "S is a collection of states, plus p is a distribution of beginning states (s0)"
- "a series of actions A."
- "$T(st+1|st,at)$ transition dynamics that translate a state action pair at time t onto a distribution of states at time $t+1$."
- "$R(st,at,st+1)$ is an immediate/instantaneous reward function."
- "A discount factor $\gamma \in [0,1]$, with smaller values emphasizing quick rewards [11]."

15.8 Model-Based RL

While talking about model-based RL it is a commonly known that expressive the transition dynamics p(st+1|st, at) makes things more convenient. Models are what these dynamics are. Model-based approaches are algorithms which study the transition dynamics that determine which state would come next after taking action in the current state, s(t), after a given action. When that happens, methods will decide how to select actions. Briefly stated, this type of algorithm study models of system dynamics and employs optimal control to select actions. From the study of optimal control, model-based RL emerges. Typically, specific problems are constructed using models like the Gaussian process and Bayesian network, and then solved using ML techniques or optimal control techniques, such as model predictive control (MPC), linear quadratic regulator (LQR), and linear-quadratic-Gaussian control [12].

15.9 Experimental Results

The final result of the project is a comprehensive and complete application which can run independently on systems of varying configurations and specifications. The application primarily consists of two environments which we can access in real time and observe.

The two environments each independently have two modes of being accessed by the user:

- The first one being the main focus of this project and that is to observe a trained AI Agent in action, repeatedly completing the objective we have set for it.
- The second one being putting the user in the shoes of the Agent and getting to see them act towards completing the same objective repeatedly.

In my honest opinion, while the second mode of access is not a hundred percent essential to this project, it acts as a reminder to us that some real-world processes are mind numbingly repetitive, boring and monotonous. Keeping that in mind, the second mode of accessing thus acts as representation of how we could actually automate and streamline such processes on a larger scale in the real world.

These two environments are joined together by a navigation menu which was also created simultaneously in the Unity Editor using inbuilt functions, scripts and parameter definitions.

The environments we have made, however primitive, serve as a placeholder for a potentially larger project that could be implemented in the real world for various applications such as warehouses management, self-driving cars, automated repair and maintenance, self-sustaining garbage management etc. These applications may sound slightly farfetched right now, but they are a very real possibility in the near future and are viable solutions to pre-existing problems.

15.9.1 Machine Learning Models Used for the Environments

Visualizing the model used for the web browser version of Netron. Netron is a model viewer for deep learning, machine learning, and neural networks. Netron supports a variety of programming languages and frameworks, including ONNX, TensorFlow Lite, Caffe, Keras, Darknet, PaddlePaddle, ncnn, MNN, Core ML, RKNN, MXNet, MindSpore Lite, TNN, Barracuda, Tengine, CNTK, TensorFlow.js, Caffe2, and UFF. Additionally, Netron offers beta support for TensorFlow, PyTorch, TorchScript, OpenVINO, Vitis AI, kmodel, Arm NN, BigDL, Chainer, Deeplearning4j, MediaPipe, ML.NET, and scikit-learn.

15.9.2 PushBlock

The model properties are as follows which includes information about the inputs and outputs. The model starts with vector observation node (shown in Figure 15.7) and end at action and concat nodes (shown in Figure 15.8).

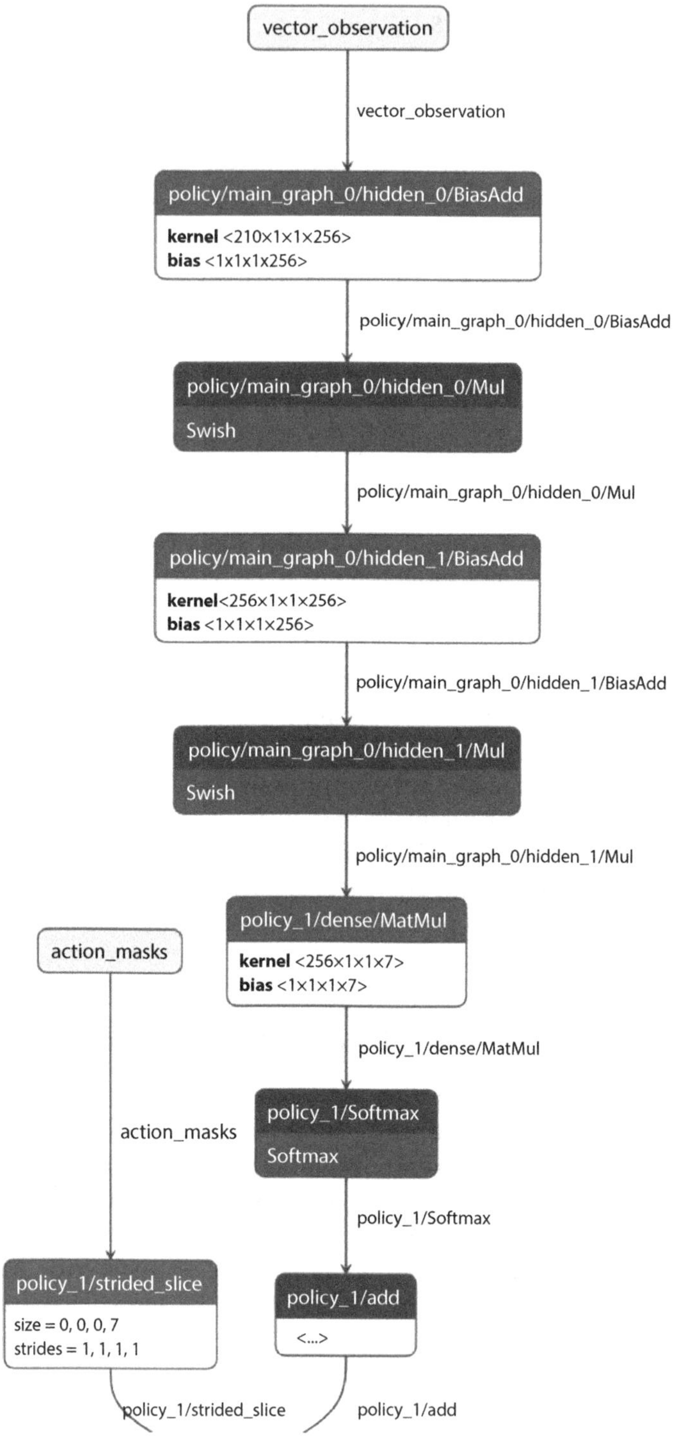

Figure 15.7 The ML model starts from the vector_observation node.

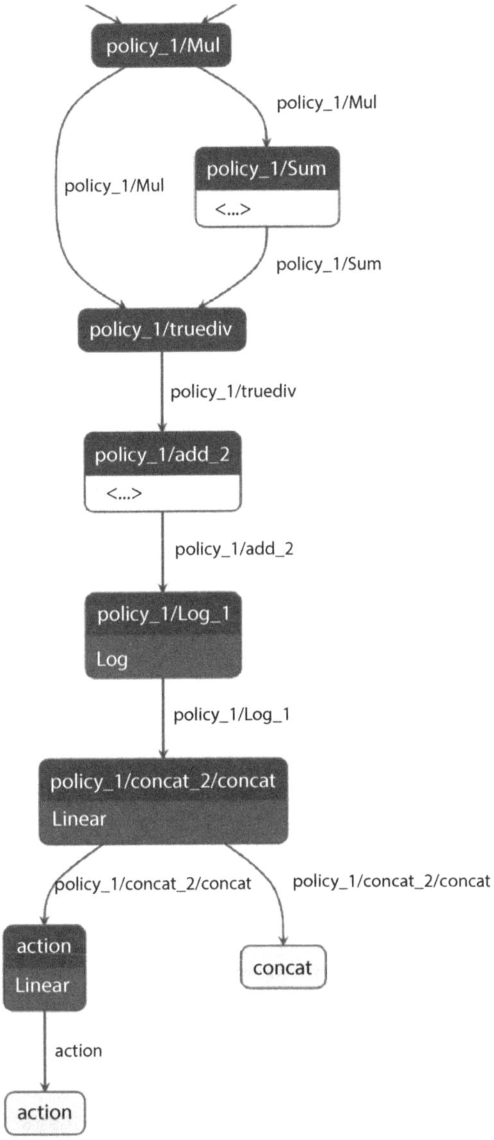

Figure 15.8 The model ends at action and concat nodes.

The policy_/strided_slice and policy_1/add nodes join themselves to the policy_1/Mul node.

15.9.3 Hallway

Now let us look at the ML model of the Hallway environment. The model properties are as follows which includes information about the inputs and outputs shown in Figures 15.9 and 15.10.

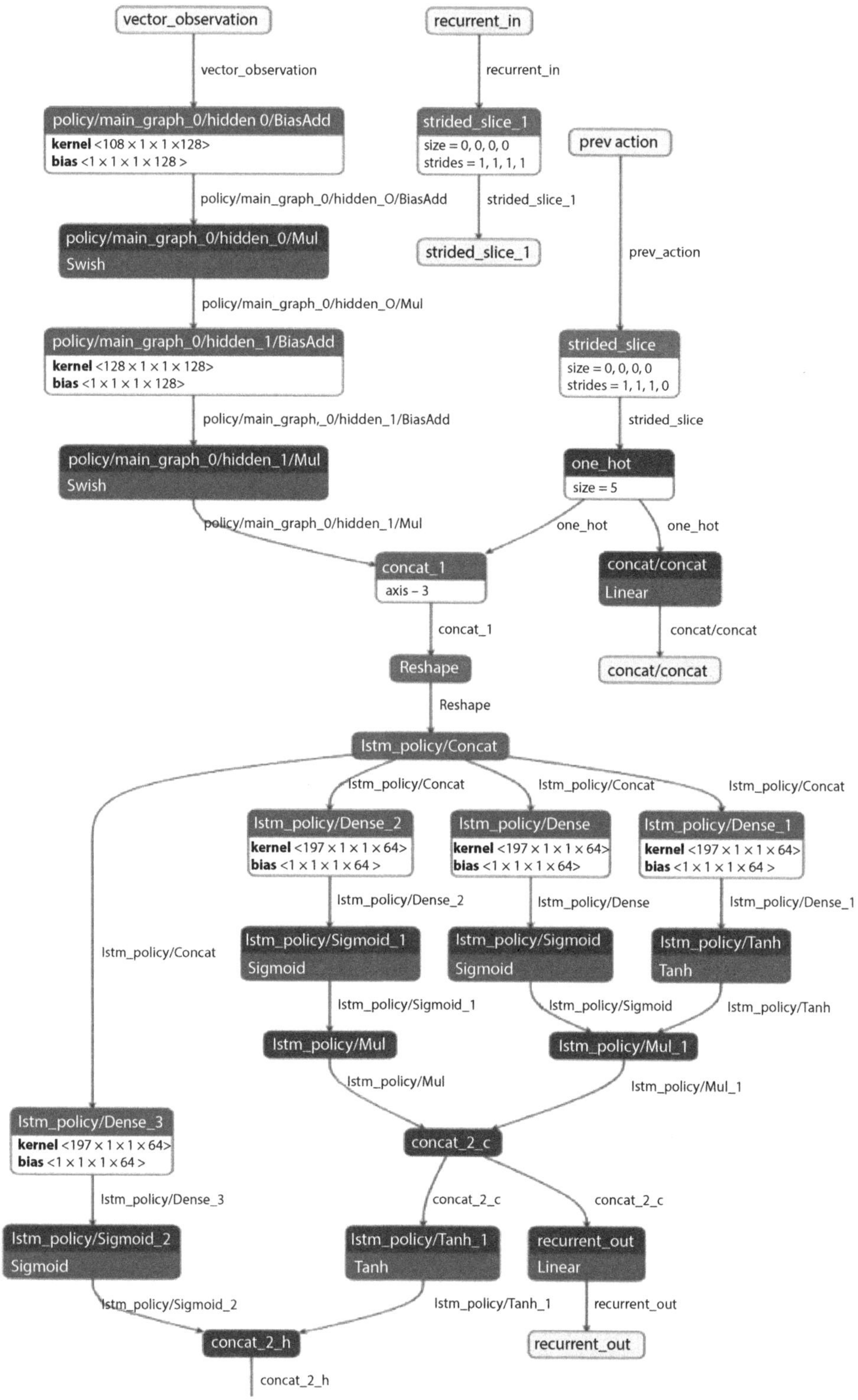

Figure 15.9 The ML model starts from vector_observation, recurrent_in and prev_action nodes.

The concat_2_h node joins into the policy_1/dense/MatMul node.

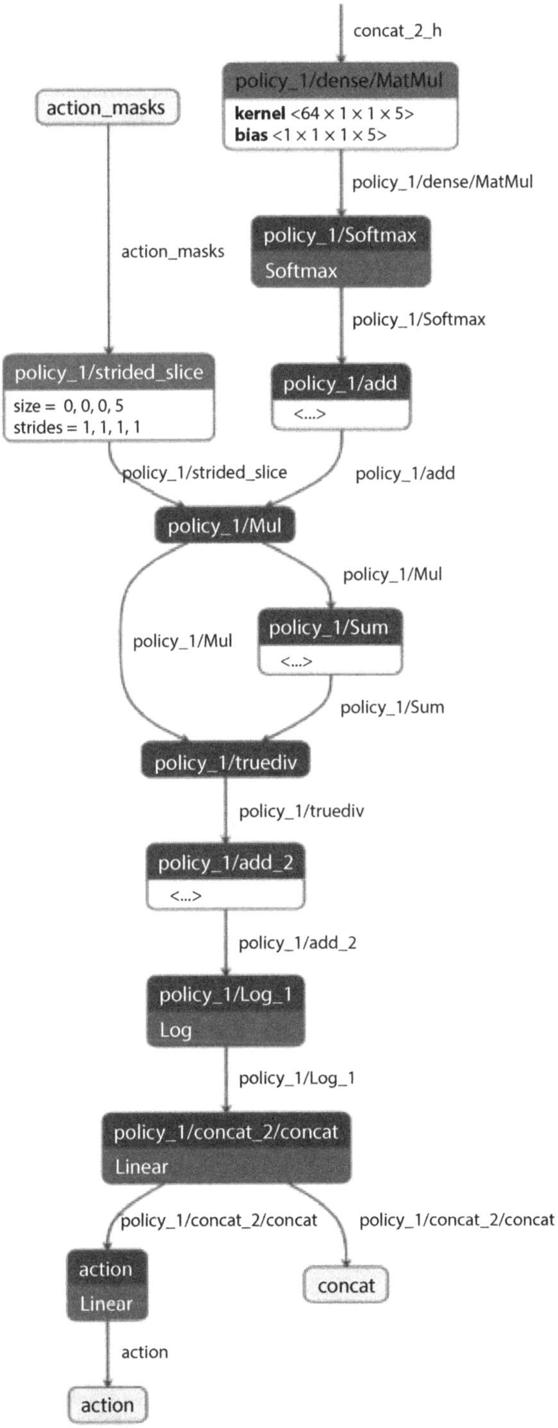

Figure 15.10 The model ends at action and concat nodes.

15.9.4 Screenshots of the PushBlock Environment

The green zone (shown in Figure 15.11) represents the target goal area, and the dark gray zone represents the movable area where the agent and target block spawn. Let us take a quick look at the moving space of the area. Here the blue stylized cube is our agent or player cube (shown in Figure 15.12), and its job is to move the white stylized cube, which is our target block, in the direction of the green zone, which is the objective zone.

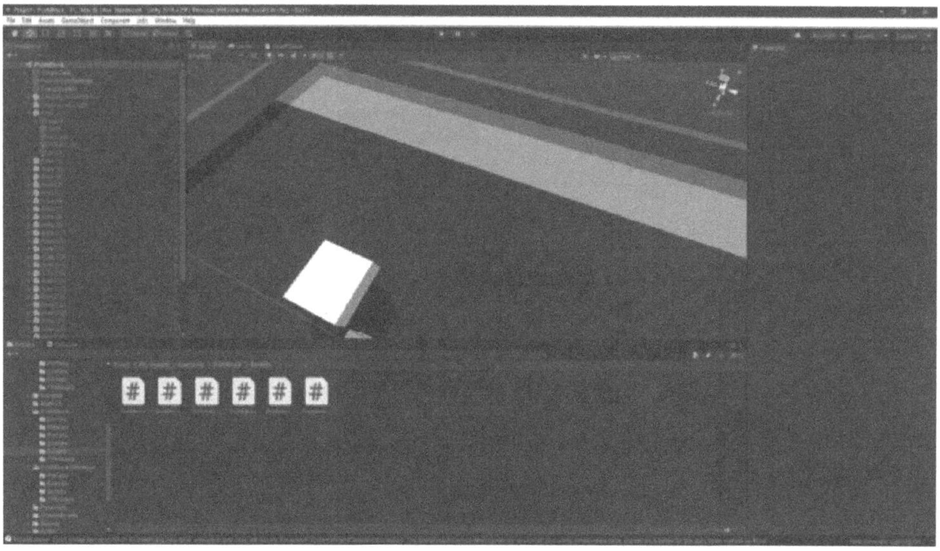

Figure 15.11 Overview of moving space of the area.

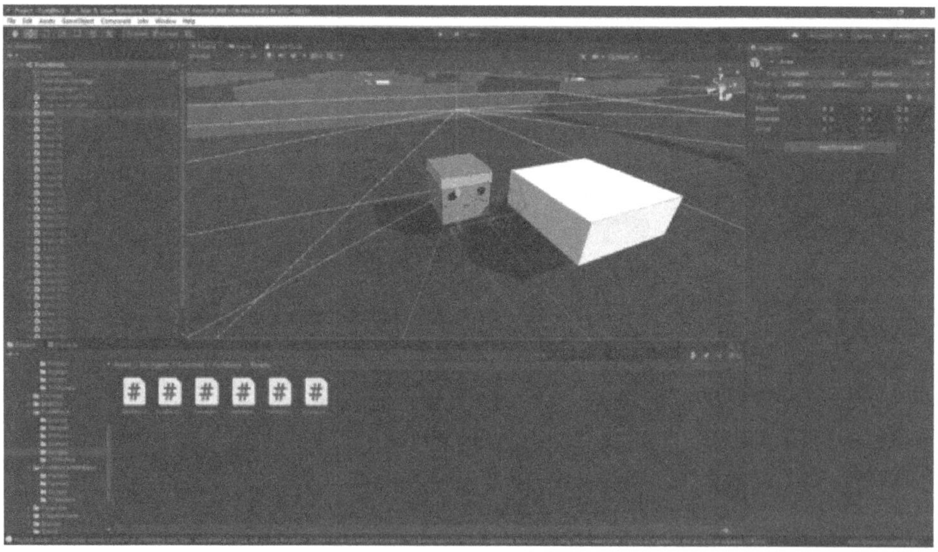

Figure 15.12 Screenshot of the agent and the object interacting.

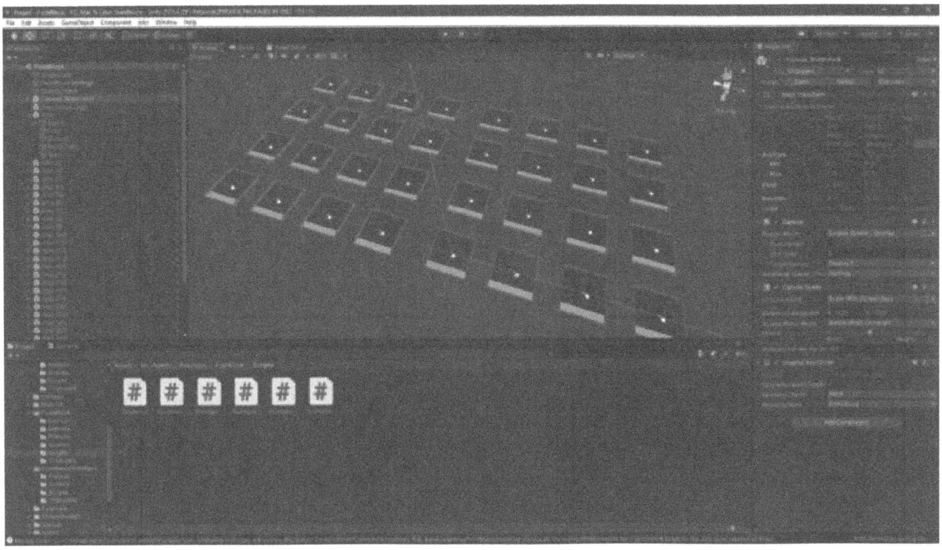

Figure 15.13 Screenshot of the multiple scenarios used for training.

There exist 32 such zones arranged in a neat 8x4 grid which each contain (shown in Figure 15.13):

- Gray zone (movable area)
- Green zone (objective area)
- Blue cube (agent or player)
- White cube (target block)

This is done so as to get a more accurate and precise depiction of the simulation that we are running so that we can get a broader picture of the possible scenarios. Running multiple scenarios simultaneously also makes the AI agent more robust and thus increases its chances of successfully completing the given task.

The player-controlled mode has a much smaller environment consisting of only one zone which resets every time the player completes the objective. This zone is completely identical to the other zones used for training our agent. Starting with camera placed directly on the player Agent, from where target block is located and proceed to push it into the green zone shown in Figure 15.14.

15.9.5 Screenshots of the Hallway Environment

Let us take a quick look at the area's moving space. Green zones stand in for the target goal areas, and the area that can move is where the agent and target block spawn. It is enclosed by light gray walls in the dark gray zone (shown in Figure 15.15).

The blue stylized cube here represents our agent or player cube, which has the task of looking at the image on white cube which represents our target image (either O or X), remembering it, and then moving towards one of the two greens zone which is the objective zone corresponding to the image seen in Figure 15.16.

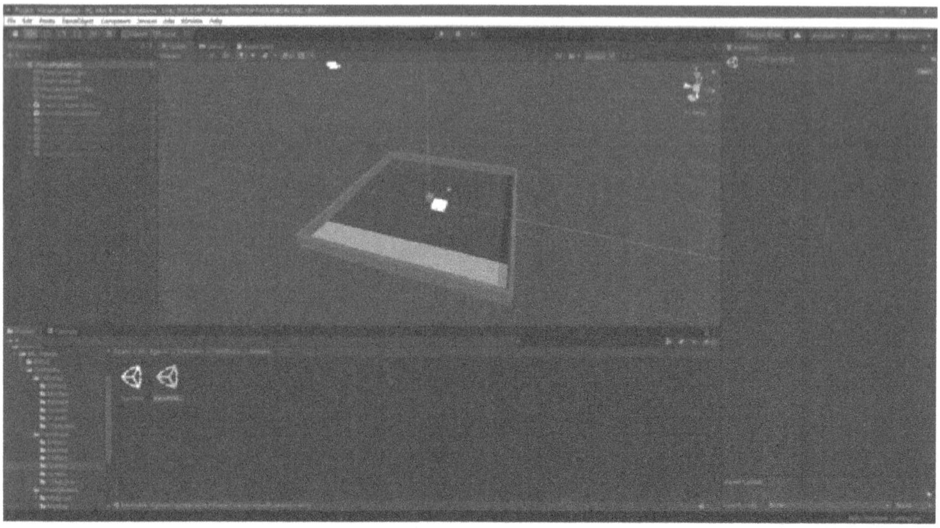

Figure 15.14 Screenshot of the player-controlled PushBlock.

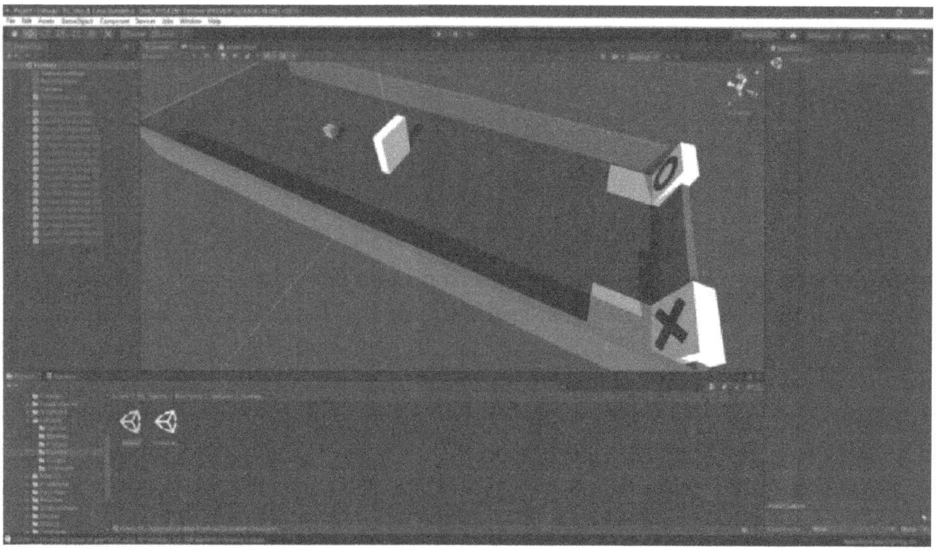

Figure 15.15 Overview of the hallway environment.

There exist 16 such zones arranged in a neat 8×2 grid which each contain (shown in Figure 15.17):

- Gray zone (movable area)
- 2 green zones (target areas)
- Blue cube (agent or player)
- White cube (target image displayer)

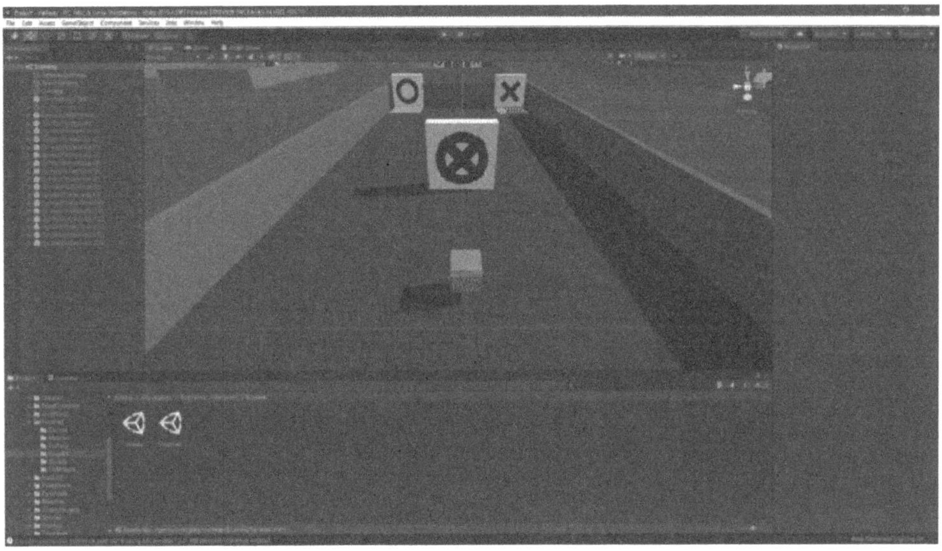

Figure 15.16 Screenshot of the agent and the target images.

Figure 15.17 Overview of training the agent in multiple scenarios.

This is done so as to get a more accurate and precise depiction of the simulation that we are running so that we can get a broader picture of the possible scenarios. Running multiple scenarios simultaneously also makes the AI agent more robust and thus increases its chances of successfully completing the given task.

The player-controlled mode has a much smaller environment consisting of only one zone which resets every time the player completes the objective. This zone is completely identical to the other zones we have used for training our agent. The camera was placed directly

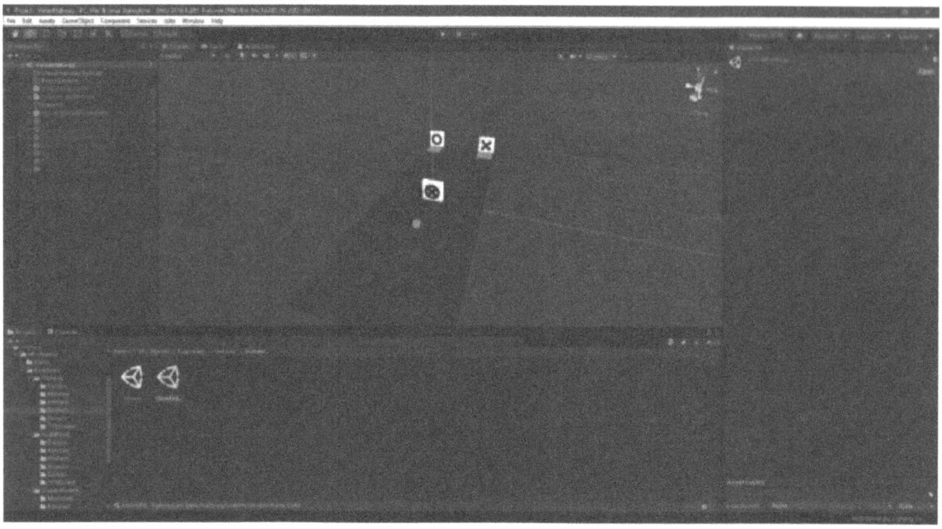

Figure 15.18 Screenshot of the player-controlled hallway.

on the player agent, from where look at the given image and navigate towards the correct answer, i.e., the correct image is shown in Figure 15.18.

15.10 Conclusion

With the aid of Python, PyTorch, and Unity Engine, the suggested model aims to create an intelligent agent that can be trained to perform tasks in a Unity 3D environment via reinforcement learning. Sanitation, engineering, biomedicine, farming, and other fields of application would all benefit greatly from an intelligent agent that could navigate and carry out activities in a variety of environments under a variety of conditions with little to no user input. The task assigned to the agent would be successfully completed, and the workload on the human population would be significantly reduced. This kind of technology may have countless uses in a wide range of global industries.

The world benefits from this study, and it is desperately needed to raise the standard of living. The advantages and cost savings are ground-breaking, but this research is still very much in its infancy, and much more work is needed before something like this can be implemented in the real world.

This project serves as an introduction to what will eventually happen in the real world, which is that artificial intelligence will become more capable and take control of our menial tasks and perform them more effectively than any human could ever hope.

References

1. Kaelbling, L.P., Littman, M.L., Moore, A.W., Reinforcement Learning: A Survey. *J. Artif. Intell. Res.*, 4, 237–285, May 1996, doi: 10.1613/JAIR.301.
2. Sutton, R.S. and Barto, A.G., *Reinforcement Learning: An Introduction*, MIT Press, 1998.

3. Blackman, S., Beginning 3D Game Development with Unity, in: *Begin. 3D Game Dev. with Unity*, 2011, doi: 10.1007/978-1-4302-3423-4.
4. Van Otterlo, M. and Wiering, M., Reinforcement learning and markov decision processes. *Adapt. Learn. Optim.*, 12, 3–42, 2012, doi: 10.1007/978-3-642-27645-3_1.
5. Boakes, R.A., Performance on Learning to Associate a Stimulus with Positive Reinforcement, in: *Operant. Interact*, pp. 67–101, Sep. 2021, doi: 10.4324/9781003150404-4.
6. Fonseca-Júnior, A.R. and Hunziker, M.H.L., Negative reinforcement of behavioral variability without previous training with positive reinforcement. *Behav. Processes*, 204, 104797, Jan. 2023, doi: 10.1016/J.BEPROC.2022.104797.
7. Strehl, A.L., Lihong, L., Wiewiora, E., Langford, J., Littman, M.L., PAC model-free reinforcement learning. *ACM Int. Conf. Proceeding Ser.*, vol. 148, pp. 881–888, 2006, doi: 10.1145/1143844.1143955.
8. Yarats, D., Zhang, A., Kostrikov, I., Amos, B., Pineau, J., Fergus, R., Improving Sample Efficiency in Model-Free Reinforcement Learning from Images. *Proc. AAAI Conf. Artif. Intell.*, 35, 12, 10674–10681, May 2021, doi: 10.1609/AAAI.V35I12.17276.
9. Burkart, N. and Huber, M.F., A Survey on the Explainability of Supervised Machine Learning. *J. Artif. Intell. Res.*, 70, 245–317, Jan. 2021, doi: 10.1613/JAIR.1.12228.
10. Alloghani, M., Al-Jumeily, D., Mustafina, J., Hussain, A., Aljaaf, A.J., Supervised and unsupervised learning for data science, pp. 3–21, 2020, doi: 10.1007/978-3-030-22475-2_1.
11. Plaat, A., Kosters, W., & Preuss, M., High-accuracy model-based reinforcement learning, a survey. *Artif. Intell. Rev.*, 56, 9, 9541–9573, 2023.
12. Jang, B., Kim, M., Harerimana, G., Kim, J.W., Q-Learning Algorithms: A Comprehensive Classification and Applications. *IEEE Access*, 7, 133653–133667, 2019, doi: 10.1109/ACCESS.2019.2941229.
13. Zhao, D., Wang, H., Shao, K., Zhu, Y., Deep reinforcement learning with experience replay based on SARSA. *2016 IEEE Symp. Ser. Comput. Intell. SSCI 2016*, Feb. 2017, doi: 10.1109/SSCI.2016.7849837.
14. Wiering, M.A., QV (λ) -learning : A New On-policy Reinforcement Learning Algorithm. *Proc. of the 7th Eur. Reinf. Learn. (ed. D Leone)*, pp. 17–18, 2005.
15. Jiang, H., Gui, R., Chen, Z., Wu, L., Dang, J., Zhou, J., An improved sarsa(λ) reinforcement learning algorithm for wireless communication systems. *IEEE Access*, 7, 115418–115427, 2019, doi: 10.1109/ACCESS.2019.2935255.
16. Roderick, M., MacGlashan, J., & Tellex, S. (2017). Implementing the Deep Q-Network. arXiv e-prints, arXiv-1711.
17. Liu, B., Cai, Q., Yang, Z., Wang, Z., Neural Trust Region/Proximal Policy Optimization Attains Globally Optimal Policy. *Adv. Neural Inf. Process. Syst.*, 32, 2019.
18. Karis, B., & Games, E., Real shading in unreal engine 4. *Proc. Physically Based Shading Theory Practice*, 4, 3, 1, 2013.
19. dan Rosad, S., Unity 4 Fundamentals: Get Started at Making Games with Unity. *Routledge*, 5, 3, 248–253, 2020.
20. Juliani, A., Berges, V. P., Teng, E., Cohen, A., Harper, J., Elion, C., ... & Lange, D. (2018). Unity: A general platform for intelligent agents. arXiv preprint arXiv:1809.02627.
21. François-Lavet, V., Henderson, P., Islam, R., Bellemare, M. G., & Pineau, J., An introduction to deep reinforcement learning. *Foundations and Trends® in Machine Learning*, 11, 3-4, 219–354, 2018.

16

Emerging Technologies in Healthcare Systems

Ravi Kumar Sachdeva[1], Priyanka Bathla[2*], Samriti Vij[1], Dishika[1], Madhur Jain[1], Lokesh Kumar[1], G. S. Pradeep Ghantasala[3†] and Rakesh Ahuja[1]

[1]*Chitkara University Institute of Engineering and Technology, Chitkara University, Punjab, India*
[2]*Chandigarh University, Mohali, Punjab, India*
[3]*Department of Computer Science and Engineering, Alliance College of Engineering and Design, Alliance University, Bengaluru, India*

Abstract
This book chapter provides an overview of emerging trends and opportunities in healthcare research. The objective of the chapter is to explore the impact of emerging healthcare technologies in optimizing patient care. It focuses on addressing challenges and setting future directions for healthcare research. The chapter discusses several subjects, such as the growing significance of patient-centered care and personalized medicine, as well as the application of artificial intelligence (AI), machine learning, and big data analytics to enhance patient outcomes. Examples include the use of AI for drug discovery, machine learning algorithms for predictive analytics, and Internet of Things (IoT) applications related to healthcare enabled by big data analytics. Additionally, it looks at how wearables, telehealth, and apps for mobile health might change the way that healthcare is provided and make care more accessible. The chapter also emphasizes the significance of covering social aspects of health to improve health equity and lessen health disparities as well as the necessity of interdisciplinary collaboration. Finally, the chapter discusses challenges and recommendations related to healthcare research. Overall, the book chapter touches on emerging directions in healthcare research standards, including benchmarks, challenges, and recommendations for advanced technologies, offering a reflective analysis of recent progress and envisioning future mentoring in healthcare.

Keywords: Artificial intelligence, machine learning, personalized medicine, immunotherapy, regenerative medicine, telemedicine

16.1 Introduction

Health research affects society significantly by providing vital data on sickness patterns and causes, treatment efficacy, public health efforts, functional abilities, care models, and healthcare expenditures and utilization. Different research approaches present various points of view. Delivering high-quality, cost-effective healthcare services is becoming increasingly

**Corresponding author*: priyankabathla85@gmail.com
†*Corresponding author*: ggspradeep@gmail.com

Shubham Mahajan, Pethuru Raj and Amit Kant Pandit (eds.) Deep Reinforcement Learning and Its Industrial Use Cases: AI for Real-World Applications, (375–394)

difficult, but current research directions in healthcare have the potential to solve this issue. Developments in health information technology are facilitating a shift in health research that may allow previously unthinkable investigations, yielding novel insights into health and disease. Artificial intelligence (AI) has contributed to the advancement of many areas, so it is not unexpected that it has the potential to change medicine. AI can recognize patterns and trends that humans cannot, resulting in major advancements in illness diagnosis and healthcare [1]. The prospects for the future of healthcare are boundless, thanks to constant discoveries and developments [2].

This chapter concentrates on several important fields of healthcare, such as personalized medicine, immunotherapy, regenerative medicine, etc. The chapter first discusses personalized medicine. Beginning with a focused examination of personalized medicine and patient-centered care, the chapter highlights the importance of tailoring medical interventions to individual characteristics. Next, it discusses the use of AI and machine learning (ML) in the healthcare sector, showcasing their applications in optimizing patient outcomes. The exploration extends to the realm of immunotherapy, and its implications for the future of medical interventions. A dedicated section on regenerative medicine follows, emphasizing the groundbreaking potential of harnessing the body's own regenerative capabilities for therapeutic purposes. The chapter seamlessly transitions to the transformative influence of wearables, telehealth, and mobile health applications on healthcare accessibility. Highlighting the imperative of addressing health disparities, the chapter delves into the social aspects of health, advocating for a comprehensive approach to improve health equity. After that, the chapter discusses future directions in healthcare research standards. Furthermore, it discusses standards, challenges, and recommendations for advanced levels of healthcare technologies and the status of healthcare technologies in developing and underdeveloped countries. The last section compares recent progress and future mentoring in healthcare using technology.

16.2 Personalized Medicine

According to pharmacokinetic, pharmacodynamic, and pharmacogenomic data, personalized medicine simply refers to the prescription of certain treatments that are best suited for a given individual. A vast and quickly developing area of medicine, personalized medicine, draws on the specific clinical, genetic, chromosomal, and environmental data of each patient. An integrated, organized, and scientifically proven strategy for customizing patient treatment throughout the transition from health to illness is personalized medicine. These objectives require overcoming several challenges. The most significant genetic markers for clinical use, reducing the side effects of gene-based medicines and performing clinical research to find genetic variations associated with therapeutic response, are a few examples of scientific difficulties in this area. The information in Figure 16.1 relates to personalized medicine's diagnosis, treatment, and prevention.

Although the fact that DNAs from various cells are identical, the way that genes function in different organs varies. Although many cancerous tumors may share comparable DNA, they exhibit diverse patterns of gene expression. Researchers can study the gene expression profiles of hundreds of genes at once and separate a melanoma gene expression record from typical profiling thanks to technologies like genomic microarray analysis. Cohort-based

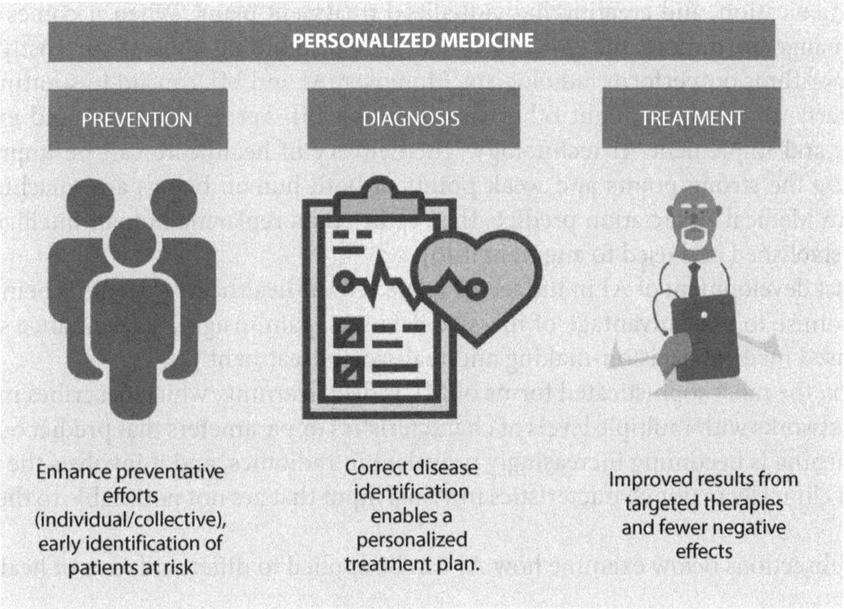

Figure 16.1 Prevention, diagnosis, and treatment of personalized medicine.

observational studies, in which particular genetic diversity is not taken into consideration and the majority of the results are established at the societal level, have guided routine medical care for decades. Contemporary personalized medicine considers a patient's genetic makeup and medical history before prescribing a course of action [3].

A straightforward but essential method for providing information on personal health risks is family health history (FHH). A strong FHH can approximate biological and genomic hazard information and incorporate it into medical safety since it reflects the complex interaction of hereditary, environmental, and behavioral variables. FHH evaluations would help to identify those who are more susceptible to disease, allowing for proactive and preventative actions such as dietary adjustments, health checks, testing, and early intervention when necessary.

Three key elements are required to successfully integrate FHH into public health:

- accessible, standardized collecting techniques;
- access for health care professionals; and
- clinical advice for analysis and usage.

Presently, a participant's medical providers may obtain FHH data that is insufficient, is challenging to understand, or contains vastly different information [4].

16.3 AI and ML in Healthcare Sector

Various studies have demonstrated that AI has the potential to perform crucial tasks in the sector related to healthcare, such as clinical decision support, assisting in medical diagnosis, finding medication, and creating individualized treatment plans. When it comes to diagnosing malignant tumors and guiding researchers in building cohorts for costly clinical trials, algorithms outperform radiologists. Moreover, AI and ML can aid in spotting trends and patterns that people might otherwise overlook [5]. Yet it is difficult and expensive to create and implement AI technology. The delivery of healthcare can be improved by combining the strong points and weak points of both human beings and machines. The American Medical Association predicts that, rather than replacing human intelligence, AI will be established and used to augment it [6].

The fast development of AI in the sector related to the healthcare systems is being driven by its promise to take advantage of massive data sets, gain insight, and enhance scientific proof-based medical decision-making and real-worth treatment [7].

Among the most sophisticated forms of ML is deep learning, which describes models of neural networks with multiple levels of characteristics or parameters that predict outcomes. Deep learning is becoming increasingly prevalent in radionics, and it involves the identification of clinically major characteristics in visual input that are not noticeable to the human sense.

The subsections below examine how AI can be applied to different facets of healthcare.

16.3.1 AI in Medical Diagnosis

AI is frequently used in medical analysis [8]. In healthcare, it helps with illness detection. It also helps to address the issue of misdiagnosis by using automatic AI and medical screening [9]. It is frequently used to help with different types of cancer diagnosis using ultrasound or X-ray imaging. In this case, supervised learning classification helps to determine whether an individual is diagnosed with cancer, by using a discrete binary label [10]. An example of the utilization of ML in medicine is the forecasting of Alzheimer's disease [11]. There are a few issues to consider when utilizing ML to detect diseases. First off, AI cannot take the position of a doctor. While ML can assist in forecasting the probability of a disease, it cannot replace a professional in all aspects of their duties. [12].

The Internet of Medical Things (IoMT) integrates network-connected biomedical hardware and software to enhance human wellness, including heart monitoring, cancer scare, and diabetes control. The contribution of AI is to support sophisticated robotic procedures created for cutting-edge biomedical applications. IoMT provides healthcare monitoring without the need for human interaction. Smart wrists, electronic fabrics and clothing, smartphone-integrated gadgets, and sports wristwatches for activity and fitness tracking all fall under the category of wearable personal health monitoring devices. Body sensors connected to the individual's body, woven into clothing, or surgically implanted in people are some other kinds of sensors that can be used. Smart sensor technology is applied to examine the data before it is uploaded to the cloud. The body sensors communicate with the output recipient over the cloud [13].

16.3.2 Drug Discovery

One of the most significant translational science endeavors that can improve human well-being is the discovery and production of new drugs. Through increasing productivity in the production of therapeutic drugs AI has proven its value in the healthcare sector.

Table 16.1 Significant tools for drug discovery.

Sr. no.	Name	Description
1.	DeepChem	A multilayer perceptron (MLP) model that employs an AI system built on Python to identify potential candidates for drug development
2.	DeepTox	Software that estimates the toxicity of 12,000 different medications
3.	DeepNueralNetQSAR	A Python-based system with computational power to identify a substance's chemical activity
4.	Organic	A chemical synthesis tool that helps to create compounds with the desired properties
5.	PotentialNet	Predicts ligand binding affinity using neural networks
6.	Hit Dexter	Uses ML to anticipate compounds that may respond to biochemical assays
7.	DeltaVina	Uses a scoring system to determine the drug–ligand binding capacity
8.	Nueral Graph Fingerprint	Aids in predicting the characteristics of new compounds
9.	AlphaFold	Protein 3D structure prediction
10.	Chemputer	Aids in the standard reporting of chemical synthesis procedures

AI is already being used in several stages of drug development, from drug design to drug screening, and has already demonstrated its potential. Table 16.1 lists various significant tools that are useful in the search for new drugs [14].

16.3.3 Personalized Treatment Plans

Personalized healthcare (PH) is a healthcare method that is patient-centric and its aim is to advance the existing health system. It emphasizes the patient data amassed through IoT sensor devices, wearable, electronic health records, portable devices, internet-based information, and social networking sites. To improve the processes for early diagnosis, and progression of the disease, PH uses AI methods to analyze the obtained dataset. The development of analytical models in this context frequently uses ML techniques. Numerous supports for clinical decision systems and applications for healthcare services use these models [15]. These models identify the medical condition and patterns of behavior of the patient by analyzing data collected from various sources, including sensor devices. Clinical decision support and healthcare applications suggest lifestyle modifications, specialized care, and patient care plans based on those patterns. To bolster lifestyle recommendations, the care plan creation process might also involve doctors and carers [16].

16.3.4 Pattern Matching or Trend Detection

Early diagnosis and treatment can drastically alter one's perspective of the world in the healthcare industry by lowering the mortality rate brought on by numerous chronic diseases [17]. There are specialized exams that can detect the disease while it is asymptomatic—that is, when it has neither symptoms nor signs. Using statistical theories, ML creates models from data sets. Using statistical and predictive models, computers can reliably clean, curate, analyze, and forecast the outcomes of certainties (or uncertainties). Modern machine learning and pattern recognition techniques have made it possible to perform precise and non-intrusive disease detection and monitoring using a variety of data sources, including speech, language, movement, gait, handwriting, video, neural activity (EEG, electroencephalography), and others [18].

16.4 Immunotherapy

Cancer is the second most common cause of death worldwide. The majority of patients, particularly those with primary tumors can be treated with a mixture of surgeries, radioactivity, and chemotherapy. Immunotherapy, laser, targeted therapy, and other methods are also available [19]. Immunotherapy is a type of cancer treatment in which chemicals produced in a lab or by the body are used to strengthen the body's defenses and help the body destroy cancer-causing cells. Before employing the immune system's components to treat the disease, cancer immunotherapy aims to change the patient's immune system. Figure 16.2 depicts a simple representation of cancer.

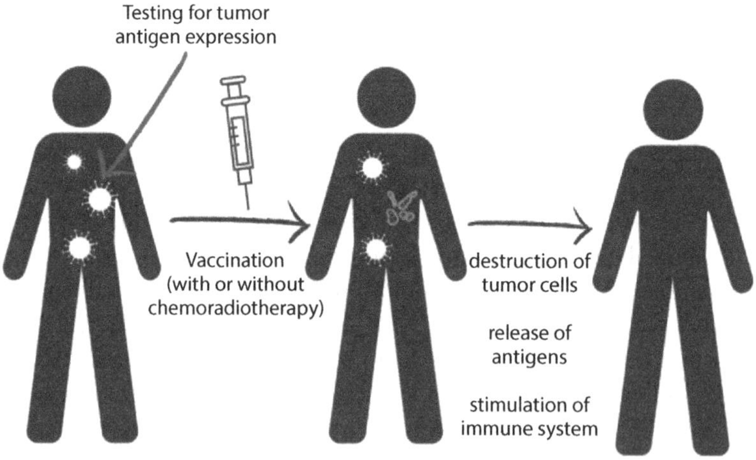

Figure 16.2 Simple representation of cancer.

Immunotherapy comes in a variety of ways, such as the following:

- Inhibitors of immune checkpoints and monoclonal antibodies
- General immunotherapies
- Viral oncolytic treatment
- T-cell treatment
- Vaccines for cancer

16.4.1 Monoclonal Antibodies

A monoclonal antibody is a protein made in a lab that can bind to targeted agents in the body, like antigens found on the exterior of cancerous cells.

Monoclonal antibodies are permitted for curing diseases involving many organ systems, such as the cardiovascular, respiratory, haematological, renal, immunological, and oncological systems. They are approved to treat orphan diseases or indications, such as multiple sclerosis, asthma, rheumatoid arthritis, and breast cancer, which affect millions of people. They can also treat cancers and various sclerosis, which impact hundreds of people [20]. Since antibodies are proteins, sometimes giving them causes a sensitive reaction. The potential side effects include, but are not limited to, sickness, cold, fatigue, headache, vomiting, diarrhea, decreased blood pressure, sores, etc.

16.4.2 Checkpoint Inhibitors

An immunity system's common function is immunological checkpoints. They are there to protect healthy cells in the body from being harmed by an overreaction of the immune system. Proteins on the outermost layer of immune cells known as T cells recognize and bind partner proteins on different cells, such as certain tumour cells, to activate immunological checkpoints. These proteins are known as immune checkpoint proteins. Once the partner proteins and checkpoint bind, the T cells get an "off" signal. As a result, the immune system may have a harder time eliminating the cancer [21].

16.4.3 CAR-T Cell Therapy

A successful immunotherapy that goals cancers at both the cellular and inherited levels is required due to the complicated behaviour of cancers and the participation of various hereditary and cellular variables. During CAR-T cell therapy, T cells extracted from the blood of the individual are altered in vitro to create synthetic receptors specific to a given tumour antigen. CAR-T cell therapy is an option for treating certain lymphomas, leukemias, and multiple myeloma [22].

16.5 Regenerative Medicine

It is an exciting study field that involves growing and replacing damaged or diseased cells, tissues, or organs. This includes techniques such as stem cell therapy and tissue engineering [23]. With this, affected body parts can be recovered to get things to work back to normal. Potential applications include the treatment of congenital abnormalities as well as the replacement or repair of tissues and organs injured by disease, aging, or trauma. The potential for curing both chronic illnesses and acute insults, treatments for various cancers, and more is backed by encouraging preclinical and current clinical results [24]. Figure 16.3 represents two approaches, i.e., cell therapy and tissue engineering, to the use of stem cells.

This covers operations, surgical devices like artificial joints, and complex biomaterial scaffolds. Also, it has to do with tissue engineering and is inspired by medical practices like bone marrow and organ transplants.

Each of these treatments leaves behind residual effects that may prevent the patient from being regarded as being healthful naturally concerning the ailment they were treated for, but there is no specific moment at which they stop evolving into fully-fledged regenerative medicine.

Figure 16.4 represents the ingredients of regenerative medicine tissue engineering: human tissues, cellular therapy, stimulation of endogenous repair, gene therapy, xenotransplantation, and biologic-device combination products. Due to its potential to circumvent direct transplantation's limitations of a lack of donors and immune complications, regenerative

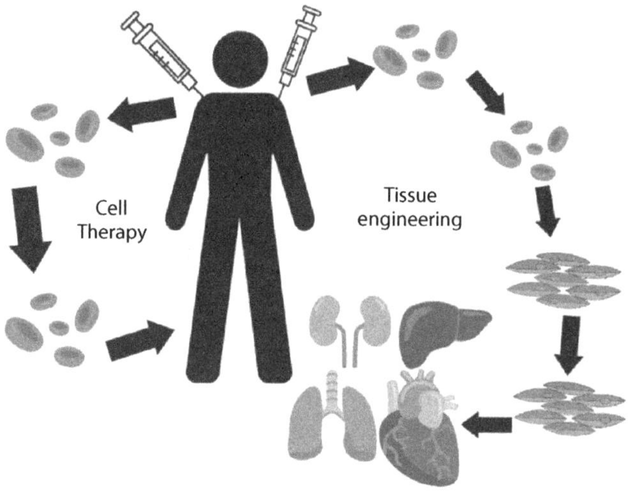

Figure 16.3 Two approaches to the use of stem cells.

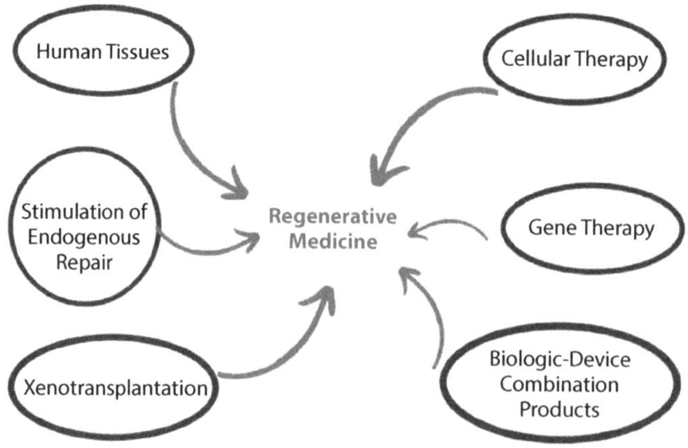

Figure 16.4 Regenerative medicine tissue engineering and molecular biology.

medicine has attracted a lot of attention recently. The rapid development of living materials—which are made up of living biological agents and can be created in conjunction with synthetic materials to satisfy the application requirements of regenerative medicine—is required by the field's ongoing advancement [25].

16.6 Digital Health (Use of Technology in Healthcare)

16.6.1 Wearable Devices

Step count, pulse rate, and other health-related metrics are all quietly measured by smart watches and other wearable tech. In essence, smart technology collects user health data and transmits it to a clinic or doctor. Additionally, it helps clinicians completely understand their patients. Additionally, these gadgets have some safety features that can help the user in an emergency, like alerting someone when the user's heart rate rises and calling an ambulance in the event of a mishap. In numerous times, smart technology has prevented fatalities. Monitoring our health is feasible thanks to wearable technology. The future of healthcare will be shaped by them [26]. Figure 16.5 shows various innovations in digital health: mHealth, wearables, clinical trials, wellness, telehealth care, health analytics, and visualization. Wearable technology also has some drawbacks.

- Privacy concern
- They may be costly.
- Uncomfortable
- They might not be correct [27].

Figure 16.5 Digital health innovations.

16.6.2 Telemedicine

People can remotely consult a doctor using telemedicine while relaxing in their own homes. Online appointments are used here, and a doctor virtually joins the patient from his clinic. This technique is currently quite popular. Many healthcare facilities and professionals now offer this service. With this, a person does not need to organize child care, take a leave of absence from work, or drive to the doctor's office or clinic and wait for his appointment to arrive. It worked well during the COVID-19 epidemic.

Also, numerous online resources let patients self-diagnose a variety of conditions, including depression and mental illnesses. They ask the patient regarding their health and give them the diagnostic test results. Many platforms provide counseling to patients with depression or any mental illness [28]. Figure 16.6 depicts different components of digital health.

16.6.3 Electronic Health Records

The patient records in the database are known as electronic health records. In essence, it is a digital replacement for paper records. It facilitates the doctor's comprehension of the patient's issues. It includes all the patient's medical information, such as test results, prescriptions currently being taken, past medical histories, and other information pertinent to the patient's health. It enables doctors to retrieve patient information and medical history quickly and easily from the database. It improves patient care and outcomes. The health sector very quickly adopted this practice. This has the potential to increase access to healthcare and improve patient engagement [29]. It also has some shortcomings, such as the following:

- Issues with cybersecurity
- Frequent updating is necessary
- What happens if technology fails? (power outage, loss of Internet connectivity)

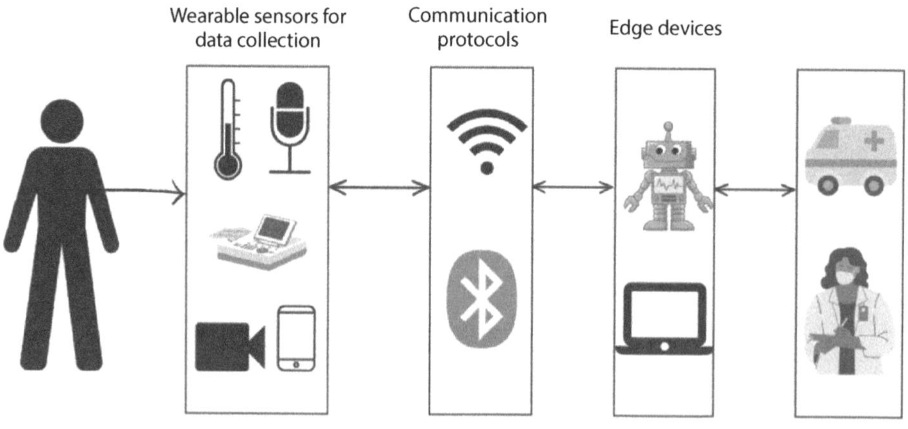

Figure 16.6 Digital health: Telemedicine with the use of wearable sensors.

16.7 Health Inequity

16.7.1 Health Disparity

If the word "disparity" is searched in a dictionary, it is likely to be defined as nothing more than a difference, a variation, or possibly inequality. However, the phrase "health disparity", coined in the US in the 1990s, was not intended to include every instance of racial or ethnic disparity in health. Rather, the intention was to highlight a specific kind of variation, i.e., lower healthiness among socially and economically disadvantaged individuals within all racial and ethnic groupings as well as among members of disadvantaged racial/ethnic groups.

Health inequalities are not always differences in health. A few instances of health inequalities that do not correspond to health disparities include the lower health of older people compared to younger people and the higher prevalence of arm injuries among professional tennis players compared to the general population.

Ambiguity over the definitions of health disparities and health equity could allow for the misuse of scarce resources. For instance, if these phrases are not clearly defined, socially and economically privileged groups may appropriate them and push for funding to support the health needs of their privileged group [30].

16.7.2 Health Equity

Health equity refers to the goal of reducing and eventually destroying inequities in health and its factors, such as socioeconomic determinants. Aiming for the highest level of health for each person while giving particular attention to those who are more likely to become ill due to their socioeconomic circumstances is the pursuit of health equity.

16.8 Future Directions in Healthcare Research

Overall, the future of healthcare research is promising and exciting, with many opportunities for discoveries and innovations. As technology advances and our understanding of diseases and treatment improves, we can expect to see many more breakthroughs in the years to come. Following are different directions in which we see the breakthroughs in healthcare:

- Remote surveillance: Physicians would gain deeper comprehension of their patient's health with the assistance of customized hardware and software that will read data from medical cards in real time [31].
- Healthcare accessories: There are many products available that can monitor patients' daily activities in real-time and store the data. These tools give patients knowledge of their regular activities. They can also help to prevent emergencies because the doctor would immediately obtain the patient's information [32].
- Asset management: IoT can help by providing controllers and functionality for various pieces of critical medical equipment. The device is vital to the course of treatment, thus any flaw in it could be catastrophic. Once connected,

the employees will be able to easily monitor their work. Additionally, real-time fault detection reduces the likelihood of unintentional therapy [33].

- Better reporting and oversight: Real-time monitoring utilizing loT devices can save lives in circumstances where there is a medical emergency, such as an asthma attack or heart failure. The connected device can collect important data regarding a patient's health and immediately send it to the physician.
- Communication from end to end: The patient care process can be automated with the use of loT and healthcare mobility technologies. It creates medical service interoperability, machine-to-machine connection, data mobility, and information exchange are all made possible, increasing efficiency. Thanks to numerous connectivity protocols in the gadgets, hospital workers can recognize patients who are displaying early signs of illness.
- Analysis of statistics: Through fast data collection, reporting, and analysis, loT devices eliminate the requirement for data storage. Medical practitioners will be able to focus on the important information they need to handle the patient as a result. The decision-making process for doctors will be accelerated by the data-driven conclusions.
- Monitoring and alerts: In situations where there is a risk to one's life, prompt warnings may be required. IoT facilitates It is feasible for medical equipment to collect crucial information and transmit it to clinicians in real time.
- Reduce costs: Wearable technology along with other connected devices will allow patients to consult with physicians from the comfort of their homes. The number of visits required for various tests and examinations will reduce. Everyday time and financial savings will benefit patients.
- Management of medicines: Smart wireless pill bottles will make it easier to keep track of a patient's prescription schedule. This will help people who forget to take their medications on time. Additionally, the IoT-enabled pharmaceutical management systems will provide doctors with information so they can treat a patient's health [34].

16.9 Challenges and Recommendations for Advanced Level of Modern Healthcare Technologies

Healthcare systems are now more effective than they were before because of recent developments in computers and wireless communications. Without any human assistance from medical personnel, modern healthcare technologies can automatically monitor and manage a patient's various health issues. Also, the whole patient monitoring and treatment process is enhanced via means of IoT, body area networks, and implanted medical devices in healthcare systems [35]. Modern healthcare technologies include:

- Portable monitors
- Wearable devices
- Electronic health records
- Automated IV pumps

- Telehealth and apps
- Smart bio patches
- Smart hearing aids

Figure 16.7 shows different components of modern healthcare technologies. Worldwide interest in and usage of telehealth for medical diagnosis, treatment, and rehabilitation has grown since it was created and successfully implemented in clinical settings. Offering remote healthcare apps and services has become easier with the development of advanced signal processing methods, high-speed communications, and current developments in cloud computing. In this environment, affordable medical diagnosis, and care options in addition to healthcare services are crucial and must be extensively implemented.

16.9.1 Challenges

The asymmetry between healthcare facilities, hospitals, and user-ends continues to pose a serious obstacle to the broad use of this technology, particularly in poor nations. Due to the complexity of the software and hardware used in these systems, healthcare systems need to balance security, privacy, and therapy since any security or privacy breach might hurt patients' treatments and general health conditions. The healthcare industry's technology and apps have several design flaws and inadequate security measures that make them vulnerable. The healthcare domain is, in fact, rapidly experiencing security difficulties and dangers [36]. Following are the major challenges being faced by modern healthcare technologies

- AI-enhanced software: Technology and artificial intelligence have dominated every industry, boosting the productivity and efficiency of every activity. To handle the growing data, the health industry also urgently needs more AI-integrated software tools. When these data sets start to grow quickly, better tools are needed for managing and storing administrative data, clinical notes, and medical records.

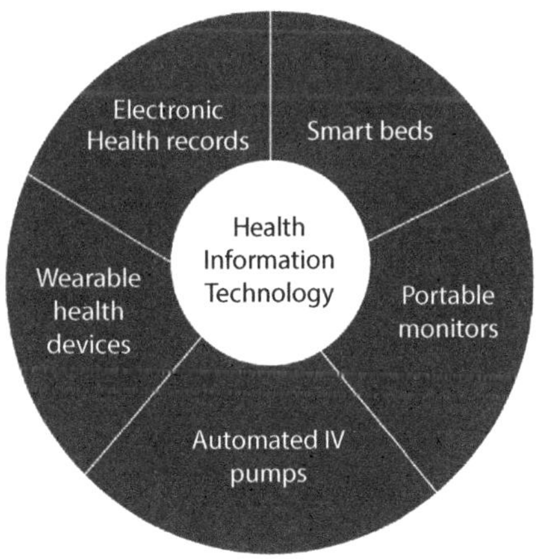

Figure 16.7 Modern healthcare technologies.

- Expensive healthcare: The growing cost is another urgent problem in this sector. The growing expense of healthcare hinders patients from having normal health exams and maintaining up with follow-ups. The outcome is managing, preventing, and monitoring health-related problems is more difficult.
- Cybersecurity: Cyberattacks and data breaches are two downsides of the digital age that require rapid and efficient security measures to combat. Hospitals that engage in these deliberate assaults face the risk of harming their image and usually incur significant penalties. As a result, healthcare organizations must invest in cybersecurity, which includes installing multi-factor authentication and potent firewalls.
- Better patient experience: The medical infrastructure is now better and more up-to-date thanks to reforms made to the healthcare industry. Additionally, patients have higher expectations of the treatment they receive. They want services to be straightforward and easy to access, like getting appointments, getting records, scheduling testing, and paying medical bills. Hospitals can draw more patients and expand their patient base with a simplified user experience.

16.9.2 Recommendations

Some actions that may be made to enhance healthcare include:

- Creating a healthy physical environment
- Infrastructural development
- Decentralizing public health services
- Improving the doctor–patient ratio [37]

16.10 Healthcare Sector in Developing and Underdeveloped Countries

16.10.1 Healthcare Sector in Developing Countries

Developing nations are among those most severely impacted by the pandemic and impacted by its aftereffects on a wide scale. The actual number of casualties in humanity will depend on the capacity to limit the transmission of disease and the usefulness of health services. The availability of medications, medical supplies, and personal protective equipment is still a problem in many countries.

The healthcare system in various developed countries is rife with difficulties and problems. Although Ghana's health system is regarded as one of the most competitive and strong in West Africa, it lags behind other developing nations in providing timely and high-quality care [38].

Health systems in Sub-Saharan Africa will not be able to handle the significant rise in cancer rates that is predicted for the area during the years that follow. To focus reform efforts, more knowledge regarding cancer treatment in this area is required. The only place in Cameroon which has an estimated population of 18.8 million, where patients can obtain chemotherapy is Yaounde General Hospital (YGH). Patients' experiences at this facility serve as an effective model for outlining cancer care in this area [39].

World Economic Forum has estimated that by the year 2040, worldwide healthcare expenditures will be around US$25 trillion. Particularly in the developing world, technological advancements in a variety of fields, like blockchain, mobile technology, AI, the IoT, and big data analytics are expected to have an impact on how healthcare and other services are provided [40]. One of the industries with the most IoT applications enabled by big data analytics, for instance, is healthcare. According to one prediction, 15% of all IoT applications worldwide will be related to healthcare by 2020 [41].

16.10.2 Healthcare Sector in Underdeveloped Countries

For those living in rural parts of less developed nations, having access to healthcare facilities is essential for good medical care. The lack of all-season roads in many regions reduces accessibility and makes access dependent on the weather. To make things more accessible, ideal designs of healthcare facilities have been suggested using location-allocation models, however, these models are predicated on the supposition that the underlying transportation network is static and always accessible [42].

Many issues with the availability and quality of electricity in developing nations endanger the lives of patients, the infrastructure of the healthcare system, and healthcare professionals. Grid electricity, non-renewable energy sources like diesel or gas, and mobile health clinics have all been tried as solutions, but none of them have been successful. In reaction, there have been several changes in the previous ten years regarding access to dependable energy and healthcare, particularly in isolated and rural areas [43].

Various strategies were used by these countries: in Cambodia, nongovernmental organizations (NGOs) were hired to provide primary healthcare; in Argentina, food was distributed through neighborhood mothers' committees; in Ghana and Zambia, treated bed nets were distributed through immunization campaigns; in Nepal, marginalized populations were involved; in India, government hospitals established quality improvement initiatives; and in Brazil, antenatal and child care were rearranged to prioritize serving impoverished communities. Out of the eleven national studies, six indicated that the strategies helped the poor; the other four produced inconsistent results, and only one nation did not help the poor. There were some differences in the methods. The government, NGOs, the commercial private sector, or mixtures of these were employed as implementing organizations. Accessibility might be increased by contracting between agencies, reorganizing the way services are delivered, empowering local communities, or a mix of these strategies and actors. Some of these contained poor-specific programs, while others aimed to broaden coverage for everyone [44].

16.11 Comparison of Recent Progress and Future Mentoring in Healthcare Using Technology

In recent years, there has been a complete revolution in the way that cancer patients are managed as several molecular alterations have been identified as contributing elements to the development and spread of cancer. Historically, this biomarker-based strategy has moved from fundamental research to clinical validation. As a result, rather than focusing on the anatomic location of origin of a particular tumor, attention has progressively shifted to biomarkers. Large groups of patients with tumors that display a specific molecular characteristic have been effectively treated as a result of this change with a single targeted drug. High-throughput technologies have been used to identify several possible targets for novel drugs, and many substances have recently received approval or are currently being studied [45].

Huge amounts of electronic health data are produced daily due to the IoT's quick spread. Improvements in data security are required. The integration of blockchain technology and smart healthcare may address the problems of traditional healthcare in terms of data exchange, data security, and privacy protection. Blockchain technology uses an encrypted data record to link secure data chunks [46]. Figure 16.8 represents different aspects of smart healthcare in the modern world.

Similarly, in immunotherapy, with an incidence of two to three per 100,000, adults having the most prevalent and dangerous primary brain tumor are those with glioblastoma multiforme (GBM). Although GBM shares a lot of the basic mechanisms that other solid tumors use to induce aggression, it has a few unique characteristics. Metastasis is a sentinel event in the development of most solid tumors and an indicator of chronic illness. On the other hand, multifocal GBM is uncommon in that it is unclear from the illness pattern whether recurring *de novo* tumor formation or disease dissemination is occurring. Additionally, it has been suggested that metastasis can occur outside the central nervous system, but this is still under investigation [47].

Moreover, cardiac tissue engineering is a quickly expanding research over the past ten years, a field that has allowed the construction of objects that resemble the native human heart. It has huge promise for use in disease modeling, drug screening, and cardiac

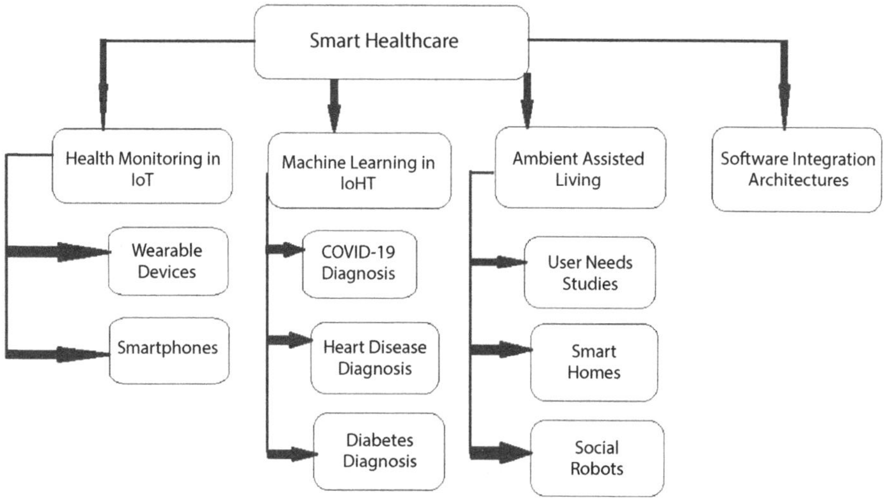

Figure 16.8 Smart healthcare in the modern world.

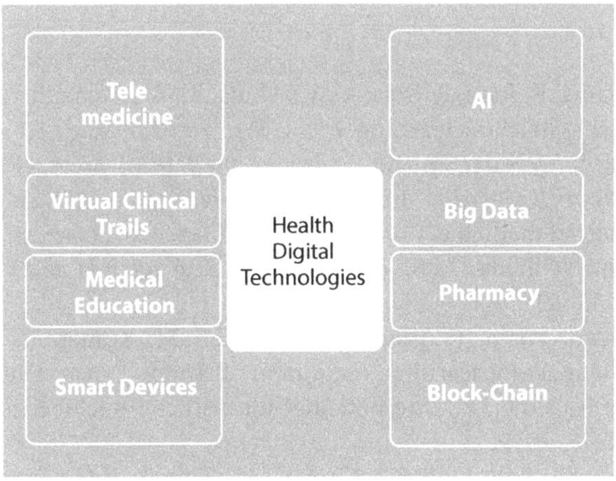

Figure 16.9 Tools used for digital health in medical diagnosis.

regenerating therapies toward personalized medicine, as was already mentioned. Soon, personalized bioengineering technologies and medical treatments are anticipated to become a reality, but continued efforts are needed to guarantee the precision, applicability, repeatability, and durability of these designed technologies and constructions [48]. Figure 16.9 depicts different tools for digital health in medical diagnosis.

Radiology is a key area for application based on AI in healthcare. Image-reading machines [magnetic resonance imaging, computed tomography (CT) scan, and ultrasound] that can train themselves in their early inspection for effective planning of prognosis and therapy can significantly increase the effectiveness of available medical resources. Image acquisition and reconstruction can take the benefit from AI. For example, An ML-based image reconstruction tool for CT scanners was recently released by GE Healthcare and Canon Medical Systems. Reducing radiation exposure while retaining image quality was the aim. Another example is the product for MR imaging process optimization offered by the US company Subtle Medical. The first certified AI products are their SubtlePETTM and SubtleMRTM systems [49].

16.12 Conclusion

In conclusion, future directions in healthcare research are exciting and hold great promise for improving standards in patient care and outcomes. Technological advancements like artificial intelligence, personalized medicine, and telemedicine will play a significant part in the future of healthcare. Additionally, interdisciplinary collaboration and a patient-centered approach will continue to be critical for successful healthcare research. Furthermore, there is a need to focus on addressing health disparities and improving healthcare access for underserved populations. With continued investment and innovation, the future of healthcare research holds great potential for improving the health of people around the world.

References

1. Sachdeva, R.K., Bathla, P., Rani, P., Solanki, V., Ahuja, R., A systematic method for diagnosis of hepatitis disease using machine learning. *Innov. Syst. Software Eng.*, 19, 1–10, 2023.
2. Gostin, L.O., Levit, L.A., Nass, S.J. (Eds.), *Beyond the HIPAA privacy rule: enhancing privacy, improving health through research*, The National Academies Press, Washington, D.C., 2009.
3. Jain, K.K., Personalized medicine. *Curr. Opin. Mol. Ther.*, 4, 6, 548–558, 2002.
4. Hoffman, M.A. and Williams, M.S., Electronic medical records and personalized medicine. *Hum. Genet.*, 130, 33–39, 2011.
5. Verma, K., Bhardwaj, S., Arya, R., Islam, U.L., Bhushan, M., Kumar, A., Samant, P., Latest tools for data mining and machine learning. *Int. J. Innov. Technol. Explor. Eng.*, 8, 9S, 18–23, 2019.
6. Davenport, T. and Kalakota, R., The potential for artificial intelligence in healthcare. *Future Healthc. J.*, 6, 2, 94–98, 2019.
7. Bhardwaj, S., Jain, S., Trivedi, N.K., Kumar, A., Tiwari, R.G., Intelligent Heart Disease Prediction System Using Data Mining Modeling Techniques, in: *Soft Computing: Theories and Applications. Lecture Notes in Networks and Systems*, R. Kumar, C.W. Ahn, T.K. Sharma, O.P. Verma, A. Agarwal (Eds.), p. 425, Springer, Singapore, 2022.
8. Sachdeva, R.K., Garg, T., Khaira, G.S., Mitrav, D., Ahuja, R., A Systematic Method for Lung Cancer Classification, in: *2022 10th International Conference on Reliability, Infocom Technologies and Optimization (Trends and Future Directions) (ICRITO)*, IEEE, pp. 1–5, 2022.
9. Ishak, W.H.W. and Siraj, F., Artificial intelligence in medical application: An exploration. *Health Inf. Europe J.*, 16, 1–9, 2002.
10. Saxena, G., Determining success factors for projects with supervised machine learning. *ScienceOpen Preprints*, 2–13, 2021.
11. Tanveer, M., Richhariya, B., Khan, R.U., Rashid, A.H., Khanna, P., Prasad, M., Lin, C.T., Machine learning techniques for the diagnosis of Alzheimer's disease: A review. *ACM Trans. Multimed. Comput. Commun. Appl. (TOMM)*, 16, 1s, 1–35, 2020.
12. Shaheen, M.Y., Adoption of machine learning for medical diagnosis. *ScienceOpen Preprints*, 2021.
13. Manickam, P., Mariappan, S.A., Murugesan, S.M., Hansda, S., Kaushik, A., Shinde, R., Thipperudraswamy, S.P., Artificial Intelligence (AI) and Internet of Medical Things (IoMT) Assisted Biomedical Systems for Intelligent Healthcare. *Biosensors*, 12, 8, 562, 2022.
14. Chan, H.S., Shan, H., Dahoun, T., Vogel, H., Yuan, S., Advancing drug discovery via artificial intelligence. *Trends Pharmacol. Sci.*, 40, 8, 592–604, 2019.
15. Rani, P., Lamba, R., Sachdeva, R.K., Kumar, R., Bathla, P., Big Data Analytics: Integrating Machine Learning with Big Data Using Hadoop and Mahout, in: *Intelligent Systems and Smart Infrastructure: Proceedings of ICISSI*, vol. 2022, p. 366, 2023.
16. Ahamed, F. and Farid, F., Applying internet of things and machine learning for personalized healthcare: Issues and challenges, in: *2018 International Conference on Machine Learning and Data Engineering (iCMLDE)*, IEEE, pp. 19–21, 2018.
17. Ramesh, T.R., Lilhore, U.K., Poongodi, M., Simaiya, S., Kaur, A., Hamdi, M., Predictive analysis of heart diseases with machine learning approaches. *Malays. J. Comput. Sci.*, 132–148, 2022.
18. Haque, F., Reaz, M.B., Chowdhury, M.E., Shapiai, M., II, Malik, R.A., Alhatou, M., Bhuiyan, M.A.S., A Machine Learning-Based Severity Prediction Tool for the Michigan Neuropathy Screening Instrument. *Diagnostics*, 13, 2, 264, 2023.
19. Schuster, M., Nechansky, A., Kircheis, R., Cancer immunotherapy. *Biotechnol. J.: Healthcare Nutr. Technol.*, 1, 2, 138–147, 2006.
20. Singh, S., Tank, N.K., Dwivedi, P., Charan, J., Kaur, R., Sidhu, P., Chugh, V.K., Monoclonal antibodies: a review. *Curr. Clin. Pharmacol.*, 13, 2, 85–99, 2018.

21. Hargadon, K.M., Johnson, C.E., Williams, C.J., Immune checkpoint blockade therapy for cancer: an overview of FDA-approved immune checkpoint inhibitors. *Int. Immunopharmacol.*, 62, 29–39, 2018.
22. Mohanty, R., Chowdhury, C.R., Arega, S., Sen, P., Ganguly, P., Ganguly, N., CAR T cell therapy: A new era for cancer treatment. *Oncol. Rep.*, 42, 6, 2183–2195, 2019.
23. Mason, C. and Dunnill, P., A brief definition of regenerative medicine. *Regener. Med.*, 3, 1, 1–5, 2008.
24. Mao, A.S. and Mooney, D.J., Regenerative medicine: Current therapies and future directions. *Proc. Natl. Acad. Sci.*, 112, 47, 14452–14459, 2015.
25. Yu, Y., Wang, Q., Wang, C., Shang, L., Living materials for regenerative medicine. *Eng. Regener.*, 2, 96–104, 2021.
26. Glaros, C. and Fotiadis, D.I., Wearable devices in healthcare, in: *Intelligent paradigms for healthcare enterprises*, pp. 237–264, Springer, Berlin, Heidelberg, 2005.
27. Iqbal, S.M., Mahgoub, I., Du, E., Leavitt, M.A., Asghar, W., Advances in healthcare wearable devices. *NPJ Flexible Electron.*, 5, 1, 9, 2021.
28. Craig, J. and Petterson, V., Introduction to the practice of telemedicine. *J. Telemed. Telecare*, 11, 1, 3–9, 2005.
29. Menachemi, N. and Collum, T.H., Benefits and drawbacks of electronic health record systems. *Risk Manage. Healthcare Policy*, 4, 47-55, 2011.
30. Braveman, P., What are health disparities and health equity? We need to be clear. *Public Health Rep.*, 129, 5–8, 2014.
31. Atreja, A., Otobo, E., Ramireddy, K., Deorocki, A., Remote patient monitoring in IBD: current state and future directions. *Curr. Gastroenterol. Rep.*, 20, 1–10, 2018.
32. Pandey, A., Asthana, A.K., Tiwari, R., Kumar, L., Das, A., Madan, M., Physician accessories: doctor, what you carry is every patient's worry? *Indian J. Pathol. Microbiol.*, 53, 4, 711, 2010.
33. Peterson, S.B., The future of asset management, in: *Engineering Asset Management: Proceedings of the 1st World Congress on Engineering Asset Management (WCEAM)*, Springer London, pp. 460–472, 2006.
34. Jost, T.S., Oversight of the quality of medical care: Regulation, management, or the market. *Ariz. L. Rev.*, 37, 825, 1995.
35. Meingast, M., Roosta, T., Sastry, S., Security and privacy issues with health care information technology, in: *2006, the International Conference of the IEEE Engineering in Medicine and Biology Society*, IEEE, pp. 5453–5458, 2006, August.
36. Newaz, A., II, Sikder, A.K., Rahman, M.A., Uluagac, A.S., A survey on security and privacy issues in modern healthcare systems: Attacks and defenses. *ACM Trans. Comput. Healthcare*, 2, 3, 1–44, 2021.
37. Raftery, J., Review of NICE's recommendations, 1999-2005. *BMJ*, 332, 7552, 1266–1268, 2006.
38. Bedeley, R. and Palvia, P., A Study of the Issues of E-Health Care in Developing Countries: The Case of Ghana, in: *AMCIS*, pp. 1–12, 2014.
39. Price, A.J., Ndom, P., Atenguena, E., Mambou Nouemssi, J.P., Ryder, R.W., Cancer care challenges in developing countries. *Cancer*, 118, 14, 3627–3635, 2012.
40. Agarwal, A.K., Tiwari, R.G., Kaushal, R.K., Kumar, N., A systematic analysis of applications of blockchain in healthcare, in: *2021 6th International Conference on Signal Processing, Computing and Control (ISPCC)*, IEEE, pp. 413–417, 2021.
41. Sharma, R. and Kshetri, N., Digital healthcare: Historical development, applications, and future research directions. *Int. J. Inf. Manage.*, 53, 102105, 2020.
42. Murawski, L. and Church, R.L., Improving accessibility to rural health services: The maximal covering network improvement problem. *Socio-Econ. Plan. Sci.*, 43, 2, 102–110, 2009.

43. Soto, E.A., Hernandez-Guzman, A., Vizcarrondo-Ortega, A., McNealey, A., Bosman, L.B., Solar Energy Implementation for Health-Care Facilities in Developing and Underdeveloped Countries: Overview, Opportunities, and Challenges. *Energies*, 15, 22, 8602, 2022.
44. Peters, D.H., Garg, A., Bloom, G., Walker, D.G., Brieger, W.R., Hafizur Rahman, M., Poverty and access to health care in developing countries. *Ann. N. Y. Acad. Sci.*, 1136, 1, 161–171, 2008.
45. Gambardella, V., Tarazona, N., Cejalvo, J.M., Lombardi, P., Huerta, M., Roselló, S., Cervantes, A., Personalized medicine: recent progress in cancer therapy. *Cancers*, 12, 4, 1009, 2020.
46. Nasr, M., Islam, M.M., Shehata, S., Karray, F., Quintana, Y., Smart healthcare in the age of AI: recent advances, challenges, and future prospects. *IEEE Access*, 9, 145248–145270, 2021.
47. Jackson, C.M., Lim, M., Drake, C.G., Immunotherapy for Brain Cancer: Recent Progress and Future Promise Immunotherapy for Brain Cancer. *Clin. Cancer Res.*, 20, 14, 3651–3659, 2014.
48. Häneke, T. and Sahara, M., Progress in Bioengineering Strategies for Heart Regenerative Medicine. *Int. J. Mol. Sci.*, 23, 7, 3482, 2022.
49. Senbekov, M., Saliev, T., Bukeyeva, Z., Almabayeva, A., Zhanaliyeva, M., Aitenova, N., Toishibekov Y. Fakhradiyev, I. (2020). The recent progress and applications of digital technologies in healthcare: a review. *Int. J. Telemed. Appl.*, Volume 2020, Article ID 8830200, 2020.

Index

Printed in the USA/Agawam, MA
September 17, 2026
910661.173